SO-BRS-495

PROBABILISTIC REASONING IN EXPERT SYSTEMS

Probabilistic Reasoning in a Causal Network

PROBABILISTIC REASONING IN EXPERT SYSTEMS

Theory and Algorithms

RICHARD E. NEAPOLITAN
Northeastern Illinois University
Chicago, Illinois

A Wiley-Interscience Publication

JOHN WILEY & SONS, INC.

New York • Chichester • Brisbane • Toronto • Singapore

006.33
N 353

Copyright © 1990 by John Wiley & Sons, Inc.

All rights reserved. Published simultaneously in Canada.

Reproduction or translation of any part of this work
beyond that permitted by Section 107 or 108 of the
1976 United States Copyright Act without the permission
of the copyright owner is unlawful. Requests for
permission or further information should be addressed to
the Permissions Department, John Wiley & Sons, Inc.

Library of Congress Cataloging in Publication Data:
Neapolitan, Richard E.
  Probabilistic reasoning in expert systems: theory and algorithms
  /Richard E. Neapolitan.
      p.   cm.
  "A Wiley-Interscience publication."
  Bibliography: p.
  Includes index.
  ISBN 0-471-61840-3
  1. Expert systems (Computer science)   2. Probabilities.
I. Title.
QA76.76.E95N43   1989
006.3'3–dc19   89-16423
                                                              CIP

Printed in the United States of America

10 9 8 7 6 5 4 3 2 1

To my mother,
without whom I would have ill-fitting clothes

PREFACE

An *expert system* is a computer program which can make reasonable judgments in a complex area. The intent is that the program has the capability to perform a task ordinarily performed by an expert. In some domains, such as the determination of the family of a given plant from a description of its characteristics, the analysis may proceed entirely in logical steps. However, in many other domains, such as medical applications, the relationships are often inexact and the conclusions are uncertain. For example, the physician may not be absolutely certain of the presence of a certain symptom. Moreover, the presence of that symptom might indicate the presence of a disease but not imply conclusively that the disease is present. Reasoning with such imprecise information and relationships, the physician often makes conclusions of which he is not absolutely certain. To reach similar conclusions the expert system must have the capability to reason under uncertainty. This book is both an introduction to the use of probability to represent uncertainty in expert systems, and a guide on how to use probabilistic techniques to create expert systems which must perform uncertain reasoning.

Although many researchers now advocate the use of probability theory in systems which reason under uncertainty, a few years ago many thought it to be inadequate. An early well-known expert system, MYCIN [Buchanan & Shortliffe, 1984], represented uncertainty by assigning real numbers, called certainty factors, to relationships or rules. The method for combining these numbers was based on considerations of how the expert reasons. Other techniques used to reason under uncertainty in expert systems include fuzzy set theory and fuzzy logic [Yager, Ovchinnikov, Yong, & Nguyen, 1987] and the Mathematical Theory of Evidence [Shafer, 1976]. In PROSPECTOR, an expert system for evaluating the mineral potential of large geographical areas [Duda, Hart, Nilsson, & Sutherland, 1978], another approach was taken. Since probability has a long history of representing uncertainty, the developers of PROSPECTOR chose to base their uncertain inference technique on probabilistic grounds.

A certain amount of controversy eventually arose as to whether the probability calculus is an appropriate tool for reasoning under uncertainty in expert systems and furthermore as to whether it is the only appropriate tool. I do not attempt to enter into this controversy. Rather my goal is to show that the probability calculus is appropriate for a certain class of problems in expert systems and to present the recent startling advances in the use of probability theory in expert systems in a coherent textbook which can be used both in the classroom and by the practitioner.

This text is written from the perspective of a mathematician, with the emphasis being on the development of theorems and algorithms. Although I did not see his text until I was 80% completed with this one, I feel that my text complements Judea Pearl's *Probabilistic Reasoning in Intelligent Systems* [Pearl, 1988], which is more from the point of view of a cognitive scientist.

The organization is as follows. Expert systems and uncertain reasoning techniques, which are not based explicitly on probability theory, are briefly reviewed in chapter 1. Since every effort is made to keep the material self-contained, the necessary background in probability theory and graph theory is covered in chapters 2 and 3. Furthermore, in chapter 2, the philosophical foundations of probability theory are reviewed and it is shown why the probability calculus is a fitting tool for a certain class of uncertain reasoning in expert systems. In chapter 4, I briefly describe a probabilistic method used in early rule-based expert systems such as PROSPECTOR to illuminate the many problems encountered when one attempts to represent and update uncertainty in rule-based expert systems. Furthermore, in that chapter, I show why the rule-based framework is often not suitable for representing uncertain knowledge and I introduce the more natural representation of causal (belief) networks. Chapter 5 covers the theoretical foundations of causal networks; chapter 6 describes Pearl's [1986a] method of belief (probability) propagation and fusion in causal networks; and chapter 7 discusses Lauritzen and Spiegelhalter's [1988] method for propagating probabilities in trees of cliques. In chapter 8, I leave the single probability propagation problem to discuss abductive inference. Abductive inference is concerned with finding the most probable set of values of a certain subset of variables called the explanation set. Chapter 9 covers the use of causal networks to represent and solve problems in decision theory. Since it is possible to be uncertain about probabilistic assessments themselves, a single point probability for a proposition can be an inadequate expression of uncertainty. Chapter 10 describes ways of coping with this problem. Furthermore, in chapter 10, I discuss the possible variability in probability values due to remaining sources of information and I show how to augment the probabilities obtained from an expert with information in a data base and how to obtain probabilities solely from a data base.

In the chapters which contain particularly difficult theory (in particular, chapters 5 through 10), I've tried as much as possible to separate the theory from the application. Throughout the text, a reader interested primarily in applications can read the theorems only for an understanding of their statements and skip the proofs.

The content of this text, including the development of all the theory, was taught as a two-quarter graduate-level course in probabilistic reasoning at the University of Chicago. There were approximately five weeks left in which we discussed special topics and worked on programming projects. Therefore it seems that the text is most appropriate for a one-semester course. If time is limited, chapters 1 and 2 may be skimmed. The student should possess the background in graph theory obtained from a data structures course and a background in discrete probability theory. However, these topics are reviewed in chapters 2 and 3. A knowledge of continuous mathematics is assumed only in the appendix of chapter 2 and in part of chapter 10. Both these topics can be skipped without affecting the understanding of the rest of the book.

Although I reviewed several hundred papers before writing this book, I actually based the content on a small percentage of those papers. This is for several reasons. First, often many papers culminated in a particular effort. Since this is a text and not a collection of papers, I've included only the culminating efforts. Second, again since my goal was to write a book which could be used as a text and as a guide to the practitioner, I felt that it was more important to cover some topics rigorously and completely rather than give a smidgen of everything. Therefore, although hopefully most of the major topics are covered, particular papers or techniques may be omitted. Since the field is very young and new discoveries are being made daily, I've undoubtedly omitted something important. However, as has often been said, if everyone waited until he were satisfied with his knowledge before writing a book, the libraries would be empty.

As to the use of personal pronouns in this text, I originally used masculine and feminine pronouns in alternating chapters. However, I found that this procedure was confusing some readers. I therefore switched to using only the masculine pronouns. These pronouns are employed without regard to sex; rather they are only meant to indicate that the antecedent has an indefinite gender.

Finally, I emphasize that the relationships and probabilities in the examples in this text are either fictitious or based loosely on fact. They are NOT meant to be used in any production expert system.

It is a pleasure to acknowledge the many individuals who have provided invaluable assistance in the writing of this book. I thank the researchers throughout the world who have been kind enough to send me their papers, answer my tedious questions, and review my work. I am particularly grateful to Gregory Cooper, John Lemmer, Judea Pearl, Yun Peng, James Reggia, Ross Shachter, and David Spiegelhalter for explaining their ideas and patiently answering my queries. The following people kindly reviewed this text and provided many useful suggestions: Martha Evens, Donald LaBudde, and Ross Shachter. Three of my students, Luigi Di Santo, Neil Hurwitz, and Mitchell Marks, also provided helpful comments. I am thankful to Sandy Zabell for explaining subjectivistic theory to my frequentistic mind, to my son, Richard, for our many conversations concerning the assignment of subjective probabilities, and to David Malament for our conversations concerning the philosophy of quantum mechanics. I thank Sue Coyle for creating the illustrations in this text and Gordon Lamb for granting me the sabbatical which gave me time to complete my writing. The following individuals also provided assistance through correspondence and conversations: James Foster, John Gilbert, Howard Karloff, James Kenevan, Stuart Kurtz, and Mihalis Yannakakis. Finally, this book could not have been written had it not been for my wife, Ro, who mowed the lawn.

RICHARD E. NEAPOLITAN

Chicago, Illinois

CONTENTS

CHAPTER 1

INTRODUCTION

An *expert system* is a computer program capable of making judgments or giving assistance in a complex area. The tasks the system performs are ordinarily performed only by humans. Some well-known expert systems include XCON [McDermott, 1982], a system for configuring Digital Equipment Corporation's VAX computer; DENDRAL [Lindsay, Buchanan, Feigenbaum, & Lederberg, 1980], a system for analyzing mass spectrograms in chemistry; ACRONYM [Brooks, 1981], a vision supporting system; INTERNIST/CADUCEUS [Pople, 1982], a system for diagnosis in internal medicine; CASNET [Kulikowski & Weiss, 1982], a system for dealing with glaucoma; PROSPECTOR [Duda, Hart, Nilsson, & Sutherland, 1978], a system which helps geologists evaluate the mineral potential of exploration sites; and, probably the most well-known expert system of all, MYCIN [Buchanan & Shortliffe, 1984], a system for diagnosing bacterial infections and prescribing treatments for them.

A common characteristic of expert systems is the separation of the knowledge base, the inference engine, and the data available in a specific application. The *knowledge base* includes assumptions about the particular domain, while the *inference engine* is a control mechanism which uses the knowledge base to solve a particular problem. This separation of control from knowledge led to the creation of kernel systems such as EMYCIN [Buchanan & Shortliffe, 1984]. Such systems are actually programming languages for creating expert systems; they contain the control structure (inference engine), but not the knowledge. The knowledge is supplied when the system is used to create a new expert system. The concepts of "knowledge base" and "inference engine" are discussed briefly in section 1.1. For a complete introduction to expert systems, the reader is referred to Jackson [1986].

Many expert systems are not capable of supplying decisions or judgments with absolute certainty. For example, when a physician diagnoses a bacterial infection, he is often uncertain of many of the conclusions leading to his diagnosis. A particular item of evidence may suggest a conclusion or lend credence to the conclusion, but not imply it for certain. Thus it is necessary for the MYCIN system to "reason under uncertainty." In recent years there have been startling advances in the explicit use of probability theory to handle uncertain reasoning. We have seen probabilistic methods evolve from a collection of ad hoc seat-of-the-pants procedures to a valid mathematical discipline. This text is devoted to the theoretical development of that discipline and to a discussion of the algorithms for implementing probabilistic methods. We begin discussing probability theory and its application to expert systems in chapter 2. Before that, in section 1.4, we introduce uncertain reasoning in expert systems and briefly summarize uncertain reasoning techniques which were developed before these advances in probabilistic methods and which are not based explicitly on probability theory.

A minimal knowledge of the basic notations of set theory and of mathematical logic is necessary in order to read section 1.4 and the remainder of this text. Therefore, before introducing uncertain reasoning in section 1.4, these notations are briefly reviewed in sections 1.2 and 1.3.

1.1. RULE-BASED EXPERT SYSTEMS

The best-known expert systems are rule-based expert systems. Causal (belief) networks, which are closely related to rule-based systems, will be the primary focus of this text. In chapter 7, we will even see that some systems, which are not explicitly called causal networks, can be considered as such. As will be discussed in chapter 4, rule-based systems are appropriate for representing categorical knowledge; however, causal networks are much better for representing uncertain knowledge. In this section, we briefly review expert systems concepts by discussing the rule-based approach.

In rule-based systems, the *knowledge base* is in the form of IF–THEN rules. We will focus on a particular example, taken from Thompson and Thompson [1985], to illustrate these rules.

Suppose a botanist is asked, on the phone, to identify the botanical family of a plant. He cannot see the plant, so he must ask the caller questions in order to obtain the facts necessary to an identification. Plants are classified according to type, class, and family. Within a type there are many classes, and within a class many families. A portion of a tree structure, which contains this classification scheme, is contained in Figure 1.1. A leaf of the tree contains a particular family and the description of the family. To determine the family of the particular plant at hand, the botanist could describe a cypress to the caller. He could first ask if the stem was woody. If the answer were yes, he could ask if the position was upright. If the answer were no, he would know that the family was not cypress. He could then go on to pines and ask questions concerning pines. Although effective, such a procedure would be a bit foolish. All trees have upright positions. Thus, once he had ascertained that the position was not upright, he would know that the family was not in the type tree. A reasonable botanist would first ask questions which narrowed down the particular type. For instance, if he learned that the stem was woody and the position was upright, he would ask if there was one main trunk. If the answer to this question were also yes, he would then know that the type was tree. Next he would ask questions which would determine the class of the plant, thereby further narrowing the possibilities. A portion of a decision tree which mimics such reasoning is depicted in Figure 1.2. The entire decision tree would be an expert system for determining the family of a plant.

In a complex domain, an expert does not have his knowledge explicitly organized in a decision tree. Rather it appears that his knowledge is locally organized among closely connected vertices. For example, the botanist would know that

IF   stem is woody
AND   position is upright
AND   there is one main trunk
THEN   type is tree.

This is an item of knowledge which is expressed in an IF–THEN rule. For a plant that he knew to be a tree, he would know that if the leaves were broad and flat, then the plant was a gymnosperm. That is, he would know

IF   type is tree
AND   leaves are broad and flat
THEN   class is gymnosperm.

These local relationships are the types of knowledge which a knowledge engineer (one whose specialty is the creation of expert systems) can extract from an expert. The collection of all these items of knowledge, or rules, is called a knowledge base. The knowledge base, which contains the knowledge in the decision tree in Figure 1.2, is in Table 1.1. This is only a portion of a knowledge base for an expert system which determines the family of a plant. For a particular rule, the IF portion is called the antecedent of the rule and the THEN portion is called the conclusion. Each individual clause in the antecedent is called a premise.

Notice that a knowledge base contains no method for applying the rules. It is simply a collection of separate items of knowledge. The mechanism which applies the rules to solve a problem is the *inference engine*. The particular inference engine which uses these rules to solve the botanist's problem, in the same fashion as the decision tree in Figure 1.2, is called backward chaining.

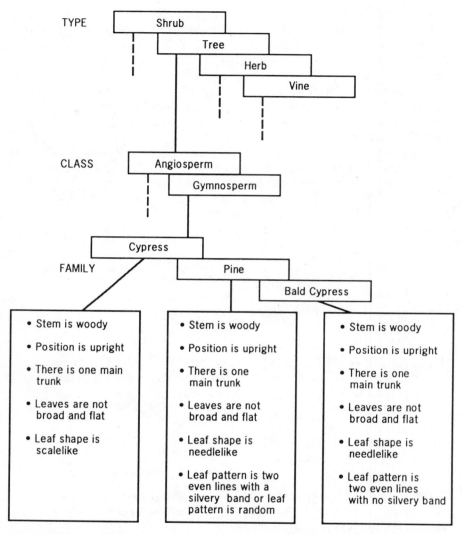

FIGURE 1.1 A portion of a classification tree for plants. (After Thompson & Thompson [1985].)

1.1.1. Backward Chaining

The caller is interested in the family of the plant. To determine the family we cycle through the rules until we find one whose conclusion determines a family. Rule 1 in Table 1.1 is such a rule. If both premises in the antecedent are true, then we know that the family is cypress. Therefore we try to determine whether the premises are true. How can we accomplish this? By using the rules. To determine whether the class is gymnosperm, we again cycle through

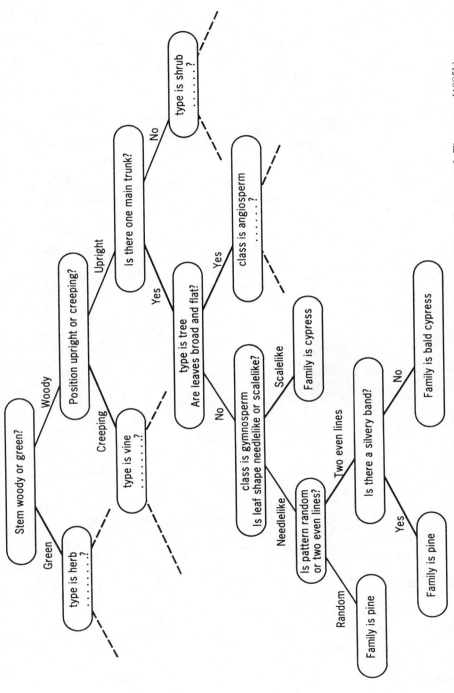

FIGURE 1.2   A portion of a decision tree for identifying the family of a plant. (After Thompson & Thompson [1985].)

5

**TABLE 1.1 A Subset of the Rules in an Expert
System for Determining the Family of a Plant**[a]

1.	IF	class is gymnosperm
	AND	leaf shape is scalelike
	THEN	family is cypress.
2.	IF	class is gymnosperm
	AND	leaf shape is needlelike
	AND	pattern is random
	THEN	family is pine.
3.	IF	class is gymnosperm
	AND	leaf shape is needlelike
	AND	pattern is two even lines
	AND	silvery band is yes
	THEN	family is pine.
4.	IF	class is gymnosperm
	AND	leaf shape is needlelike
	AND	pattern is three even lines
	AND	silvery band is no
	THEN	family is bald cypress.
5.	IF	type is tree
	AND	broad and flat is yes
	THEN	class is angiosperm.
6.	IF	type is tree
	AND	broad and flat is no
	THEN	class is gymnosperm.
7.	IF	stem is green
	THEN	type is herb.
8.	IF	stem is woody
	AND	position is creeping
	THEN	type is vine.
9.	IF	stem is woody
	AND	position is upright
	AND	one main trunk is yes
	THEN	type is tree.
10.	IF	stem is woody
	AND	position is upright
	AND	one main trunk is no
	THEN	type is shrub.

[a] Based on example from Thompson and Thompson [1985].

the rules looking for one whose conclusion determines a class. Rule 5 is such a rule. If rule 5 can determine that the class is angiosperm, then we know to give up on rule 1 and go on to another rule to determine the family. Again, we must determine if the premises in rule 5 are true. To determine whether the type is tree, we again cycle through the rules until we get to rule 9. Rule 9 will tell us that the type is tree if all the premises in rule 9 are true. Again we cycle through the rules looking for one whose conclusion determines the property of the stem. There is no such rule. This means that the user (in this case the caller) should be able to determine whether the current premise is true or false. Thus we ask the user whether the stem is woody. Notice that this is the same question we asked first when using the decision tree in Figure 1.2. If he answers yes, we ask him about the next premise in the rule. If he answers no, we look for another rule which determines a type. Suppose the user answers yes to all the premises in rule 9. Then we conclude that the type is a tree, and we go back to rule 5 to determine whether the other premise in rule 5 is true. If we determine that the leaves are broad and flat, we conclude that the class is angiosperm. We then go back to rule 1 and, noticing that the first premise is false, give up on that rule and go on to rule 2 to try to determine the family. We thus proceed in this manner until one of the rules eventually determines the family or we learn that the user cannot supply enough information to determine the family. This procedure is called backward chaining because of the way it backs up from rules, which contain as their conclusion the information desired by the user, to rules containing premises, which the user is asked to verify.

Following is high-level pseudocode for backward chaining. As we chain back through the rules, we save the current rule in a rule stack. In a given clause, the quality in the clause is called an attribute. For example, the attribute in "class is gymnosperm" is "class." The attribute, whose value is requested by the user, is our initial goal. Each time we chain back to a new rule, the attribute in the conclusion of that rule becomes our new goal. As we chain back, we save the current goal in a goal stack. Once we learn the value of an attribute, we save that fact in a set of true assertions so that we do not bother to try to determine that value again. For example, if the user says that the "stem is woody," we save that assertion in the true assertion set.

BACKWARD CHAINING

Var Goalstack: a stack of attributes;
      Rulestack: a stack of rules;
      Assertion_List: a set of assertions;

Function Backchain (Goal: an attribute);

Var Assertion, True_Assertion: assertions;
      Rule: a rule;
      All_Premises_True: Boolean;

```
Begin
    push (Goal, Goalstack);
    Assertion := Null;
    Assertion := the assertion in Assertion_List with Goal as its Attribute;
    if Assertion = Null then {we found no assertion}
        begin
            Rule := first rule;
            while (Assertion = Null) and more rules do
                begin
                    if Goal = Attribute in Rule's conclusion then
                        begin
                            Push(Rule, Rulestack);
                            All_Premises_True := false;
                            Premise := first premise in Rule's antecedent;
                            repeat
                                Goal := Attribute in Premise;
                                True_Assertion := Backchain(Goal);
                                if Premise = True_Assertion then
                                    begin
                                        Premise := next premise in Rule's antecedent;
                                        if Premise = Null then All_Premises_True := true
                                    end
                            until All_Premises_True or (Premise ≠ True_Assertion);
                            pop from the Rulestack;
                            if All_Premises_True then Assertion := Rule's Conclusion;
                        end;
                    Rule := next rule
                end;
            if (Assertion = Null) then {rules could not determine the value}
                begin
                    prompt user for the value of Goal;
                    read (Value);
                    if Value ≠ Null then Assertion := Goal "is" Value {user knows}
                    else Assertion := Goal "is Unknown"
                end;
            add Assertion to Assertion_List;
        end;
    pop from the Goalstack;
    if the Goalstack is empty then
        begin
            if Assertion = Goal "is unknown" then
                write ("You cannot answer enough questions for us to determine
                        the value of your goal")
            else write (Assertion)
        end;
```

    Backchain := Assertion
end.

In the main program, the user is asked for his goal, and "Backchain" is called with the value of that goal as its argument.

This procedure can be enhanced considerably. First, we should impose certain restrictions on the rules. One restriction is that the knowledge base should not contain rules which are circular. That is, if A, B, and C represent assertions, we should not have the following three rules:

IF A, THEN B IF B, THEN C IF C, THEN A.

Clearly, such a set of rules simply means that A, B, and C are all always true simultaneously and therefore they can be represented by a single assertion in our knowledge base. The existence of such a set of rules in our knowledge base could cause the backward chaining program to run in an infinite loop. Therefore an error check for cyclic rules could be built into the program. Such a check is quite simple; we need only note whether a goal is already in the goal stack when we are about to add it to that stack.

Another restriction on the rules is that there should not be contradictory rules. For example, we should not have the following two rules:

IF A, THEN B IF A, THEN NOT B.

A third rule restriction is best illustrated with an example. Suppose that we have the following two rules:

IF A AND B AND C, THEN D IF A AND B, THEN D.

The set of premises in the second rule is said to subsume the set of premises in the first rule. That is, if both A and B are true, we can conclude D regardless of the value of C. Therefore the first rule is superfluous and should be eliminated.

There exist other possible enhancements besides rule restrictions. One such enhancement is that we should allow the user to inquire as to why a question is being asked. For instance, if he wants to know why we are asking if the stem is woody, the system could look at the top rule in the rule stack (say rule 9 in Table 1.1) and reply that it is because we need to know the type of the plant. If the user then wanted to know why we needed to know the type, the system could look at the next rule in the rule stack (say rule 5) and reply that it is because we need to know the class of the plant. The system could continue down the stack as long as the user kept asking questions.

Computationally, backward chaining is clearly very inefficient compared to a program which simply implements the decision tree in Figure 1.2. However, recall that our goal is to separate the control structure from the knowledge base. The reason for this is to permit items of knowledge to be added and deleted freely as the expert system is being constructed. This would not be possible with a hard-coded decision tree. In practice the fully developed knowledge base is often converted to a decision tree before actual use of the expert system.

1.1.2. Forward Chaining

Suppose the botanist received a letter which contained a description of the plant and his job was to deduce as much as possible about the plant from the description. For example, the letter might say that the stem is woody, the position is upright, there is one main trunk, and the leaves are not broad and flat. He certainly would not go through the same questioning procedures. Instead he would apply his knowledge to deduce as much as possible. Since the stem is woody, the position is upright, and there is one main trunk, he would conclude that the type is tree. Then, since the leaves are not broad and flat, he would conclude that the class is gymnosperm. This is all he could conclude from the facts in the letter. We can accomplish the same thing by applying a different inference engine to the knowledge base. That inference engine is called forward chaining. It is much more straightforward than backward chaining. We simply put all our true assertions in an assertion list. Then we cycle through the rules starting with the first one. If all a rule's premises are in the assertion list, then the conclusion must be true and we add the conclusion to the assertion list. Since a conclusion of one rule can be a premise in another rule, we must start over at the first rule each time we add a new conclusion. We proceed until we reach the end of the rules.

The above procedure can be streamlined considerably by sorting the rules before performing the forward chaining. The sorting scheme is the following: If rule A's conclusion is a premise in rule B's antecedent then we place rule A before rule B. Assuming we are not engaged in circular reasoning, this is always possible. It is clear that with the rules sorted in this order, there is no need to return to the first rule when a conclusion is added to the true assertion list.

1.1.3. Conflict Resolution

A forward chaining inference engine can be substantially more complex than one that simply cycles through the rules making all possible deductions. We will focus on an example from Winston [1984] to illustrate just a few of the possibilities. The set of rules in Table 1.2 is a portion of a knowledge base for bagging groceries. The rules look at the data base of groceries in Table 1.3 to determine which item to bag next. The "bagged?" entry is set to "yes" when an item is bagged. We will now bag the groceries by forward chaining through the rules just as an inference engine would do.

Large items are bagged first, medium items bagged next, and small items bagged last. We assure that large items are bagged first by initially

TABLE 1.2   A Subset of the Rules in an Expert System for Bagging Groceries[a]

---

1. IF   step is Bag_large_items
    AND   there is a large item to be bagged
    AND   there is a bag with < 6 large items
    THEN   put the large item in the bag.

2. IF   step is Bag_large_items
    AND   there is a large item to be bagged
    AND   there is a large bottle to be bagged
    AND   there is a bag with < 6 items
    THEN   put the bottle in the bag.

3. IF   step is Bag_large_items
    AND   there is a large item to be bagged
    THEN   start a fresh bag.

4. IF   step is Bag_large_items
    THEN   step is Bag_medium_items.

5. IF   step is Bag_medium_items
    AND   there is a medium item to be bagged
    AND   there is a bag with < 10 medium items
    AND   that bag contains 0 large items
    AND   the medium item is frozen
    AND   the medium item is not in an insulated bag
    THEN   put the medium item in an insulated bag.

6. IF   step is Bag_medium_items
    AND   there is a medium item to be bagged
    AND   there is a bag with < 10 medium items
    AND   that bag contains 0 large items
    THEN   put the medium item in the bag.

7. IF   step is Bag_medium_items
    AND   there is a medium item to be bagged
    THEN   start a fresh bag.

8. IF   step is Bag_medium_items
    THEN   step is Bag_small_items.

9. IF   step is Bag_small_items
    AND   there is a small item to be bagged
    AND   there is a bag which is not full
    THEN   put the small item in the bag.

10. IF   step is Bag_small_items
    AND   there is a small item to be bagged
    THEN   start a fresh bag.

11. IF   step is Bag_small_items
    THEN   halt.

---

[a] Based on example from Winston [1984].

TABLE 1.3   A Data Base of Items to Be Bagged

Item	Container	Size	Frozen?	Bagged?
Soda	Bottle	Large	No	No
Bread	Bag	Medium	No	No
Ice cream	Carton	Medium	Yes	No
Detergent	Box	Large	No	No
Eggs	Carton	Small	No	No
Popsicles	Insulated bag	Medium	Yes	No

a Based on example from Winston [1984].

setting

Step:=Bag_large_items.

Next we look at the rules. Rules 1 and 2 in Table 1.2 both require the truth of the assertion "there is a bag with < 6 large items." Since we have not yet gotten any bags, this assertion is not true. Rule 3 requires that "step is Bag_large_items" and that "there is a large item to be bagged." Since both these conditions are true, the rule triggers, and we start a fresh bag, Bag 1. We then return to the start of the rules, and again look for a rule whose premises are all true. Notice that the premises in rules 1–4 are all true. In such a situation we say that there is a conflict as to which rule to trigger first. Simple forward chaining would just trigger these rules in sequence. However, the rule we want to trigger is rule 2. That is, we want to bag large bottles first, since a smart bagger always puts bottles on the bottom. The correct rule is chosen by incorporating into the inference engine a type of conflict resolution called specificity ordering:

Definition 1.1. Specificity Ordering. If the set of premises of one rule are a superset of the set of premises of another rule, then the first rule is triggered on the assumption that it is more specialized to the current situation.

Thus rule 2 triggers because all the premises in the other three rules are subsets of the premises in rule 2. We therefore place the bottle of soda in Bag1 and mark that the soda has been bagged. We look at the rules again, noticing that the premises in rules 1, 3, and 4 are all true. This time rule 1 is the rule that triggers because the other rules" premises are subsets of the premises in rule 1. Therefore we place the detergent in Bag1 and mark that the detergent has been bagged. Looking at the rules again, we see that only the premises in rule 4 are true. This rule acts to change the value of the variable "step to Bag_medium_items" now that all the large items have been bagged. By using the variable step, we have separated the rules into disjoint subsets; only the rules in a given subset are active at any particular time. This is an example of conflict resolution called context limiting.

Definition 1.2. Context Limiting. Separate the rules into disjoint subsets. Only the rules in one subset are active at any particular time. The context is changed to a new context by a rule in the current context.

The rules for medium items will now execute until there are no more medium items. Notice that we have written the rules so that a medium item is never placed in a bag with a large item. After the medium items are all bagged, the context is changed to the set of rules for bagging small items. Note that we can place a small item in any available bag. After all the small items are bagged, rule 11 halts execution. It is left as an exercise to walk through the steps which bag the remaining groceries.

XCON [McDermott, 1982] is an expert system which operates in a fashion similar to our grocery bagger. This system configures Digital Equipment Corporation's VAX computers.

There are other conflict resolution strategies. We mention just two more commonly used ones:

Definition 1.3. Recency Ordering. The rule which has triggered most recently has the highest priority, or the rule which has triggered least recently has the highest priority. The choice clearly depends on the needs of the particular system.

Definition 1.4. Priority Ordering. Order the rules according to the priority with which they should trigger.

Since this is a text in probabilistic methods, we have been very sketchy in our introduction to expert systems. Again, a reader desirous of a more substantial treatment is referred to Jackson [1986].

PROBLEMS 1.1

1. Finish bagging the groceries in Table 1.3.

2. Create a rule base for a mini-classification expert system. For example, the system could determine the type of an animal or a mineral based on its description, or it could determine the drink for a patron of a bar based on his particular requirements.

3. Create a rule base for a mini-diagnostic expert system. For example, the system could determine the problem with an automobile or a video cassette recorder.

4. Create a rule base for a mini-configuration expert system. The system could be an enhancement of the grocery bagger. For example, consideration could be given to placing crushable items such as potato chips on the top.

COMPUTER PROBLEM 1.1

Using your favorite programming language, write routines which perform backward chaining and forward chaining. A particular expert system may require both of these routines. For example, based on the user's goal, the system chains backward until a point where it asks the user a question. When the user enters a response, the system chains forward to determine what other deductions can be made from the new fact. Implement one of the expert systems for which a rule base was created in Problems 2, 3, or 4. Use forward chaining, backward chaining, or both, depending on the needs of your system. Incorporate conflict resolution strategies in your forward chaining routine as needed. The final program should have an expert interface and an end-user interface. The expert can add or delete rules, while the end user can enter true propositions.

1.2. A BRIEF REVIEW OF THE NOTATION OF MATHEMATICAL LOGIC

Logic deals with propositions (sentences) which are assumed to be true or false. We can define them loosely as follows:

Definition 1.5. A proposition (sentence, statement) is an entity which is either true or false.

Examples of propositions include

All men are mortal.
The stem of this particular plant is green.
This particular patient has a cancerous tumor.

We will ordinarily denote propositions by capital letters near the beginning of the alphabet, such as A, B, and C. The actual truth or falsity of a proposition is of no concern to the logician. He is only interested in whether the truth of some propositions logically imply the truth of others. In expert systems we are also interested in such implications. For example, the truth of the proposition "the stem is woody" implies the truth of the proposition "the plant is an herb" due to one of our rules in the previous section. Of course, in expert system applications we also are interested in the truth of propositions. That is, we really want to know whether a particular patient has a cancerous tumor.

The elementary logical operations on propositions are the following:

Definition 1.6. If A is a proposition, then $\neg A$ is the negation of the proposition.

For example, if A is "this particular patient has a cancerous tumor," then $\neg A$ is "this particular patient does not have a cancerous tumor." The relationships between the truths of propositions can be characterized in truth tables. The following is a truth table for A and $\neg A$:

A	$\neg A$
T	F
F	T

where T stands for true and F for false.

Definition 1.7. If A and B are two propositions, the conjunction of A and B is the proposition obtained by putting the word "and" between A and B. It is denoted by $A \wedge B$, and has the following relationship to A and B:

A	B	$A \wedge B$
T	T	T
F	T	F
T	F	F
F	F	F

We see that $A \wedge B$ is defined to be true if both A and B are true, as we would expect.

Definition 1.8. If A and B are propositions, the proposition obtained by putting the word "or" between A and B is called the disjunction of A and B. It is denoted by $A \vee B$ and has the following relationship to A and B:

A	B	$A \vee B$
T	T	T
F	T	T
T	F	T
F	F	F

We see that $A \vee B$ is defined to be true if either A is true or B is true or both A and B are true. This is called an inclusive "or." An exclusive "or" would be defined to be true if one and only one of A and B was true.

The last important operation on propositions is the conditonal: "IF A, THEN B." We are not concerned here with its formal meaning in logic. The rules in expert systems are clearly conditional statements. In the notation introduced in this section, rule 2 in Table 1.1 is as follows:

IF   class is gymnosperm
∧   leaf shape is needlelike
∧   pattern is random
THEN   family is pine.

In expert systems, which are not involved in uncertain reasoning, we simply assume that our rules are true. Then, if all the premises in the antecedent are true, we conclude that the conclusion is true.

1.3. A BRIEF REVIEW OF THE NOTATION OF SET THEORY

We define a set informally as follows:

Definition 1.9. A set is a collection of objects (elements).

For example, the set of all digits includes the objects

$$0, 1, 2, 3, 4, 5, 6, 7, 8, 9.$$

If a patient has a cold, hay fever, or the flu, the set of possible diseases for the patient includes the objects

Cold, hay fever, flu.

In this text, we will ordinarily denote sets by capital letters and by script capital letters. We will enclose the objects in a set with braces. Thus, if S is the set of all even digits, then

$$S = \{0, 2, 4, 6, 8\}.$$

We will often use a shorthand description of the objects in a set. For example, the set of all digits could be represented by

$$T = \{x \text{ such that } x \text{ is an integer and } 0 \leq x \leq 9\}.$$

We will use the symbol $\in$ to denote that an object is in a set and the symbol $\notin$ to denote that an object is not in a set. Thus if S is the set of all even digits, then

$$2 \in S \quad \text{and} \quad 3 \notin S.$$

Definition 1.10. Let A and B be two sets. If every object in A is also in B, then A is said to be a subset of B, in symbols $A \subseteq B$. Moreover, if there is some object in B which is not in A, then A is said to be a proper subset of B, in symbols $A \subset B$.

If $A = \{1,3,5\}$ and $B = \{1,2,3,5,6\}$,   then $A \subset B$.

Definition 1.11. Let A and B be sets. The intersection of A and B is defined to be the set of all objects which are in both A and B. It is denoted by $A \cap B$.

If $A = \{2,3,5\}$ and $B = \{2,3,6\}$,   then $A \cap B = \{2,3\}$.

Definition 1.12. Let A and B be sets. The union of A and B is defined to be the set of all objects which are in either A or B. It is denoted by $A \cup B$.

If $A = \{2,3,5\}$ and $B = \{2,3,6\}$,   then $A \cup B = \{2,3,5,6\}$.

Definition 1.13. The set which contains no objects is called the empty set. It will be denoted by $\emptyset$ and by NULL.

If $A = \{1,3,4\}$ and $B = \{2,5,6\}$, then A and B have no objects in common and therefore $A \cap B = \emptyset$. In this case A and B are said to be disjoint sets.

Definition 1.14. Let A and B be sets. The relative complement of A with respect to B is defined to be all those objects which are in B but are not in A, and is denoted by $B - A$.

If $A = \{2,3,5\}$ and $B = \{2,3,6\}$,   then $B - A = \{6\}$.

Definition 1.15. The universal set is defined to be the set consisting of all objects under consideration.

Clearly, the universal set is relative to a specific problem or application. If U is the universal set and A is any other set, then $A \subseteq U$.

Definition 1.16. Let U be the universal set, and A any set in U. The absolute complement (or simply complement) of A is defined to be all those elements which are in U but not in A. It is denoted by $\overline{A}$.

If U is the set of all digits and $A = \{1,3,5,7,8,9\}$, then the absolute complement of A is $\overline{A} = \{0,2,4,6\}$.

1.4. INTRODUCTION TO UNCERTAIN REASONING AND NONPROBABILISTIC METHODS

In the expert systems considered in section 1.1, we are always absolutely certain of the logical implications of the rules. For example, a botanist knows for certain that if a stem is green, then the class of the plant is an herb. Further-more, we made the assumption that the user either knows for certain that an assertion is true or false or that he supplies no useful information. For exam-ple, if the user says that the stem is woody, we take that assertion as being absolutely true. If he cannot determine for certain whether the stem is woody or green, we assume that he can supply no useful information. This is an ide-alized situation that may work in some expert systems. However, in others we are quite interested in how certain the user is of assertions. We also must ac-knowledge that in some cases the truth of certain premises may be suggestive of the truth of a conclusion, but not imply it conclusively. Thus many expert systems must do more than make simple logical deductions; they must be able to "reason under uncertainty." Szolovits and Pauker [1978] explain why such reasoning is necessary in medicine:

> Why are categorical decisions not sufficient for all of medicine? Because the world is too complex! Although many decisions may be made straightforwardly, many others are too difficult to be prescribed in any simple manner. While many factors may enter into a decision, when those factors may themselves be uncer-tain, when some factors may become unimportant depending on other factors, and when there is a significant cost associated with gathering information that may not actually be required for the decision, then the rigidity of the flow chart makes it an inappropriate decision-making instrument.

In this section we briefly review uncertain reasoning methods, which were developed before the advances in the use of probability theory and which are not explicitly based on probability theory. Occasionally in this section, we refer to probability theory for the sake of comparison. It is therefore assumed that the reader has a minimal background in discrete probability theory.

1.4.1. MYCIN and Certainty Factors

Suppose a culture taken from a patient is analyzed. A typical MYCIN rule which uses the information in this analysis is:

    IF   the organism grows in clumps
    AND   the organism grows in chains
    AND   the organism grows in pairs
    THEN   the organism is streptococcus.

If the verity of the three premises implied the conclusion for certain and if we were able to determine for certain whether the three premises were true, there would be no uncertainty. The conclusion would either be made or would not be made. However, it turns out that the verity of these three premises usually implies that the organism is streptococcus, but not always. A physician, when analyzing these results, would note that it is highly likely that the or-ganism is streptococcus based on this evidence, but he would leave open the possibility that it may not be. To incorporate such judgments into an expert system, the MYCIN system attaches real numbers called certainty factors to rules. Each rule is assigned a value between -1 and $+1$, where positive values indicate that the verity of the premises should increase our belief in the conclusion and negative values indicate the verity of the premises should decrease our belief in the conclusion. For example, if the above rule had a certainty factor of .8, this would mean that the verity of the three premises should increase to a large degree our belief that the organism is streptococcus. If the certainty factor were .1, this would mean that the verity of the three premises should increase our belief that the organism is streptococcus, but to a small degree. If the certainty factor were $-.8$, this would mean that the verity of the three premises should decrease to a large degree our belief that the organism is streptococcus. There is nothing special about the values 1 and -1. They only represent relative measures. Thus we could just as well have used the values 100 and -100, or the values 100 and 0. The values 1 and -1 are simply convenient. As noted by Heckerman [1986], a common misconception is that certainty factors represent measures of absolute belief. Rather, as we have shown, they are meant to represent changes in belief.

Often the user is not absolutely certain of the assertions he supplies to the system. For example, the analysis of the culture might indicate that it is very likely that the organism grows in chains, be indicative that the organism may grow in pairs, and be almost conclusive that it grows in clumps. The user might therefore supply the following measures of belief with these assertions:

.8 that the organism grows in chains
.3 that the organism grows in pairs
.95 that the organism grows in clumps.

These measures of belief are also between -1 and 1, where 1 represents absolute belief and -1 represents absolute disbelief. MYCIN takes this uncertainty in the premises into account when triggering the rule as follows:

$$\text{Belief in premises} = \max[0, \min(.8, .3, .95)] = .3$$

$$\text{Change in belief in conclusion} = \text{certainty factor} \times \text{belief in premises}$$

$$= (.8) \times (.3) = .24.$$

Thus our change in belief that the organism is streptococcus is only equal to .24. The reason that we take the maximum of 0 and the minimum of the beliefs in the premises is that the rule only has to do with evidence for the presence of streptococcus. The idea is that the rule can only tend to increase the belief in the conclusion, not decrease it. For example, if the belief that the organism growing in chains were $-.9$ and we did not take the maximum with 0, we would conclude that the change in belief that the organism is streptococcus is $-.72$. In other words, the rule would tend to make us believe that the organism is not streptococcus. However, this is not what the rule is meant to indicate.

The absence of the organism growing in chains is not supposed to indicate that the organism is not streptococcus. If indeed this were the case, there would be a separate rule indicating that fact.

Suppose next that another rule triggers which individually increases the belief that the organism is streptococcus with value .6. The MYCIN system must determine the total change in belief that the organism is streptococcus based on these two items of evidence. This is accomplished in the following way. If B_1 is the change in belief obtained from the first rule, B_2 is the change in belief obtained from the second rule, and B_{12} is the combined change in belief obtained from the two rules, then

$$B_{12} = \begin{cases} B_1 + B_2(1 - |B_1|) & B_1 \text{ and } B_2 \text{ both positive or both negative} \\ \dfrac{B_1 + B_2}{1 - \min(|B_1|, |B_2|)} & \text{one of } B_1 \text{ and } B_2 \text{ positive, other negative.} \end{cases}$$

In this particular example, the resultant change in belief that the organism is streptococcus is given by

$$B_{12} = .24 + .6(1 - .24) = .696.$$

Notice that the combined change in belief is greater then the change in belief due to each rule individually. If a third rule implies a change in belief that the organism is streptococcus equal to B_3, B_3 is combined with B_{12} using the same formula. It is a simple matter to show that this operation is commutative and associative, and that therefore the order in which evidence is applied is immaterial. The bottom expression in the formula for B_{12} is for the case in which one item of evidence confirms the conclusion while the other disconfirms the conclusion.

The calculus for manipulating certainty factors has many desirable properties which we would expect from an uncertain reasoning technique. It takes into account both the uncertainties in the implications and the user's uncertainty in the evidence; it can combine changes in belief obtained from several rules; it can combine positive changes in belief with negative changes in belief. Regardless, certainty factors have been the target of some criticism. For one thing, originally there was no *operational definition* of a certainty factor. That is, the definition of a certainty factor does not prescribe a method for determining a certainty factor. Without an operational definition there is no way of knowing whether two experts mean different things when they assign different certainty factors. For example, suppose expert 1 assigns a certainty factor of .6 to a particular rule, whereas expert 2 assigns a certainty factor of .8 to that same rule. They could mean the same thing. We have no way of knowing without an operational definition of a certainty factor. As we shall see in chapter 2, every approach to probability theory does include an operational definition. Certainty factors were eventually given a probabilistic interpretation [Heckerman, 1986]. However, other problems remained. We shall discuss these problems in the next chapter after presenting the foundations of probability theory.

Before ending this section, we note that the certainty factor model also includes a method for sequential updating, that is, a method for propagating changes in belief in the case where the conclusion of one rule can be a premise in another rule. For a complete discussion of the MYCIN expert system and certainty factors, the interested reader is referred to Buchanan and Shortliffe [1984].

1.4.2. Nonnumerical Methods for Reasoning Under Uncertainty

Since this text is devoted to the representation of uncertainty by real numbers, we mention these nonnumerical methods very briefly. The most notable are the use of nonmonotonic logic [Moore, 1985] and the theory of endorsements [Cohen, 1985]. In nonmonotonic logic there is assumed to be a world of n variables. At any one time, based on the available evidence, it is believed that the variables have specific values. The values of the variables may have to be revised when new information is received. As noted by Bhatnagar and Kanal [1986], this "formalism can represent an uncertain world model only if predicates with uncertain interpretations can be used."

In the theory of endorsements, each hypothesis has associated with it a body of endorsements based on the evidence. The uncertainty is maintained in the relative endorsements of competing hypotheses. There is, however, no general scheme for comparing two bodies of endorsements. They can be pairwise ranked, creating a partial order, but this does little to help compare two large bodies of endorsements.

Both of the above methods have their places in expert systems. However, here we are concerned with determining the relative likelihood of competing hypotheses. For example, we want to know how much more likely it is that the organism is streptococcus than that it is some other bacterium. Thus we associate real numbers with uncertainty.

Next we discuss the uncertain reasoning mechanism which is the closest relative of probability theory (it is, in fact, an extension of probability theory [Horvitz, Heckerman, & Langlotz, 1986]).

1.4.3. The Dempster–Shafer Theory of Evidence

We will illustrate the Dempster–Shafer theory with an example taken from Spiegelhalter [1986a]. A complete introduction can be found in Shafer [1976]. Let D stand for a particular conclusion. For example, D could stand for the conclusion "the organism is streptococcus." Suppose one rule implies D with strength .8, while another rule implies D with strength .9. In the Dempster–Shafer theory of evidence, these strengths are called basic probability assignments (bpa). They are denoted by a function m. We have

$$m_1(D) = .8 \qquad \text{and} \qquad m_2(D) = .9.$$

If we let θ be the set of all possibilities (in this simple example, θ contains only D and $\neg D$), the Dempster–Shafer theory assigns a bpa to every subset of θ (excluding the empty set), and the sum of all the bpa's must equal 1. For example, we must have that

$$m_1(\{D\}) + m_1(\{\neg D\}) + m_1(\{D, \neg D\}) = 1.$$

This is very similar to a probability assignment except that it is not necessary that the sum of the bpa's assigned to the members of θ be equal to 1. For example, the assignments of

$$m_1(\{D\}) = .8, \qquad m_1(\{\neg D\}) = 0, \qquad \text{and} \qquad m_1(\{D, \neg D\}) = .2$$

mean that the rule commits .8 of our belief to D and none of our belief to $\neg D$, and that the remaining belief is uncommitted. In probability theory it would be necessary that $m_1(\{D\})$ and $m_1(\{\neg D\})$ sum to 1. Using the Dempster–Shafer method we can leave some of the "probability" unassigned. Thus the rule is evidence for D but not evidence against it.

Before proceeding with this example, we note that θ can contain more elements that just a proposition and its negation. In general, θ must contain a set of mutually exclusive and exhaustive possibilities. For example, if a patient has either hepatitis, cirrhosis, or pancreatic cancer, and he has only one of these, and we let

$$H = \text{hepatitis}, \qquad C = \text{cirrhosis}, \qquad \text{and} \qquad P = \text{pancreatic cancer},$$

then

$$\theta = \{H, C, P\}$$

and the set of subsets of θ is equal to

$$\{\{H\}, \{C\}, \{P\}, \{H, C\}, \{H, P\}, \{C, P\}, \{H, C, P\}\}.$$

We must assign a bpa to each of these subsets.

Returning now to our original example, for a given subset x, the $\text{Bel}(x)$ is the sum of all the belief committed to the possibilities in x. For example,

$$\text{Bel}_1(\{D, \neg D\}) = m_1(\{D\}) + m_1(\{\neg D\}) + m_1(\{D, \neg D\}) = .8 + 0 + .2 = 1.$$

It is always the case that the $\text{Bel}(\theta)$ is equal to 1. For individual members of θ (in this case, D and $\neg D$), Bel and m are equal. Thus

$$\text{Bel}_1(\{D\}) = m_1(\{D\}) = .8 \qquad \text{and} \qquad \text{Bel}_1(\{\neg D\}) = m_1(\{\neg D\}) = 0.$$

The plausibility, $\text{Plaus}(x)$, of a subset x is equal to 1 minus the Bel in the complement of the set. Therefore

$$\text{Plaus}(\{D\}) = 1 - \text{Bel}(\{\neg D\}) = 1 - 0 = 1.$$

Thus we have a belief interval for each subset of θ. The belief interval for $\{D\}$ is given by

$$(\text{Bel}(\{D\}), \text{Plaus}(\{D\})) = (.8, 1).$$

The lower bound is a measure of how much belief we have committed to $\{D\}$, whereas the upper bound is a measure of how much belief we have committed to $\{\neg D\}$. The width of the interval is a measure of the uncertainty in our beliefs. For example, if the belief interval for $\{D\}$ were $(.8, .8)$ we would be absolutely certain that we believed $\{D\}$ to degree .8 and $\{\neg D\}$ to degree .2.

Let us now look at how beliefs obtained from two separate rules are combined. Suppose

$$m_1(\{D\}) = .8 \qquad m_1(\{\neg D\}) = 0 \qquad m_1(\{D, \neg D\}) = .2$$
$$m_2(\{D\}) = .9 \qquad m_2(\{\neg D\}) = 0 \qquad m_2(\{D, \neg D\}) = .1.$$

Then m_{12} is obtained as follows:

		m_2			
		$\{D\}$ .9		$\{D, \neg D\}$ .1	
m_1	$\{D\}$	.8 $\{D\}$	.72	$\{D\}$	.08
	$\{D, \neg D\}$	.2 $\{D\}$	.18	$\{D, \neg D\}$	.02.

The idea is that the product of the original separate beliefs committed to two subsets is the new belief committed to the intersection of those subsets. For example, m_1 commits .8 belief to $\{D\}$ and m_2 commits .1 belief to $\{D, \neg D\}$. The only way for one of the possibilities in both of these sets to occur is for one of the possibilities in their intersection, namely $\{D\}$, to occur. Thus the combination of these two beliefs, that is, $.8 \times .1$, is all assigned to $\{D\}$. It is a simple matter to show that, with m_{12} obtained in this manner, the sum of the new bpa's is equal to 1. The new belief committed to D is obtained by summing all the components committed to D:

$$m_{12}(\{D\}) = .72 + .18 + .08 = .98 \qquad m_{12}(\{\neg D\}) = 0$$
$$\text{Bel}_{12}(\{D\}) = m_{12}(\{D\}) = .98 \qquad \text{Bel}_{12}(\{\neg D\}) = m_{12}(\{\neg D\}) = 0$$
$$\text{Plaus}_{12}(\{D\}) = 1 - 0 = 1. \qquad \text{Plaus}_{12}(\{\neg D\}) = 1 - .98 = .02.$$

The new belief interval for D is now $[.98, 1]$. The two confirming rules for D yield greater belief in D than either rule separately.

Next suppose that one of the rules confirms D, while the other disconfirms D. That is, we have the following situation:

$$m_1(\{D\}) = .8 \qquad m_1(\{\neg D\}) = 0 \qquad m_1(\{D, \neg D\}) = .2$$
$$m_2(\{D\}) = 0 \qquad m_2(\{\neg D\}) = .9 \qquad m_2(\{D, \neg D\}) = .1.$$

Then m_{12} is obtained as follows:

		m_2			
		$\{\neg D\}$ .9		$\{D, \neg D\}$ .1	
m_1	$\{D\}$	.8 $\varnothing$	.72	$\{D\}$	.08
	$\{D, \neg D\}$	.2 $\{\neg D\}$	.18	$\{D, \neg D\}$	.02.

In this case, .72 of our belief is committed to the empty set. Since there are no possibilities in this set, the beliefs in our other sets must be normalized to 1. This normalization yields

$$m_2$$

		$\{\neg D\}$	.9	$\{D,\neg D\}$	.1
m_1	$\{D\}$	.8 $\emptyset$	0	$\{D\}$	.29
	$\{D,\neg D\}$	.2 $\{\neg D\}$	.64	$\{D,\neg D\}$	.07.

The new beliefs committed to D and $\neg D$ are obtained as follows:

$$m_{12}(\{D\}) = .29 \qquad\qquad m_{12}(\{\neg D\}) = .64$$

$$\text{Bel}_{12}(\{D\}) = m_{12}(\{D\}) = .29 \qquad \text{Bel}_{12}(\{\neg D\}) = m_{12}(\{D\}) = .64$$

$$\text{Plaus}_{12}(\{D\}) = 1 - .64 = .36 \qquad \text{Plaus}_{12}(\{\neg D\}) = 1 - .29 = .71.$$

The new belief interval for D is now $[.29, .36]$. Notice that not only has our belief in D gone down in light of the evidence against D, but also our uncommitted belief has gone down.

As originally developed, the Dempster–Shafer theory of evidence does not include an operational definition of a bpa. However, as we shall see in the next chapter, such a definition follows once we realize that the Dempster–Shafer theory is an extension of probability theory.

1.4.4. Fuzzy Set Theory and Fuzzy Logic

Fuzzy set theory and fuzzy logic [Yager, et al., 1987] address a fundamentally different class of problems then that addressed by either of the two previous methods or by probability theory. The other methods and probability theory deal with propositions which are definitely either true or false. We are simply uncertain as to whether they are true. For example, the bacterium in a given culture is either streptococcus or it is not. We are uncertain as to which is true, but exactly one is true. The real number we attach to the proposition "the organism is streptococcus" is a measure of how much we believe that it is a true proposition in this particular case. Fuzzy set theory, on the other hand, deals with propositions which have vague meaning. For example, a physician might supply the following rule:

IF   the growth is quite large

THEN there is a good chance the tumor is cancerous.

The rule itself is concerned with a proposition which is definitely either true or false; that is, the growth either is or is not cancerous. Thus when we replace the phrase "good chance" by a real number, say .8, we are dealing with uncertainty as considered by probability theory, the Dempster–Shafer theory, and certainty factors. However, although the proposition "the growth is quite large" is not precise, it is not an expression of uncertainty. We assume that we have measured the growth and know exactly how large it is. The question is how much this growth's particular size fits the physician's vague description of "quite large."

Zadeh [Yager, et al., 1987] addresses such problems with his fuzzy set theory. Fuzzy set theory associates a real number between 0 and 1 with the membership of a particular element in a set. For example, let $\mathcal{X}$ be the set of all integers which represent the concept "several." Are 3, 4, and 5 definitely in $\mathcal{X}$ and all other numbers definitely not in $\mathcal{X}$? According to fuzzy set theory the answer is no. Rather the set memberships might be represented as follows:

$$.1 \mid 1 \quad .3 \mid 2 \quad .8 \mid 3 \quad .9 \mid 4 \quad .6 \mid 5 \quad .4 \mid 6 \quad .3 \mid 7 \quad .1 \mid 8,$$

where the integer is on the right and its fuzzy set membership in $\mathcal{X}$ is on the left. This number is not a representation of uncertainty as considered by the other methods. We can never find out for certain whether 5 is in $\mathcal{X}$. Rather it is always partially in $\mathcal{X}$. The same is true for the tumor which we have accurately measured. The measurement obtained has a definite fuzzy set membership in the set of tumor sizes which are "quite large."

Some may argue that fuzzy sets can be given a probabilistic interpretation. For example, if we asked 10,000 people if 5 represented "several," the membership of 5 in $\mathcal{X}$ would be the relative frequency with which they answered yes. The argument in fuzzy set theory is that they could not answer definitely yes or no. The following example illustrates these concepts more concretely.

Suppose we had a kennel in which there were 100 dogs, of which exactly 50 were full-blooded beagles. If we picked a dog at random from the kennel, the probability of that dog being a full-blooded beagle would be .5. The dog either is or is not a full-blooded beagle. Once we picked the dog and inspected it, we would know for certain. On the other hand, if one parent of a particular dog was a full-blooded beagle and the other was a full-blooded collie, then that dog's fuzzy set membership in the set of beagles would be .5. There is no uncertainty which we can eventually resolve. We know for certain that its fuzzy set membership is .5. Thus fuzzy set theory addresses a fundamentally different class of problems than probability theory, certainty factors, and the Dempster–Shafer theory of evidence.

1.5. THE ORGANIZATION OF THIS TEXT

In this text we are concerned with propositions which are definitely either true or false and with the association of real numbers with the propositions to represent our relative uncertainty in their verity. For example, when we analyze a culture, either the streptococcus organism is or is not present. The real number represents our uncertainty as to its presence. When a patient enters a clinic, he either does or does not have a cancerous tumor. The real number represents our uncertainty. When we explore a site containing mineral deposits, a specific mineral of interest either is or is not present. The real number again represents our uncertainty as to its presence. In general, there are more alternatives of interest than just a proposition and its negation. For example, an organism may be one of many possible varieties, but only one. Given an uncertain situation in which we know that exactly one of a set of propositions is true, we are concerned with the determination of a method for assigning real numbers to represent our uncertainty in these propositions, and of a calculus for manipulating these real numbers (and therefore change our uncertainty) as evidence is accumulated. In chapter 2, we review the foundations of probability theory and show that the probability calculus is appropriate for our aims. Chapter 2 also contains a discussion of the appropriateness of the disciplines introduced in this chapter.

In chapter 3, we discuss the graph theory which is necessary to the rigorous probability-based methods (i.e., those which are based on causal networks). These methods are discussed in chapters 5 through 10. In chapter 4, we show why the rule-based approach and the inference engines described in this chapter are, in many cases, inadequate for handling uncertain reasoning. Furthermore, in chapter 4, we introduce causal networks and show that, in these cases, they are an appropriate framework for representing uncertain knowledge.

CHAPTER 2

PROBABILISTIC CONSIDERATIONS

This chapter is not an introduction to probability theory as found in the first few chapters of a standard text on probability theory. Although the basics of discrete probability theory are contained here, our main purpose is to show that probability theory is an appropriate discipline for representing and updating uncertainty in expert sytems. To accomplish this, we must look at the history and the philosophical foundations of probability theory. At times it may seem that we are digressing somewhat from our main purpose; however, the position is taken here that it is not possible to determine the suitability of applying probability theory in a particular instance without an understanding of the foundations of probability theory. Of course, our treatment must be sketchy. For a complete introduction to the philosophical foundations of probability theory, the interested reader is referred to Fine [1973] or Kyburg [1970]. A basic introduction to probability theory can be found in Feller [1968], while a more mathematically sophisticated introduction is in Ash [1970]. A philosophically and mathematically rigorous discussion of random sequences and the limiting frequency approach to probability can be found in van Lambalgen [1987].

In the dictionary, the word "probability" is defined in terms of the word "likelihood," while the word "likelihood" is defined in terms of the word "probability." If George told his 3-year-old son that it will probably rain today, and the child asked what he meant by "probably," George may respond that he meant it is likely that it will rain. If the child asked what he meant by "likely," hopefully George would not respond that he meant that it will probably rain. Rather he would offer several explanations, possibly including "the chances are good," "there is only a slight possibility that it will not rain," etc. Thus the child would eventually come to an understanding of the word "probable." In a similar way, we have all come to such an understanding. If, on the other hand, George told his son that there is a .87 probability of it raining today, George would have a far more difficult time explaining what he meant to the child or to anyone else. Indeed the meaning of this statement is a difficult philosophical question, relating to the development of a theory of probability itself. Lewis [1962] has said that he could not explain probability to anyone who did not already possess a primordial sense of probability.

As discussed in chapter 1, in this text we are concerned with the assignment of such numerical likelihoods or measures of uncertainty to statements, and with the manipulation of these likelihoods to obtain the likelihoods of other statements. The question is not only what these numbers mean, but also where they are obtained and what the rules are for manipulating them. Mathematical probability theory contains specific rules or axioms for manipulating numerical probabilities. These rules constitute the calculus of probability theory. We must determine whether they are a valid method for manipulating the numerical likelihoods which are useful in expert system applications. To determine this, we first review several of the major theories of probability in this chapter. We discuss the classical approach to probability in the first section, the limiting frequency approaches in the second section, and the subjectivistic approach in the third section. In section 2.4 we show that the nonextreme frequentist is not much different from the nonextreme subjectivist. After this review, in section 2.5 we discuss the appropriateness of the use of the probability calculus for manipulating numerical likelihoods in expert systems. The last two sections contain some final notions on probability. Namely, section 2.6 discusses an extension of the classical approach called the maximum entropy formalism, while section 2.7 briefly investigates whether there are "absolute" probabilities.

2.1. THE CLASSICAL APPROACH TO PROBABILITY

The classical approach is a term used to denote most efforts to deal with probability up to and including the time of Laplace (1749–1827). Laplace and his contemporaries did not consider their efforts an "approach"; it is a term applied retrospectively. This approach remains the most common way of dealing with probability. At its heart we find Laplace's classical definition of probability [Laplace, 1951]:

The (Classical) Definition of Probability

The theory of chance consists in reducing all the events of some kind to a certain number of cases equally possible, that is to say, such as we may be equally undecided about in regard to their existence, and in determining the number of cases favorable to the event whose probability is sought. The ratio of this number to that of all the cases possible is the measure of the probability.

Probabilities are assigned within the framework of the classical approach by using the *principle of indifference* (a term first used by Keynes [1948]). The fundamental idea in this principle is that "alternatives are always to be judged equiprobable if we have no reason to expect or prefer one over the other" [Weatherford, 1982].

Example 2.1. Suppose we have a standard poker deck of cards and are interested in the probability that the top card is a spade. There are 52 total cards, each of which we are equally undecided as to its being the top card. Furthermore, there are 13 cards which are spades. According to the principle of indifference, the probability that the top card is a spade is therefore 13 divided by 52, or .25.

We are all used to assigning such probabilities. If George told his son that the probability of the top card being a spade is .25, he would have an easy time explaining how he obtained the number. Although the 3-year-old may not understand, anyone with a minimal background in arithmetic would. Of much greater importance is the significance of the number. Even if George's 10-year-old daughter did understand the arithmetic, her response might be "so what?" That is, what should she do with this number? George might respond that the chances are 1 out of 4 that a spade will be picked; therefore we should bet 3 dollars to 1 dollar that a spade will not turn up. If she asked why we should bet like this, he might explain that since the chances are 1 out of 4 that a spade will turn up, if we repeat this situation many times the spade will turn up very close to one-quarter of the time. Therefore, if we always bet 3 to 1 against the spade turning up, and we repeat the situation x times, we will approximately win $(1)(3/4)x$ dollars and lose $(3)(1/4)x$ dollars and thus break even. Such explanations are often at the heart of attaching significance to probability values. We'll return to these considerations later. We now wish to derive the axioms of probability theory from Laplace's definition. First we formalize the matter a bit with some definitions.

Definition 2.1. Let an experiment which has a set of mutually exclusive and exhaustive outcomes (alternatives) be given. That set of outcomes is called the sample space and is denoted by Ω.

We will formally define "mutually exclusive and exhaustive" for sets later. Here it simply means that at least one of the outcomes must occur and it is not possible for two of them to occur. In the example above, the experiment is the picking of the top card from a deck, and the sample space Ω is the set of the 52 different outcomes. The sample space does not necessarily have to be a set of undecomposable alternatives. If we are only interested in the suit of the top card, the sample space could be the set of four possible suits.

Definition 2.2. Let $\mathcal{F}$ be a set of subsets of Ω such that

1. $\Omega \in \mathcal{F}$
2. E_1 and $E_2 \in \mathcal{F}$ implies $E_1 \cup E_2 \in \mathcal{F}$
3. $E \in \mathcal{F}$ implies $\overline{E} \in \mathcal{F}$

Then $\mathcal{F}$ is called a set of events relative to Ω (in mathematics such an $\mathcal{F}$ is called an algebra).

Since we have been mathematically rigorous in the definition of a set of events, the above definition is more complicated than the actual concept of an event. In the case of picking the top card from a deck, the usual set of events $\mathcal{F}$ is the set of all subsets of Ω. For example, if

$E_1 = \{$two of hearts picked, three of hearts picked, five of spades picked$\}$,

then $E_1 \in \mathcal{F}$. An event E is simply a set of possible outcomes (sample points) of the experiment. The event E actually means that one of the outcomes in E occurs. Thus there is an equivalence between specifying an event by a proposition and specifying it by a set of outcomes. For example, if E_1 is the proposition "a spade turns up," then E_1 contains the 13 possible outcomes which produce a spade. If E_2 is the proposition "a king turns up," then E_2 contains the four possible outcomes which produce a king. The definition does not require that we always consider all the subsets of Ω, but it does require that if we want both E_1 and E_2 to be in the set of events $\mathcal{F}$, then $\overline{E}_1$, $\overline{E}_2$, and $E_1 \cup E_2$ must also be in $\mathcal{F}$. In the present context, $E_1 \cup E_2$ contains 16 possible outcomes (since the king of spades is in both sets) and is the event that either a spade or a king turns up. Since the complement of a set in $\mathcal{F}$ must be in $\mathcal{F}$, it is easy to show that $E_1 \cap E_2$ must also be in $\mathcal{F}$. In the present context, this event contains one outcome (the king of spades) and is the event that both a spade and a king turn up.

At this point we adopt a nonstandard convention. In the coming chapters of this text we will use the union symbol $\cup$, the intersection symbol $\cap$, and the symbol for the empty set $\varnothing$ in a different context. We have already seen that an event can be specified in a proposition. Therefore, since $E_1 \cup E_2$ actually means that either E_1 or E_2 occurs and $E_1 \cap E_2$ means that both E_1 and E_2 occur, we will replace the symbol $\cup$ by the logic symbol for disjunction $\vee$, and the symbol $\cap$ by the logic symbol for conjunction $\wedge$. The set which contains no sample points will be denoted by NULL. Thus

$E_1 \vee E_2$ is the event containing the outcomes in either E_1 or E_2

$E_1 \wedge E_2$ is the event containing the outcomes in both E_1 and E_2

NULL is the event containing no sample points.

Some [Cox, 1979] have developed probability theory entirely within the context of propositions representing events instead of sets. Mathematicians prefer the set theoretic development, as presented here, because it leads to a more straightforward development of mathematical probability theory.

Next we formally define the probability of an event within the framework of the classical approach:

Definition 2.3. Classical Definition of Probability. For each event $E \in \mathcal{F}$, there corresponds a real number $P(E)$, called the probability of E. This number is obtained by dividing the number of equipossible alternatives favorable to E by the total number of equipossible alternatives.

As we shall see in the following sections, the other approaches to probability theory have different *operational definitions* of the probability of an event.

We can now prove some properties of probabilities based on the above definition. These properties are the axioms of probability theory and the definition of conditional probability. That is, in a probability text, they, not the classical definition of probability, are taken as given. Again, our goal is to determine the appropriateness of these axioms and the definition of conditional probability to applications in expert systems. Thus we cannot take them as given.

Theorem 2.1. Let Ω be a finite set of sample points, $\mathcal{F}$ a set of events relative to Ω, and, for each $E \in \mathcal{F}$, $P(E)$ the probability of E according to the classical definition of probability. Then

1. $P(E) \geq 0$ for $E \in \mathcal{F}$
2. $P(\Omega) = 1$
3. If E_1 and E_2 are disjoint subsets of $\mathcal{F}$, then $P(E_1 \vee E_2) = P(E_1) + P(E_2)$.

Proof. Let n be the number of equipossible outcomes in Ω.

1. If m is the number of equipossible outcomes in E, then, according to the classical definition,

$$P(E) = \frac{m}{n} \geq 0.$$

2. According to the classical definition,

$$P(\Omega) = \frac{n}{n} = 1.$$

3. Let E_1 and E_2 be disjoint events, let m be the number of equipossible outcomes in E_1, and let k be the number of equipossible outcomes in E_2. Then, since E_1 and E_2 are disjoint, $k + m$ is the number of equipossible outcomes in $E_1 \vee E_2$. Thus, according to the classical definition,

$$P(E_1) + P(E_2) = \frac{m}{n} + \frac{k}{n} = \frac{m+k}{n} = P(E_1 + E_2). \quad \square$$

Example 2.2. Let the experiment again be the drawing of the top card, E_1 the event that a heart is drawn, and E_2 the event that a spade is drawn. Then $P(E_1) = .25$ and $P(E_2) = .25$, and, by Theorem 2.1,

$$P(E_1 \vee E_2) = .25 + .25 = .5.$$

Example 2.3. Result 3 in Theorem 2.1 only holds if the events are mutually exclusive. For example, if E_1 is the event of drawing a spade and E_2 is the event of drawing a king, then $P(E_1 \vee E_2)$ is equal to 16/52, not 17/52. The king of spades would be counted twice if we mistakenly applied Theorem 2.1.

The proof of Theorem 2.1 is trivial. Yet, excluding for the moment the concept of conditional probability, most of the rich mathematics in probability theory is based on the three results in that theorem. These results are the axioms of probability theory. Before proceeding, we formally define a probability space (this definition was originally due to Kolmogorov [1950]):

Definition 2.4. Let Ω be a set of sample points, $\mathcal{F}$ a set of events relative to Ω, and P a function which assigns a unique real number to each $E \in \mathcal{F}$. Suppose P satisfies the following axioms:

1. $P(E) \geq 0$ for $E \in \mathcal{F}$
2. $P(\Omega) = 1$
3. If E_1 and E_2 are disjoint subsets of $\mathcal{F}$ then $P(E_1 \vee E_2) = P(E_1) + P(E_2)$.

Then the triple $(\Omega, \mathcal{F}, P)$ is called a probability space and P is called a probability measure on $\mathcal{F}$.

In many probability texts, condition 3 in the above definition is replaced by infinitely countable additivity in order to develop a richer mathematical theory. However, this assumption has not been convincingly justified either physically or philosophically. Since we are interested here in the application of probability theory, we will only consider finite additivity.

As mentioned previously, this definition is simply stated at the beginning of a probability text. However, when we apply probability theory, we must determine whether the definition is justified. Due to Theorem 2.1, it appears that it is justified in problems involving simple combinatorics when we apply the principle of indifference. If one's only experience with probability theory is with problems involving combinatorics, it may seem that this is all there is. However, we shall see shortly that the situation is far more complicated. First, however, we define the concept of conditional probability.

Suppose that immediately before the top card is drawn from the deck, George's sneaky friend, Clyde, peaks at the top card. However, he peaks so quickly that although he is able to discern for certain that it is a king, he obtains no clue as to the suit. When the top card is drawn, would he compute the probability of the king of spades to be 1/52? Hopefully, he would not. For Clyde, there are only four equipossible alternatives and therefore for him the probability of the top card being the king of spades is 1/4, while for George, who is not privy to Clyde's information, it is still 1/52. Since we have two probabilities for the same event, we may ask, which is the real probability of choosing the king of spades? The question of real or *objective* probabilities has concerned philosophers from the start. Laplace believed in a deterministic universe. He felt that if a demon knew the position and momentum of every particle in the universe, then he could predict all future events. Thus there are no real probabilities in the sense that there are no absolute probabilities. Probability only exists relative to partial information or knowledge. If one had sufficient information about an experiment, all probabilities would be 0 or 1. For example, if Clyde had a longer peek and had seen that the top card was indeed the king of spades, the probability of this event would be 1. In the light of the theory of quantum mechanics, some have abandoned the notion of determinism. We will return to this issue briefly in section 2.7. The point here is that probabilities exist only relative to information. When a probability space is created, it must be created relative to some information. For example, in the experiment of drawing the top card from the deck, if the only information were that it was one of the 52 possible cards, we would create the probability space in which the probability of a king of spades was 1/52.

We see that all probabilities are conditional on information. When a probability space is created, we call the probabilities which are conditional on the initial information *a priori* or *prior* probabilities (here we are not using the literal meaning of "a priori"; i.e., to be independent of experience. The literal meaning is used in many of the philosophical discussions of probability). Probabilities based on additional information are called conditional probabilities. For example, when George creates his probability space based on considering all 52 cards equipossible, the a priori probability of selecting the king of spades is 1/52. The conditional probability of drawing the king of spades given that the top card is a king is then 1/4. Of course, a conditional probability in one space is an a priori probability in a space based on additional information. We have the following theorem based on the classical definition of probability:

Theorem 2.2. Let $(\Omega, \mathcal{F}, P)$ be a probability space created according to the classical definition of probability. Suppose $E_1 \in \mathcal{F}$ is nonempty and therefore has a positive probability. Then, if we *assume* that the alternatives in E_1 remain equipossible when it is known for certain that E_1 has occurred, the probability of E_2 given that E_1 has occurred is equal to

$$\frac{P(E_1 \wedge E_2)}{P(E_1)}.$$

Proof. Let n, m, and k be the number of sample points in Ω, E_1, and $E_1 \wedge E_2$, respectively. Then the number of equipossible alternatives based on the information that E_1 has occurred is equal to m (we have assumed that they have remained equipossible), while the number of these alternatives which are favorable to E_2 is equal to k. Therefore the probability of E_2 given that E_1 has occurred is equal to

$$\frac{k}{m} = \frac{k/n}{m/n} = \frac{P(E_1 \wedge E_2)}{P(E_1)}. \quad \square$$

Notice in the above theorem that it is necessary to assume that the alternatives in E_1 remain equipossible when it is known that E_1 has occurred. The claim that these alternatives remain equipossible does not follow from the claim that all the alternatives are originally equipossible; therefore the former claim must be an additional assumption. Based on this theorem, we have the following definition for an arbitrary probability space:

Definition 2.5. Let $(\Omega, \mathcal{F}, P)$ be a probability space and $E_1 \in \mathcal{F}$ such that $P(E_1) > 0$. Then for $E_2 \in \mathcal{F}$, the conditional probability of E_2 given E_1, which is denoted by $P(E_2 \mid E_1)$, is defined as follows:

$$P(E_2 \mid E_1) = \frac{P(E_1 \wedge E_2)}{P(E_1)}.$$

Like the definition of a probability space, the definition of conditional probability is given near the beginning of any probability text. When probability theory is applied to problems other than those involving simple combinatorics, this definition must also be justified. Before discussing these other problems and the philosophical justifications for these definitions, we obtain some important results from Definition 2.5. First, it leads immediately to another definition, namely one involving the notion of independence.

Definition 2.6. Let $(\Omega, \mathcal{F}, P)$ be a probability space and E_1 and E_2 events in $\mathcal{F}$ such that one of the following is true:

1. $P(E_1) = 0$ or $P(E_2) = 0$
2. $P(E_2 \mid E_1) = P(E_2)$.

Then E_2 is said to be independent of E_1.

The meaning of independence is clear. E_2 is independent of E_1 if knowledge that E_2 has occurred does not change the probability of E_1. It is a simple matter to show that E_1 is independent of E_2 if E_2 is independent of E_1. Thus we only say that E_1 and E_2 are independent. The following theorem holds for independent events:

Theorem 2.3. Let $(\Omega, \mathcal{F}, P)$ be a probability space and E_1 and E_2 be arbitrary events in $\mathcal{F}$. Then E_1 and E_2 are independent if and only if

$$P(E_1 \wedge E_2) = P(E_1)P(E_2).$$

Proof. The proof follows easily from the definition of independence and is left as an exercise. $\square$

Example 2.4. In the case of picking the top card from a poker deck, the probability of picking a king is 1/13 and the probablity of picking a king given that the top card is a spade is also 1/13. Thus the event of picking a king and the event of picking a spade are independent.

Definition 2.5 leads to a very powerful theorem, called Bayes' theorem in honor of its originator, Thomas Bayes (1702–1761). In order to prove that theorem we need the following definition and lemma:

Definition 2.7. Let $(\Omega, \mathcal{F}, P)$ be a probability space and $\{E_1, E_2, \ldots, E_n\}$ be a set of events such that for $i \neq j$,

$$E_i \wedge E_j = \text{NULL} \qquad \text{and} \qquad \bigvee_{i=1}^{n} E_i = \Omega.$$

Then the events in $\{E_1, E_2, \ldots, E_n\}$ are said to be mutually exclusive and exhaustive.

Lemma 2.1. Let $(\Omega, \mathcal{F}, P)$ be a probability space and $\{E_1, E_2, \ldots, E_n\}$ be a set of mutually exclusive and exhaustive events in $\mathcal{F}$ such that for $1 \leq i \leq n$, $P(E_i) > 0$. Then, for any $E \in \mathcal{F}$,

$$P(E) = \sum_{i=1}^{n} P(E \mid E_i)P(E_i).$$

Proof. Since the E_i's are exhaustive, we have that

$$E = (E \wedge E_1) \vee (E \wedge E_2) \vee \cdots \vee (E \wedge E_n).$$

Therefore, since the E_i's are mutually exclusive, by Definition 2.4 we have that

$$P(E) = P(E \wedge E_1) + P(E \wedge E_2) + \cdots + P(E \wedge E_n).$$

The proof now follows immediately from Definition 2.5. $\square$

Theorem 2.4. *Bayes' Theorem.* Let $(\Omega, \mathcal{F}, P)$ be a probability space and $\{E_1, E_2, \ldots, E_n\}$ a set of mutually exclusive and exhaustive events in $\mathcal{F}$ such that for $1 \leq i \leq n$, $P(E_i) > 0$. Then for any $E \in \mathcal{F}$ such that $P(E) > 0$, we have that for $1 \leq j \leq n$

$$P(E_j \mid E) = \frac{P(E \mid E_j)P(E_j)}{\sum_{i=1}^{n} P(E \mid E_i)P(E_i)}.$$

Proof. The proof follows easily from Definition 2.5 amd Lemma 2.1 and is left as an exercise. $\square$

If E and E' are any two events such that $P(E)$ and $P(E')$ are both positive, then the following equality follows directly from Definition 2.5:

$$P(E \mid E') = \frac{P(E' \mid E)P(E)}{P(E')}.$$

This equality is sometimes called Bayes' theorem in the literature. In this text, when we apply this equality, we will simply say that we are applying the definition of conditional probability.

Having developed the important theorems in the application of probability theory to expert systems, we now address the justification for the definitions of a probability space and conditional probability. First we return to the principle of indifference and investigate whether it should be used as the sole justification for these definitions. Even though we commonly assign probabilities using the principle of indifference, it has become perhaps the most criticized concept in all of probability theory. To show that this principle alone cannot justify the axioms of probability theory, we must review these criticisms and their significance.

One of the most common criticisms is that the application of this principle can lead to contradictory or paradoxical results. Consider the following example:

Example 2.5. Suppose that we have a mixture of wine and water and all that is known is that there is at most three times as much of one as the other. This imples that

$$1/3 \leq \frac{\text{wine}}{\text{water}} \leq 3.$$

Applying the principle of indifference to the ratio wine/water, we conclude that this ratio is uniformly distributed in the interval $[1/3, 3]$, and therefore that

$$P\left(\frac{\text{wine}}{\text{water}} \leq 2\right) = \int_{1/3}^{3} \frac{3}{8} \, dx = \frac{5}{8}.$$

However, the information also implies that

$$1/3 \leq \frac{\text{water}}{\text{wine}} \leq 3,$$

and by applying the principle of indifference to the ratio water/wine, we obtain

$$P\left(\frac{\text{water}}{\text{wine}} \geq 1/2\right) = \int_{1/2}^{3} \frac{3}{8}\,dx = \frac{15}{16}.$$

Thus we obtain two different probabilities for the same event. We could obtain yet a third probability for that event by assuming that the fraction of water is uniformly distributed in the interval $[1/4, 3/4]$. We conclude then that there is a contradiction in this application of the principle of indifference.

In order to analyze this paradox properly, we must distinguish between two possible ways of interpreting the principle of indifference. The first interpretation is that there is an experiment which uses some physical process to randomly obtain exactly one of the alternatives in the sample space, and that the physical process treats all of the alternatives equitably. That is, the process is both random and uniform. Although the concepts "uniform" and "random" are used freely, they are quite difficult to define rigorously. We will defer any attempt to define a random process rigorously until the next section. For now, we simply offer the intuitive notion that the experiment is repeatable and that separate trials of the experiment are in some sense independent. For example, it is commonly believed that processes such as coin tossing are random. As already mentioned, "uniform" means that all the alternatives are treated equitably, which thus should imply that they are equipossible. As many have noted, such a definition is circular. There is no way to avoid this circularity. However, if we impose an apparent uniform process as is done in the first interpretation of the principle of indifference, we feel that the alternatives are equipossible. For example, if we shuffle a deck of cards and then pick the top card, each card is treated equitably. Therefore we assign a probability value of $1/52$ to each card. Having assigned probabilities in this manner, we would expect that, if we repeated the process many times, each card should appear approximately $1/52$ of the time. This interpretation of the principle of indifference means that we are using the principle to obtain probability values whose existence are justified by a frequency interpretation of probability (to be discussed in the next section). The above paradox does not contradict this interpretation. For example, if we set up 100 mixtures of water and wine in which the ratio of wine to water was uniformly distributed in the interval $[1/3, 3]$ (i.e., the ratio of wine to water was $1/3$ in the first mixture, the ratio was $1/3 + (1/99)(8/3)$ in the second, and so on), we would only apply the principle of indifference to conclude that the ratio of wine to water was uniformly distributed in the interval $[1/3, 3]$ and there would be no contradiction. If we used some other physical process to set up the alternatives, we'd apply a different application of the principle of indifference. With this interpretation of the principle of indifference, we would not try to apply the principle to the information as it is stated in Example 2.5. That is, the problem is not well-formulated.

Another interpretation of the principle of indifference is that we should apply it whenever the information gives us no reason to choose one alternative over the other. This interpretation assumes that it is not necessary that there is a physical process which treats all of the alternatives equitably; rather it is only necessary that there is nothing in the information to imply that they are not treated equitably. One of the proponents of this interpretation is Jaynes [1979]. According to Jaynes, assigning probability values equitably in such a case is the least presumptive assignment of probabilities. A common criticism of this method of assigning probability values is the argument that probabilities are assigned from ignorance rather than from knowledge. In section 2.6, we will show how Jaynes addresses this criticism. We will now illustrate this interpretation of the principle of indifference with some examples.

Suppose Clyde walked into the room with a deck of cards, did not shuffle the deck, and picked the top card. Using the second interpretation of the principle of indifference, we'd still compute each card to have a probability of 1/52. This may seem reasonable; however, the situation is quite different than when the cards were shuffled. There is no uniform process which we can repeat. We could have Clyde exit and reappear with the cards. However, Clyde might always place the ace of spades on top of the deck while he is out of the room. Therefore, each alternative will not come up 1/52 of the time in repeated trials. We could argue that if we repeated the act of a person walking into the room with a deck of cards many times, each card should come up about 1/52 of the time; or, at a more abstract level, we could argue that if we repeated the information many times that there are precisely 52 alternatives, and each time there was no reason to choose one alternative over the other, and if we chose one alternative each time, then our chosen alternative would occur approximately 1/52 of the time. This latter argument is more easily illustrated with experiments which have precisely two alternatives. Suppose two people were to race, and we knew nothing as to their relative speeds. Using the second interpretation of the principle of indifference, we would assign 1/2 to the probability of each winning. Yet it could be that one individual is much faster and is almost sure to win. Proponents of the second interpretation would argue that this is not a problem; the information that one individual is faster does not have a probability value of 1/2 associated with it. Rather only the information that there are precisely two alternatives determines a probability value of 1/2. Now suppose that we repeat a situation in which the only information is that there are precisely two alternatives. That is, we watch another two-person race in which we know nothing of the relative speeds of the participants, or we toss a coin whose composition is completely unknown. We could argue that if we repeated such situations many times, then the relative frequency of an arbitrarily chosen alternative should approach 1/2. This frequency argument for the second interpretation of the principle of indifference is badly in need of justification. That is, there is little experimental evidence for the approximate convergence of the relative frequencies to the probability values. On the other hand, there is evidence for the approximate convergence of the relative frequencies in the case where a uniform process generates one of the alternatives (we will see, however, in Example 2.7 that sometimes the evidence is to the contrary). For example, coins have been repeatedly tossed and the relative frequency of heads has approached 1/2. We stress, however, that proponents of the second interpretation would not agree with the necessity of experimental evidence in the form of relative frequencies. They argue that the assigned probability values have to do with a single trial, not with repeated trials. In section 2.6, we will show that in some cases there is evidence, other than relative frequencies, for the existence of objective probability values obtained using the second interpretation of the principle of indifference.

The point here is that there are two distinct interpretations of the principle of indifference, and that the above paradox can possibly be viewed as a contradiction in this second interpretation, but it is definitely not a contradiction in the first interpretation. Defenders of the second interpretation could argue that it is not even a contradiction in that interpretation. That is, they could maintain that the fact that there exists more than one set of cases to which we can apply the principle of indifference implies that we cannot really be equally undecided about the alternatives in any of the cases, and that therefore the principle of indifference is not applicable to the problem. In other words, they could argue that the problem is not well-formulated for their interpretation either.

We consider one more paradox which might indicate that there is even a contradiction in the first interpretation of the principle of indifference, and show that this is not the case. This paradox is concerned with expected utility.

Example 2.6. Suppose that there are two envelopes containing money, and we are told only that one envelope contains twice as much money as the other. We are handed one of the envelopes and are given the opportunity to exchange it for the other. In order to determine whether this would be wise, we reason that if x is the sum in the envelope we possess, then it is equally likely that the other envelope contains $.5x$ and $2x$ dollars. Therefore the expected value of the sum in the other envelope is

$$.5(.5x) + .5(2x) = 1.25x,$$

and thus we would agree to exchange envelopes. However, based on this line of reasoning, we would have to agree to exchange the envelopes another time, and indeed to go on exchanging indefinitely.

The paradox, which results in this endless exchanging of envelopes, is resolved by precisely noting the information in the experiment and the sample points in Ω. The information is that one envelope holds an arbitrary sum of money and the other holds twice that sum, and we are given one of the envelopes. The sample points in Ω are the propositions "the sum in the other envelope is one-half the sum in our envelope" and "the sum in the other envelope is twice the sum in our envelope." These propositions are each assigned a probability of .5. Next a random variable, whose value is the amount in the other envelope, is created on Ω ("random variable" is defined in Appendix

2.A). The values assigned to random variables must be real numbers. The paradox is achieved by assigning functions of a variable (namely $2x$ and $.5x$ where x is the sum in our envelope), whose value depends on the outcome of the experiment, as values of the random variable. That is, each time we execute this experiment, there can be a different sum in our envelope and therefore x can have a different value. The problem can be solved in a different way. If we let x' be a fixed sum of money and x'' be twice that sum, we can consider the experiment where one of the envelopes contains x' dollars and the other contains x''. Based on this information, we can assign the probability value of .5 to the two propositions as before, and assign the random variable value of x' to the proposition "the sum in the other envelope is one-half the sum in our envelope" and the value x'' to the proposition "the sum in the other envelope is twice the sum in our envelope." Therefore, since $x'' = 2x'$, the expected value of the sum in the other envelope is equal to

$$.5x' + .5(2x') = 1.5x',$$

the same as the expected value of the sum in our hand. Since this is true for any value of x', we would not expect to gain anything by exchanging, regardless of the sum in each envelope. By using a variable as the value of a random variable, the paradox was achieved by treating x' and x'' as the same value.

There is apparently another paradox here. Suppose we open the envelope and find, for example, $100 in it. At that point we are equally undecided as to whether the other envelope contains $50 or $200. Based on the principle of indifference, we then compute the expected value of the sum in the other envelope to be equal to

$$.5(\$50) + .5(\$200) = \$125.$$

If we are still given the option of exchanging envelopes, it seems that we should do so. However, using the same line of reasoning, the person holding the other envelope would also wish to exchange regardless of the amount he found in his envelope. Thus both persons would always deem it wise to exchange envelopes, an apparent antinomy.

There is no contradiction in the first interpretation of the principle of indifference because the problem, as stated, is not well-formulated for that interpretation. That is, we are not given a uniform random process for generating the condition that one amount is twice or one-half of the other amount.

We could make the problem well-formulated by specifying a process as follows: Use a uniform random process to determine an amount of money. Place that amount of money in our envelope, toss a fair coin, and place twice or one-half that amount in the other envelope depending on whether the outcome of the toss is heads or tails. In this case, if we find $100 in our envelope, the expected value of the amount of money in the other envelope is given by

$$P(\text{heads}) \times \$200 + P(\text{tails}) \times \$50 = .5(\$50) + .5(\$200) = \$125.$$

Notice that the expected value is computed by applying the principle of indifference to the outcome of the coin toss. It will yield 1.25 times the amount of money in our envelope regardless of the process used to generate the money in our envelope, and therefore we would always be wise to exchange envelopes.

On the other hand, consider the individual holding the other envelope. Suppose he finds $100 in it. Should he exchange envelopes? We will analyze his decision by first considering the case where the amount of money in our envelope is uniformly generated in some interval, say [0, 1000]. Since the smaller sum is equally likely to be any value between $0 and $1000, the larger sum is equally likely to be any value between $0 and $2000, and therefore the expected value of the larger sum is equal to

$$\int_0^{2000} \frac{x}{2000} \, dx = 1000.$$

Therefore he would only be wise to exchange sums smaller than this expected value. Notice that his decision is based on applying the princple of indifference to the amount of money generated, while our decision, which is that we should always exchange envelopes, is based on applying that principle to the outcome of the coin toss. There is no paradox here. We would be wise to always exchange envelopes, while he would be wise to only exchange sums which are less than $1000.

We now return to the situation where the uniform random process generates an amount of money which is of arbitrarily large size. First, apparently there is no uniform random process which can create numbers of arbitrarily size. We can randomly generate numbers of arbitrarily large size in the following way: Construct a wheel with the digits 0 through 9 and the word STOP uniformly spaced on its perimeter, and generate digits by repeatedly spinning the wheel until STOP is hit. The numbers, however, will not be uniformly spaced in $[0, \infty)$. Rather, smaller numbers will come up more frequently. Although the above process randomly generates numbers of arbitrarily large size, it does not uniformly generate them. Another possibility would be to have someone arbitrarily pick numbers. However, since he would choose smaller numbers more than larger ones, the process is again not uniform. Even if we could somehow both uniformly and randomly generate numbers of arbitrarily large size, it is not possible to create a uniform probability distribution on the interval $[0, \infty)$. Therefore we cannot use the analysis in the previous paragraph to determine whether the other individual should exchange envelopes. If we created some nonuniform process which generated a probability distribution on $[0, \infty)$ and used that process to determine the amount of money in the smaller envelope, then we could compute the expected value of the amount in the larger envelope using the analysis in the previous paragraph, and we would conclude that the other individual should exchange envelopes if his sum is less than that expected value. If we used some nonrandom process to create numbers of arbitrarily large size uniformly (e.g., we could generate the sequence $[1, 2, 3, 4, \ldots]$), then again we could not use the analysis in the previous paragraph because we would have no probability distribution. It is left as an exercise to verify that in this case the other individual should lose in the long run if he always exchanges envelopes. Since the analysis as to whether we should exchange is based only on the coin toss, not on the way the money is generated, we would still be wise to always exchange envelopes.

An alternative process would be to determine an amount of money using a uniform random process, put that amount in one envelope, and put twice that amount in the other envelope. As a result of a fair coin toss, we receive one envelope or the other. In this case the problem is apparently well-formulated and is symmetrical. Yet it seems that both individuals would always want to exchange envelopes. We will analyze this problem by again first considering the case where the amount of money is equally likely to be any value in some interval, say $[0, 1000]$. As in the case of the process analyzed above, the expected value of the larger sum of money is then $1000, and therefore each individual would be wise to only exchange sums less than $1000. There is no paradox, since each individual would not always wish to exchange envelopes. Again, since there is no uniform probability distribution on $[0, \infty)$, this analysis does not apply to the case where we generate the amount uniformly in $[0, \infty)$. If we used some nonuniform probability distribution on $[0, \infty)$ to generate the money, then we could determine the expected value of the larger amount of money in order to decide when to exchange envelopes. It is left as an exercise to show that if we used some nonrandom process to generate sums in $[0, \infty)$ uniformly, then both individuals should approximately break even by always exchanging envelopes.

In practice, probabilistic decisions must be based on real-world constraints. For example, if the envelopes contained possible bonuses from a sporting boss, and if the employee knew that the boss was too cheap to ever give a bonus over $1000, then he should only exchange sums less than $500. If he had the additional constraint that he was flat broke, and he would lose his home if the $400 mortgage was not paid, then he should only exchange sums less than $400. We will consider probabilistic decision making more in chapter 9.

We see then that if we stay within the first interpretation of the principle of indifference, the paradoxes can be resolved. However, that interpretation cannot overcome another criticism of the principle of indifference: namely the claim that the probabilities obtained using this principle are based on logic alone [Nagel, 1939]. That is, if we expect the probability to approximate the relative frequency with which the event will occur in repeated trials of the experiment, the principle gives absolutely no proof that probabilities obtained by using it will agree with those expectations. Consider the following example:

Example 2.7. Suppose that we know that three indistinguishable balls are in four urns, but we have no information as to in which urn each is located. Clearly, there are 4^3 or 64 possible arrangements of the balls. Of these arrangements, only four have all three balls in one urn. Thus, by the principle of indifference, the probability of all three balls being in one urn is 4/64 or 1/16. This line of reasoning has been applied in physics by subdividing space into a large number of small cells and then computing the probability of indistinguishable particles occupying the cells in space. If a particle behaves in the manner indicated by the above application of the principle of indifference, the physicist says that the particle obeys Maxwell–Boltzmann statistics (this use of the word "statistic" is peculiar to physicists). The results of observing actual frequencies have shown that photons and nuclei do not obey this statistic. Rather, in this case, the relative frequency with which all three occupy the same cell is about equal to 4/20. This observed behavior is in agreement with an application of the principle of indifference to a different set of alternatives. That is, we treat only situations which our eyes can discern as equipossible. For example, if ball 1 and ball 2 are in urn 1 and ball 3 is in urn 2, our eyes could not distinguish this situation from ball 1 and ball 3 being in urn 1 and ball 2 being in urn 2 (remember the balls are indistinquishable). In this case there are only 20 equipossible alternatives, four of which are favorable to all three balls being in one urn. Particles which behave in this way are said to obey Bose–Einstein statistics.

If we repeated the experiment of randomly dropping three indistinquishable balls into four urns many times, we would expect them to distribute according to Maxwell–Boltzmann statistics. Similarly, based on pure reason, we would expect photons to behave in a similar way. However, they do not. Furthermore, electrons, neutrons, and protons obey yet another statistic, called Fermi–Dirac statistics. It is obtained by applying the principle of indifference to yet another set of alternatives. Thus, although these probability spaces can be obtained using the principle of indifference, there is no way of telling by pure reason alone which distribution agrees with observed frequencies.

In the above example, even though experimental evidence was needed to determine which probability distribution agrees with observed frequencies, the distributions could still be obtained using the principle of indifference. The final reason that we cannot use the principle of indifference as the only method of creating probability distributions (and thereby use the classical approach as the sole philosophical foundation of probability theory) is that in many applications, probabilities cannot be obtained from any application of that principle. Consider this example:

Example 2.8. The use of probability and statistics has proved to be very profitable to insurance companies in actuarial applications. For example, they would decide the insurance rate of a 30-year-old man based partly on the probability of a man entering his 31st year dying in the coming year. To obtain that probability, they would observe perhaps 10,000 men, who are in the same insurance class as the man under consideration, during their 31st year. (The insurance class is the information on which the probability space is based. An example of such a class is the class of all white-collar workers in the United

States during the 1980s.). If 20 of the observed men died in their 31st year, they would compute the probability of a new 30-year-old prospect dying in the coming year to be approximately .002.

In this example, there are not a set of equipossible alternatives from which the new prospect is selected. Some claim that the principle of indifference is still applicable because there are 10,000 men observed, and it is equally likely that any one of those men represents the new prospect. However, this line of reasoning would extend to our determining the probability of the prospect dying in the coming year to be one if we had observed only one man and he had died. Others maintain that if we took all the men who have ever lived or who ever will live, and who are in the same class as the man under consideration, and if we divide the number of such men who die during their 31st year by the number of all such men, then we would have the probability of a man, picked at random, dying in his 31st year. The probability obtained from the 10,000 men is then a statistical estimate of this true probability. Such an interpretation is actually a frequency approach to probability and will be discussed in the next section.

The following very simple example even better illustrates a probability value which cannot be obtained using the principle of indifference:

Example 2.9. Suppose we construct a coin such that one side is heavier (or we simply take a thumbtack). We know the coin to be unfair, but do not know the bias. If we tossed the coin 10,000 times and observed 6004 heads, we would suspect that there is a probability associated with heads turning up and that its value is around .6.

In this example, we may never toss the coin again. Thus we cannot consider the 10,000 tosses a sample from a larger set of alternatives. Furthermore, even if the coin were never tossed, many still feel that there is a probability value associated with the coin's composition.

If the above examples do not sufficiently demonstrate the inadequacy of the principle of indifference, a final example taken from physics should:

Example 2.10. Diffraction. Suppose we have a diaphragm with two very small slits and a screen behind the diaphragm. Attached to the screen is an apparatus capable of counting electrons which strike the screen. We then beam a stream of electrons at the two slits. Classical mechanics would imply that all the electrons would land and be counted directly behind the slits. Experimental results, on the other hand, show that electrons land all across the screen, and that although their arrival at an individual spot is random, the rate at which they arrive at a spot is fixed. Furthermore, the arrival rate patterns can be explained by considering the electron as a wave undergoing diffraction. To explain such phenomena, physicists created the theory of quantum mechanics. If an observable, such as an electron, is considered as a particle, this theory is able to obtain a probability distribution for the location of the electron, but is not able to pinpoint an exact location. This distribution is obtained by solving a partial differential equation (the Schrödinger equation).

We will return to this example when we briefly discuss *absolute* probabilities. The point here is that the probability distribution is obtained by solving a partial differential equation, not by considering a set of equipossible alternatives or even by observing a frequency of occurrences. The validity of the distributions is established by the extreme accuracy with which they predict measured energy values. We see then that these probabilities have nothing to do with the principle of indifference.

Thus some of the most important and useful applications of probability theory have nothing to do with the principle of indifference. Definitions 2.4 and 2.5 are for arbitrary probability spaces; yet we have only established their validity when the spaces are created based on the principle of indifference. To this day, some of the most popular attempts to establish their validity in these other cases are the limiting frequency approaches. These include strict frequentism as popularized by von Mises [1957], and the propensity interpretation. We discuss these approaches in the next section.

PROBLEMS 2.1

1. Prove the result in Theorems 2.3 and 2.4.

2. Suppose that James takes a routine blood test required to obtain a marriage license, and the test comes back positive for syphilis. Suppose further that the physician knows that the test has a false positive rate of 1%. The physician therefore tells James that the chance that he has syphilis is 99%. Asssume that the false positive rate is correct, that is, that the probability of a positive test given that a person does not have syphilis is .01. Additionally assume that the probability of a positive test given that a person has syphilis is .999 and that the a priori probability of a person who takes the blood test required for marriage having syphilis is .0001. Use Bayes' theorem to determine the probability of James having syphilis. Does he have reason to panic?

   Suppose next that Mary tests positive on a pregnancy test which has the same false positive rate of 1% and for which the probability of a positive test given that a person is pregnant is .999. Assume that the a priori probability of a person who takes this pregnancy test being pregnant is .3 (the value is high because ordinarily only women who feel that they may be pregnant take the pregnancy test). Determine the probability that Mary is pregnant.

   Why does it turn out that it is very improbable that James has syphilis and very probable that Mary is pregnant even though they both tested positive on equally reliable tests?

Part of this problem was taken from [Henrion, 1986]. All values in the problem are entirely fictitious.

3. Suppose amounts of money are generated according to the sequence [1, 2, 3, 4,...], the amount is placed in one envelope, a fair coin is tossed, and twice or one-half of the amount is placed in a second envelope based on the outcome of the coin toss. Show that the individual holding the second envelope should lose money in the long run if he always exchanges envelopes.

4. Suppose again that amounts of money are generated according to the sequence [1, 2, 3, 4,...], the amount is placed in one envelope, and twice that amount is placed in another envelope. As a result of a fair coin toss, we receive one envelope or the other. Show that we should approximately break even in the long run by always exchanging envelopes.

5. Suppose a uniform random process is used to generate an amount of money with some upper bound, say $1000, the amount is put in one envelope and twice the amount is put in another, and a coin is tossed to determine which envelope we receive. As discussed in the text, we would be wise to always exchange sums less than or equal to $1000. It may be thought that in many repetitions we would end up with approximately 25% more money than we would have if we never exchanged. In actuality, we would end up with approximately 50% more money. Show that this is the case.

6. Suppose your boss said that he is going to place an arbitrary amount of money in one envelope and place twice that amount in another and flip a coin to determine which envelope you receive. If you found $1000 in your envelope, would you exchange envelopes?

COMPUTER PROBLEM 2.1

Write a simulation to show that the claim in Problem 5 is correct. Write another simulation to show that if an amount of money is generated and put in our envelope, and if one-half or twice that amount is put in the other envelope depending on the outcome of a coin toss, we will increase our wealth by about 25% if we always exchange envelopes.

2.2. LIMITING FREQUENCY APPROACHES

2.2.1. Strict Frequentism

Von Mises [1957] says that numerical probability only has meaning in the case of an experiment which can be repeated; for example, the tossing of the same coin, the observance of the diffraction pattern of electrons, the observance of whether a 30-year-old man dies in the coming year, and, indeed, all of the examples in the preceding section. To von Mises, probability theory has nothing to do with the probability of the Los Angeles Lakers winning the NBA championship. Of course, people do attach numerical probabilities to such events in order to guide their betting behavior; however, to von Mises, these probabilities are outside the domain of probability theory. We will return to them in the next section.

Specifically, von Mises [1957] defines a "collective" as follows:

> This term is "the collective," and it denotes a sequence of uniform events or processes which differ by certain observable attributes, say colours, numbers, or anything else.

An example of a collective is the class of all 30-year-old men who are white-collar workers in the United States in the 1980s, along with the attribute whose possible values are the occurrence or nonoccurrence of death in their 31st year. Specifying a collective is equivalent to specifying the information in the experiment, as discussed in section 2.1. The information that the man is in his 31st year and is a white-collar worker in the United States in the 1980s determines one collective, whereas the information that he is in his 31st year and is an aborigine in New Guinea in the 1980s determines another collective. If we add the information that he has lung cancer to either of these bodies of information, yet another collective would be determined. The act of picking one member from the collective and observing the values of the attributes of interest is a single repetition of the experiment. If an experiment is repeated n times and $S^n(E)$ is the number of times an event E occurs (an attribute is observed), then von Mises defines the probability of E as follows:

$$P(E) = \lim_{n \to \infty} \frac{S^n(E)}{n}.$$

The argument is that if, for example, we tossed a biased coin 100 times it might turn up heads 67 times, after 1000 tosses there might be 681 heads, and after 10,000 tosses 6817 heads. The point is that as the number of repetitions of the experiment increases, the number of stable decimal digits in the frequency increases. Therefore there is a limiting value and that value is the probability. Of course, this cannot be proved for any experiment since we can never proceed indefinitely; however, von Mises [1957] says that it is applicable whenever we have "sufficient reason to believe that the relative frequency of the observed attribute would tend to a fixed limit if the observations were indefinitely continued." Von Mises stresses that a single experiment does not have probability values associated with it (since in a single experiment all attributes have specific values). Rather probabilities are only associated with the collective.

It is not difficult to derive the axioms of probability theory from von Mises' definition. For example, if E_1 and E_2 are two mutually exclusive events, then

$$P(E_1) + P(E_2) = \lim_{n \to \infty} \frac{S^n(E_1)}{n} + \lim_{n \to \infty} \frac{S^n(E_2)}{n}$$

$$= \lim_{n \to \infty} \frac{S^n(E_1) + S^n(E_2)}{n}$$

$$= \lim_{n \to \infty} \frac{S^n(E_1 \vee E_2)}{n}$$

$$= P(E_1 \vee E_2).$$

Furthermore, this definition covers all of the examples in the previous section and agrees with our experiences. That is, in games of chance, physics, and actuarial applications, we have observed the apparent convergence of relative frequencies. However, as we shall see, there are a number of criticisms of von Mises' approach.

Before discussing those criticisms, we note that von Mises makes another assumption besides the convergence of relative frequencies. He assumes that the class of events for which probability theory is applicable are random processes, and he defines a random process as one which, when repeated indefinitely, *definitely* generates a random sequence (here we have one rigorous definition of a random process). A random sequence, according to von Mises, is defined as follows (this definition is actually Church's [1940] paraphrase of von Mises' definition):

For the present purpose it is largely sufficient to confine attention to the case that each trial has only two possible outcomes, as with the toss of a coin adjudged as falling heads or tails, or the roll of a die adjudged as showing or not showing an ace. The *Kollectiv* may then be represented abstractly by a random sequence of 0's and 1's: in the case of the coin, for instance, we may let 1 correspond to the fall of heads and 0 to tails.

The definition of a random sequence of 0's and 1's as given by von Mises may perhaps be put in the following form:

An infinite sequence $a_1, a_2, \ldots$ of 0's and 1's is a random sequence if the two following conditions are satisfied:

(1) If $f(r)$ is the number of 1's among the first r terms of $a_1, a_2, \ldots$, then $f(r)/r$ approaches a limit p as r approaches infinity.

(2) If $a_{n_1}, a_{n_2}, \ldots$ is any infinite sub-sequence of $a_1, a_2, \ldots$, formed by deleting some of the terms of the latter sequence according to a rule which makes the deletion or retention of a_n depend only on n and $a_1, a_2, \ldots, a_{n-1}$, and if $g(r)$ is the number of 1's among the first terms of $a_{n_1}, a_{n_2}, \ldots$, then $g(r)/r$ approaches the same limit p as r approaches infinity.

The first condition in the above definition is simply the limiting relative frequency assumption as stated at the beginning of this section. The second says that we cannot improve on the original limit by *shrewdly* picking a subsequence of the original sequence. Von Mises calls this assumption the principle of the impossibility of a gambling system. That is, a gambler cannot obtain profit in the long run by using a strategy of choosing when to bet instead of betting every time. The strategy for determining whether to bet on the nth trial can only be based on the outcomes of the first $(n - 1)$ trials or on events not related to the process itself. For example, the gambler could choose to bet (on heads) every time three heads in a row come up for the given coin, or he could choose to bet on heads every time heads occurs on the toss of a different coin. However, he cannot base his choice on whether to bet on a given toss after that toss has already occurred. A subsequence of the original sequence (collective) which is obtained using this restriction is called an admissible subsequence. Using his assumption of randomness, von Mises was able to prove that separate trials of the same experiment are probabilistically independent.

Von Mises was not able to give a precise mathematical definition to the restriction on admissible subsequences. As a result, techniques were devised to incorporate mathematically the result of the nth trial into the decision as to whether include the nth trial in the subsequence. Church [1940] and Li and Vitanyi [1988] contain examples of such techniques. If the probability p is not 0 or 1, an infinite number of 1's must occur in the original sequence. Thus, by using these techniques, supposedly admissible sequences can be obtained which converge to 1 rather than to p (all the successes are simply chosen for the subsequence). Von Mises' theory was therefore criticized with the claim that the definition of the collective was inconsistent. This claim, however, is totally unjustified, since the techniques for obtaining the subsequences are contrary to von Mises' intentions. Regardless, as a result of this criticism and others, von Mises' views were, for the most part, abandoned after a conference on probability theory in Geneva in 1937. A more popular view became the propensity interpretation, discussed in the next subsection. Van Lambalgen [1987] shows that most of the other criticisms are also unjustified. Since some of these criticisms and van Lambalgen's rebuttals are beyond the scope of this book, we will only briefly consider some of the more important ones here.

Before discussing these criticisms, we note that the assumption of randomness, although today intuitively very appealing, is not based on direct experience nearly as much as is von Mises' other assumption (the convergence of limiting frequencies). That is to say, although it certainly does not appear that physically separate events affect each other, there have not been extensive tests to determine directly whether, for example, one can predict the outcome of the next toss of a coin from the previous tosses. At the time when von Mises derived his theory, the philosopher Marbe [1916] believed to the contrary. He felt that nature was endowed with a memory. That is, if a string of 15 heads come up in repeated tosses of a fair coin, the probability of tails on the next toss is increased because nature will compensate for all the previous heads. A current mathematician, Zinovy Reytblatt (mathematics department, Illinois Institute of Technology) claims that he and Serge Makarov ran experiments with coins in Yakutsk, U.S.S.R. in 1958–1959, which substantiate Marbe's belief. On the other hand, Iversen et al. [1971] conducted experiments which substantiate that the outcomes of many throws of a die do indeed generate a random sequence.

We return now to the rebuttals to the criticisms of von Mises' theory. First, the concept of a random sequence has been made mathematically rigorous. Church [1940] added the requirement that a subsequence be admissible only if the decision as to whether to include the nth element is effectively calculable from the values of the first $n - 1$ elements. That is, the decision is obtained by applying an effective algorithm to the values of the first $n - 1$ elements. The existence of random sequences in the sense specifed by Church is a consequence of a result of Doob [1936].

Church's definition does not quite solve the problem, since von Mises does allow subsequences, other than ones obtained using explicit algorithms, to be admissible. For example, in the toss of a coin, a valid subsequence is one obtained based on the results of tossing another coin. Another valid subsequence is one obtained simply by someone arbitrarily making a decision at each toss. The interested reader is referred to van Lambalgen [1987] and Li and Vitanyi [1988] for a discussion of Kolmogorov complexity and modern mathematical definitions of random sequences, which account for these other admissible subsequences. The point here is that von Mises' notion of a random sequence has been made mathematically respectable.

Another criticism of collectives is due to von Mises' statement "first the collective—then the probability" [von Mises, 1957]. Von Mises meant that "we shall not speak of probability until a collective has been defined." For example, we cannot speak of the probability of Joe Smith dying. We can only speak of probability relative to some collective of which Joe Smith is a member. An example of such a collective might be all 30-year-old white-collar workers in the United States in the 1980s. Since the collective is described as being the infinite sequence, critics claim that the statement "first the collective—then the probability" implies that the existence of probabilities depends on future events which may never occur. For example, if a coin were never tossed, there would be no probability associated with that coin coming up heads in a toss. Von Mises did not intend this interpretation. Indeed, he states [von Mises, 1957] the following:

> The probability of a 6 is a physical property of a given die and is a property analogous to its mass, specific heat, or electrical resistance.

Yet another criticism of von Mises' approach is that an infinite version of the law of the iterated logarithm is not derivable from von Mises' assumptions. However, as noted by Kolmogorov [1929], the law is only meaningful as regards finite sequences.

Thus many of the criticisms of von Mises' approach are unfounded. However, one important problem remains. The basic assumption in the approach is that a random sequence will *definitely* occur and that the relative frequency will *definitely* converge to the probability. That is, if the probability of a heads is 1/2, we will definitely not have an endless sequence of heads or any other sequence which is not a "random sequence" and which does not converge to 1/2. Such a claim is somehow contrary to the concept of probability itself. As is well-known, if p is the probability of event E, and if we assume that the axioms of probability theory hold and that separate trials of the same experiment are probabilistically independent, then it is possible to prove the weak law of large numbers, which states that given an arbitrary $\delta > 0$,

$$\lim_{n \to \infty} P\left(\left| p - \frac{S^n(E)}{n} \right| < \delta \right) = 1.$$

That is, by repeating the experiment many times, we can be very certain that the relative frequency will be close to p, but we can never be absolutely certain. Thus, no matter how large we take n, the possibility remains that the relative frequency will still be nowhere near p. The strong law of large numbers implies something stronger, but it still leaves open the possibility that, for an arbitrarily large n, the relative frequency can be far from p. These laws of convergence are all we have been able to deduce from the axioms of probability theory; yet von Mises establishes the axioms based on the assumption of a stronger type of convergence. Admittedly, all observed relative frequencies of "random" processes do apparently converge. However, if the probability that they do not converge is indeed very low, the apparent convergence of all relative frequencies could be due to our limited experiences. It seems that we should base the axioms of probability theory on a weaker type of convergence than that assumed by von Mises. Namely, we should assume convergence only in the sense of the weak law of large numbers. In Appendix 2.A we obtain such a derivation. That is, we assume only that probabilities exist and that relative frequencies approach probabilities in the sense of the weak law of large numbers. From that assumption we deduce the axioms of probability theory and the definition of conditional probability. The derivation in Appendix 2.A has nothing to do with the notion of randomness. However, once we have deduced the axioms of probability theory, we can define a random process to be one for which separate trials of the experiment are probabilistically independent. It can then be proved that "random sequences" are most probable. The principle of the impossibility of a gambling system would then be replaced by a theorem which states that it is very improbable that a gambler could achieve a profit in the long run by *shrewdly* selecting a subsequence on which to bet.

PROBLEM 2.2.1

Derive the other axioms in the definition of a probability space and the definition of conditional probability based on von Mises' definition.

2.2.2. The Propensity Interpretation

The propensity interpretation, popularized by Popper [1975, 1983], holds that probability should be thought of as a physical characteristic. As we've already noted, von Mises actually agreed with this. The real difference in this approach from that of von Mises is that this approach *assumes* the axioms of probability theory and that separate trials of the same experiment are independent and then claims to be able to derive von Mises' frequency approach. We will illustrate the argument by considering the repeated tossing of a fair coin.

If 1 stands for a heads and 0 stands for a tails, $P(\text{heads}) = 1/2$, and the coin is tossed n times, then the probability of any particular sequence of heads and tails is $(1/2)^n$ (with the assumption that separate tosses are independent). Proponents of this approach then define a new continuous probability distribution on the set of all infinite binary sequences. If we let S be the set of all infinite binary sequences starting with a particular string of 0's and 1's of length n, then $P(S)$ is defined to be $(1/2)^n$. This definition is consistent with the probabilities assigned to finite strings, and it is quite easy to show that this definition yields a probability distribution on the set of all infinite binary sequences. If we then let

$$U = \{\text{set of all infinite binary sequences with limiting}$$
$$\text{relative frequency of 1's equal to 1/2}\},$$

it is possible to show that

$$P(U) = 1.$$

This result is the strong law of large numbers. Therefore the set of infinite binary sequences which do not converge to 1/2 have probability equal to 0. Furthermore, if we let

$$W = \{\text{set of all infinite binary sequences with limiting}$$
$$\text{relative frequency of 1's equal to 1/2 and which qualify}$$
$$\text{as random sequences}\},$$

it is possible to show that

$$P(W) = 1.$$

Therefore it is also true that the set of all infinite binary sequences which are not random have probability equal to 0. It is at this point that the proponents of the propensity interpretation make an auxiliary hypothesis, which is usually not stated. That is, they assume that an event which has probability equal to 0 can be neglected as if it were an impossibility. Based on that assumption, they then conclude that a random sequence, whose relative frequency converges to the probability, will definitely occur. Thus they claim to be able to deduce von Mises' assumption.

Although the propensity interpretation is mathematically elegant, it does not appear to be philosophically or physically as sound as von Mises' approach.

First, von Mises did not agree that the set of all infinite binary sequences is really a probability space. Rather the assumption that it is a probability space is a mathematical idealization. Furthermore, even if we accept that it is a probability space with a continuous distribution, we must assume that an event with probability equal to 0 cannot occur. Yet, in the case of a continuous distribution, an event with probability equal to 0 must always end up occurring. Therefore we might as well assume that a random sequence definitely occurs in the first place and not rely on an idealized probability space.

Furthermore, the proponents of the propensity interpretation, *assuming* the axioms of probability theory and that separate trials of the same experiment are independent, are able to deduce the strong law of large numbers and the fact that a random sequence is highly probable. That is, they are able to deduce what they feel to be the physical evidence for probabilities from the axioms of probability theory (and an independence assumption). However, perhaps this evidence could be deduced from other assumptions. von Mises, on the other hand, *deduces* the axioms of probability theory from what he feels to be the physical evidence for probabilities. Thus, if we accept his arguments, the axioms of probability theory are inevitably correct.

As mentioned in the last subsection, the only problem with von Mises' approach is that it is necessary to assume that a random sequence definitely occurs and that the relative frequency definitely converges to the probability. In Appendix 2.A we remove that problem by deducing the axioms of probability theory by only assuming convergence of relative frequencies in the sense of the weak law of large numbers.

Before ending this section, we note that Kolmogorov himself was not a proponent of the propensity interpretation, as is evidenced by the following statement concerning his axioms [Kolmogorov, 1963]:

> This theory was so successful, that the problem of finding the basis of real applications of the results of the mathematical theory of probability became rather secondary to many investigators...[however] the basis of the applicability of the results of the mathematical theory of probability to real "random phenomena" must depend in some form on the *frequency concept of probability*, the unavoidable nature of which has been established by von Mises in a spirited manner.

2.2.3. The Evidence for Physical Probabilities

Much of the evidence which gives us reason to postulate the existence of probabilities is the apparent convergence of relative frequencies, which can be taken as manifestations of the laws of large numbers. Experiments with coin tossing have shown that we can be close to certain that in many tosses the relative frequency of heads will stabilize close to some value. Notice, however, that no matter how many experiments indicate that a relative frequency is approaching a limit, we can never be absolutely certain that a relative frequency will reach a limit. These experimental results are therefore least presumptuously described by hypothesizing that probabilities exist and assuming convergence only in the sense of the weak law of large numbers. The financial success of insurance companies is a testimony to the weak law of large numbers holding for the probabilities assigned in acturial cases. The predictive accuracy of the arrival rate of electrons substantiates the weak law of large numbers holding for the probabilities obtained using the Schrödinger equation. We note once again that all of these probabilities are relative to information (a collective). The probability that a man will die in his 31th year has one value if it is relative to the information that he is a 30 year old white collar worker in the United States in the 1980's, and it has a different value relative to the additional information that he has cancer. The probability of the location of the electron is relative to the information that the electron is diffracted. Whether it would ever be possible to obtain additional information which would change this probability is at the heart of the question concerning the existence of absolute probabilities (as discussed in section 2.7).

We cannot prove or know whether these probabilities are, in actuality, real any more than we can prove any physical fact:

> For example, we cannot prove mathematically that there is a physical quantity called "force." What we can do is postulate a mathematical entity called "force" that satisfies a certain differential equation.
>
> —[Ash [1970]]

The same is true for these probabilities. We hypothesize their existence for the same reason we hypothesize any theory in science; that is, because we can describe the world we observe by assuming they exist. In science we build a collection of mathematical results that provide a reasonable description of certain physical phenomena. A mathematical model, which assumes the existence of probabilities relative to information and convergence of relative frequencies to these probabilities in the sense of the weak law of large numbers, describes observable physical phenomena.

There is a difference between this view of probability and the logical or a priori views of Laplace, Carnap [1950], and Keynes [1948]. Laplace's classical approach is an a priori view of probability in that it assumes the existence of probabilities which are independent of our experience. Carnap [1950] and Keynes [1948] also developed a priori approaches to probability theory. Although we do not have space to discuss these latter two approaches here, we mention that all of these approaches not only assume that probabilities exist independent of experience, but also that we can determine them without experience. That is, we can use pure logic to determine them. We also hypothesize that probabilities exist independent of experience, but we assume that we need experience to approximate their values. This point of view is actually a propensity view of probability. However, rather than assume the axioms of probability theory, as do the proponents of the usual propensity approach, we assume the weak law of large numbers.

Both a logical approach and a propensity approach assume that probabilities have objective reality apart from an observer. However, a logical approach assumes that there exists logically correct probability values based on the information, and we do not need physical evidence to substantiate that they are correct, while a propensity approach assumes that we need physical evidence in the form of relative frequencies to substantiate the correctness of probability values. (We saw an example of the necessity of such evidence in Example 2.7.) Therefore ordinarily the probabilities associated with a propensity approach are called "physical," while the word "objective" is used for probabilities associated with a logical approach. In the remainder of this book we call both such probabilities "objective" to distinguish them from subjective probabilities (as discussed in the next section).

As previously noted, Appendix 2.A contains the derivation of the axioms of probability theory and the definition of conditional probability from the assumption that probabilities exist and that relative frequencies converge to these probabilities in accordance with the weak law of large numbers.

2.3. THE SUBJECTIVISTIC APPROACH

The previous approaches maintain that statements such as "the Los Angeles Lakers will win the NBA championship" are not in the domain of numerical probability theory. However, gamblers routinely assign probabilities to such statements, and Nevada makes an enormous amount of money by so doing. Furthermore, many practical applications contain uncertainty of this same kind. Consider the following example:

Example 2.11

A firm is setting its price for a product which is also sold by one other company. The other firm will announce its price first and then our customers will wait for our firm to set its price before choosing their orders. Some of these customers will be placing their orders early, before we make our production lot size decision. Our goal is to maximize our expected profit which depends on sales, our price, and the lot size.

—[Shachter [1988]]

There are a number of uncertain relationships in this example. The amount of "our firm's" sales depends on both the other firm's price and our price; however, we are uncertain as to the exact dependency. We are also uncertain as to the number of early orders given our total sales. There are yet other uncertainties in this problem. We will discuss this problem again in chapter 9, when we show how the techniques developed in chapters 5 through 8 can be used to recommend explicit decisions in the face of uncertainty. The point here is that the company must make a practical decision as to its price and its lot size based on uncertain relationships; its goal is to make the decision which will lead to the most profit.

The experiment consisting of the other company and our company setting our prices is not an experiment which can be repeated in the sense of a frequency approach to probability. Our company might encounter a similar situation in another year; however, clearly this is not a situation which can be repeated indefinitely. Rather the uncertainty is more like the uncertainty in the Lakers winning the NBA championship. The numerical value attached to the uncertainty is only relative to this particular situation and must be obtained from a human's careful analysis of the likelihood.

The following example shows that even in the case of a typical repeatable experiment there exists uncertainty which does not fit into the frequency approach:

Example 2.12. Suppose that Mary claimed that she could flip a coin in such a way that a heads comes up 2/3 of the time. Initially, we would doubt that she could do this, and therefore we'd bet on heads with a probability of .5. However, if in 25 tosses 17 heads came up, we would change our probability to be closer to .67. This is due to the fact that it is somewhat plausible that she could toss the coin in such a manner, and the evidence makes us start believing it. On the other hand, if Mary claimed that she could use psychokinesis to command the coin to come up heads 2/3 of the time, it would take far more tosses for us to change our probability because this claim is far less plausible.

These two examples and the situation involving the Lakers winning the NBA championship are situations containing a set of mutually exclusive and exhaustive alternatives of which only one will end up true. However, they are not repeatable experiments in the sense of a frequency approach. Our uncertainty simply represents our *belief* relative to the particular situation at hand. De Finetti [1972] derives the axioms of probability theory and the definition of conditional probability for such cases. He defines the probability $P(E)$ of an event E as the fraction of a whole unit value which one would feel is the fair amount to exchange for the promise that one would receive a whole unit value if E turns out to be true and zero units if E turns out to be false. Since one would consider either side of such an exchange fair, one should be willing to be the person relinquishing $P(E)$ units and also be willing to be the person relinquishing either 0 or 1 unit. If there are n mutually exclusive and exhaustive events E_i, and a person assigned probability $P(E_i)$ to each of them respectively, then he would agree that all n exchanges are fair and thus agree that it is fair to exchange $\sum_{i=1}^{n} P(E_i)$ units for 1 unit. Therefore, if $\sum_{i=1}^{n} P(E_i) \neq 1$, the person would have to agree to a bet that he is sure to lose. This has been called the dutch book theorem [de Finetti, 1964]. A set of probability values which do not allow a dutch book are said to be coherent.

Based on the assumption of coherency, de Finetti is able to deduce the axioms of probability theory and the definition of conditional independence.

For example, we can deduce additivity as follows: first, since any event E and its complement $\overline{E}$, are mutually exclusive and exhaustive, coherency implies that $P(E) + P(\overline{E}) = 1$. Now let E_1 and E_2 be two mutually exclusive events. Then, again by the coherency assumption,

$$P(E_1) + P(E_2) + P(\overline{E_1 \vee E_2}) = 1.$$

However, since the probability of any event plus the probability of its complement is equal to 1,

$$P(\overline{E_1 \vee E_2}) = 1 - P(E_1 \vee E_2).$$

Therefore

$$P(E_1) + P(E_2) + 1 - P(E_1 \vee E_2) = 1,$$

which implies additivity.

The deductions of the remaining axioms are left as an exercise. Since the concept of a conditional bet may not be immediately clear, we deduce the definition of conditional probability here. If E and H are two events, then the conditional probability of E given H, denoted $P(E \mid H)$, is defined as follows: Once it is learned that H occurs for certain, $P(E \mid H)$ is the fair amount one would exchange for the promise that one would receive a whole unit value if E turns out to be true and zero units if E turns out to be false. For example, one could offer a conditional bet on the Lakers winning the NBA championship given that they come in first in their conference. If they do not come in first, the bet is canceled. Given this definition we can deduce the standard definition of conditional probability as follows: One would exchange

$$P(H) \text{ units for } \begin{cases} 1 \text{ unit if } H \text{ occurs} \\ 0 \text{ unit if } H \text{ does not occur.} \end{cases}$$

Therefore one would exchange

$$P(H)P(E \mid H) \text{ units for } \begin{cases} P(E \mid H) \text{ units if } H \text{ occurs} \\ 0 \text{ units if } H \text{ does not occur.} \end{cases} \quad \text{(bet 1)}$$

Furthermore, if H does occur, one would exchange

$$P(E \mid H) \text{ units for } \begin{cases} 1 \text{ unit if } E \text{ occurs} \\ 0 \text{ units if } E \text{ does not occur.} \end{cases} \quad \text{(bet 2)}$$

Therefore, before it is known whether H occurs, one would agree to exchange the $P(E \mid H)$ which he might win in bet 1 (if H occurs) for 1 unit if E also occurs. Therefore, before it is known whether either of these events occur, one would exchange

$$P(H)P(E \mid H) \text{ units for } \begin{cases} 1 \text{ unit if both } H \text{ and } E \text{ occur} \\ 0 \text{ units if } H \text{ and } E \text{ do not both occur.} \end{cases}$$

However, since one would also exchange

$$P(E \wedge H) \text{ units for } \begin{cases} 1 \text{ unit if both } H \text{ and } E \text{ occur} \\ 0 \text{ units if } H \text{ and } E \text{ do not both occur,} \end{cases}$$

the standard definition of conditional probability follows for $P(H) > 0$.

Probabilities which represent an individual's beliefs are called subjective (in contrast to the objective probabilities offered by the previous two approaches). De Finetti [1972] claims that all probabilities are subjective. That is, any set of probabilities are based on someone's opinion. For example, when one states that the probability of the top card being the king of spades is 1/52, that is the person's opinion. Another person could have the opinion that it is 1/10. De Finetti does not argue that people mentally manipulate these probabilities using the axioms of probability theory. Rather the argument is that a rational person would have to agree that if these numbers were manipulated, it would have to be done according to those axioms.

Weatherford [1982] states that "it is the fatal flaw of the subjectivistic theory that it pretends that one person's probability is just as good as the other's and that no objective consideration can choose between them, when it is perfectly plain that one will lose his shirt *because he is wrong about the probability*." Putnam [1963] points out that we ask more of mathematical systems than mere consistency; we ask also that they truly describe the world.

These criticisms are based on an incorrect interpretation of subjectivistic theory. As shown in section 2.1, many probabilities cannot be obtained using the principle of indifference. Other probabilities, such as ones assigned to statements such as "the Los Angeles Lakers will win the NBA championship," cannot be given a frequency interpretation either. It is even an idealization to claim that there is an objective probability associated with the event that a man who is a white-collar worker in the United States in the 1980s will die in his 31st year. As noted in the previous section, we hypothesize that such a probability exists because, by so doing, we can describe our experiences. However, the notion that there really exists such a number, accurate to 100+ digits, is a bit hard to defend. Due to these limitations of the classical and frequency approaches, the subjectivists feel that neither of these approaches can be used as the axiomatic foundation of probability theory. However, since every probability does indeed represent someone's belief, subjectivists feel that the coherency argument can be used as such a foundation and that it encompasses all probabilities. Thus subjectivistic theory in itself does not deny that objective probabilities exist; rather it denies that the assumption of the existence of objective probabilities is adequate as a foundation for probability theory.

Within the framework of subjectivistic theory, an extreme subjectivist may indeed deny that objective probabilities exist. However, this does not mean that he feels that one probability is as good as another. Rather he believes that he should assign the best possible probabilities based on his available information. For example, even an extreme subjectivist would assign a probability of 1/52 to an ace of spades because that is the rational choice based on the symmetry of the situation. The well-known subjectivist Savage [1954] remarks that arguments based on the principle of indifference "typically do not find the contexts in which such agreement obtains sufficiently definable to admit of expression in a postulate." That is, the subjectivist does not deny the usefulness of the principle of indifference; he only claims that the classical definition cannot be used as the foundations of probability theory. Good [1952] also makes this point when he notes that arguments based on the principle of indifference are "suggestions for using the theory, these suggestions belonging to the technique rather than the theory itself."

Neither does the subjectivist deny the usefulness of probabilities obtained from frequencies. In repeatable experiments, such as coin tossing, many people believe that the sequence, which represents the outcome of a number of repetitions of the experiment, is *exchangeable*. That is, if we denote a heads by a 1 and a tails by a 0, they believe that these two sequences have the same probability:

$$1000100100 \quad \text{and} \quad 1110000000.$$

In addition, they believe that any other sequence containing 3 heads and 7 tails has that same probability. Based on the assumption of exchangeability in de Finetti [1931] proved the famous

De Finetti Representation Theorem. Suppose an experiment can be repeated indefinitely and that the sequence of successes in n trials is exchangeable for any n (such a sequence is said to be infinitely exchangeable). That is, the individual believes that the probability of k particular successes in n trials is the same regardless of where the k successes occur. If we let S^n be the number of successes in n trials, then

1. $p = \lim\limits_{n \to \infty} \left(\dfrac{S^n}{n} \right)$ exists with probability equal to 1.

2. If $\mu(x)$ is the probability density function for p, then

$$P(S^n = k) = \binom{n}{k} \int_0^1 x^k (1-x)^{n-k} \mu(x)\, dx.$$

If p is thought of as an objective probability, then $\mu(x)$ represents a subjective probability of the "true" value of p (being an extreme subjectivist, de Finetti does not give $\mu(x)$ this interpretation). We will discuss probabilities of probabilities much more in chapter 10. The first point here is that if an individual believes that the sequence is infinitely exchangeable, then he must believe that it is most probable that the sequence is approaching a limit. Even an extreme subjectivist such as de Finetti believes this. He only disputes the claim that limiting frequencies are objective or physical properties.

More importantly, Zabell [1988] notes that the representation theorem, along with the definition of conditional probability, implies that except in the case of very unusual density functions $\mu(x)$ (an example of such a case is when the individual says that he "knows" the point probability value for certain and no number of repetitions of the experiment will affect his belief), the individual must believe that it is highly probable that the probability of success on the $n + 1$st trial is about equal to S^n/n when n is large. Thus, if a sequence is assumed to be infinitely exchangeable and if n is large, a subjectivist uses relative frequencies to approximately represent his probabilities (in chapter 10 we will see precisely how to augment the original probabilistic assessments of an expert with information in a data base).

We see then that the subjectivist, like the frequentist, uses limiting frequencies to determine probabilities. When he does this, he is in a sense assuming something objective about the probabilities. That is, he is assuming exchangeability. However, this objectivity is part of the individual's particular application of the theory (i.e., it is part of his beliefs); it is not inherent in the theory itself.

Before closing this section, we note another well-known subjectivistic derivation of the axioms of probability theory. The physicist Cox [1946] deduced the axioms by making the commonsense assumption that $P(E_1 \wedge E_2)$ is a real valued function of $P(E_1 \mid E_2)$ and the $P(E_2)$. He then offers an elegant argument based on functional analysis to deduce the axioms.

PROBLEM 2.3

Derive the remaining axioms in the definition of a probability space based on de Finetti's definition.

2.4. THE NONEXTREME FREQUENTIST AND THE NONEXTREME SUBJECTIVIST

The extreme frequentist denies that probability theory has anything to do with statements such as "the Lakers will win the NBA championship." Yet, if forced at gunpoint to bet his life savings according to a probability which he must assign to this event, he would carefully analyze the situation to arrive at a number which he felt represents that likelihood as closely as possible. He might offer frequency arguments for this probability. For example, he might claim that there is an objective probability associated with the information which he has about this situation, and that if that information is repeated many times, the fraction of the time that a team repeats as champions will approach that probability. The number which he assigns is then his estimate of the objective probability. Alternatively, he could claim that his probability is a highly conditionalized probability based on past experiences. For example, this author lost his shirt on the 1989 Super Bowl because he bet based on the fact that in the past three super bowls the NFC team scored impressive victories in the playoffs and then proceeded to decidely defeat the AFC team in the Super Bowl. Perhaps there is some validity to these arguments. However, arguments for objective probabilities in these cases are clearly not as defensible as arguments for objective probabilities in the case of picking a card from a deck or of electron diffraction. In any event, practically we can gain nothing by offering such arguments. As noted in the last section, even the claim that there is an objective probability associated with a 30-year-old man dying in the coming year is an idealization.

Furthermore, the extreme frequentist can never know the values of his objective probabilities. He can only obtain estimates and confidence intervals. However, to him the axioms of probability theory and the definition of conditional probability only hold for the objective probabilities. Therefore he can never use those axioms or that definition to compute new probabilities from existing ones, which implies that he cannot use any of the techniques which are presented in this text.

The nonextreme frequentist is not bound to these conclusions because he realizes that he need not cling tenaciously to the tenets of his formalism. The nonextreme frequentist realizes quite well that he must make uncertain decisions in situations where uncertainties are not represented by relative frequencies. In such cases he can use the coherency argument of the subjectivists to justify the use of probability theory rather than invent far-reaching frequency claims. Furthermore, he realizes that his estimates of objective probabilities represent his beliefs, and again he can use the coherency argument to justify the use of probability theory. That is, the nonextreme frequentist can embrace the usefulness of subjectivistic theory while still maintaining that some probabilities (e.g., electron diffraction) are best conceived as being objective.

We noted in the last section that even an extreme subjectivist does not claim that the assignment of a probability of 1/10 to the ace of spades is just as good as the assignment of 1/52. However, an extreme subjectivist does deny that even the probabilities observed in electron diffraction have objective reality. The nonextreme subjectivist need not make this claim, since, as also noted in the last section, subjectivistic theory does not deny that there is an objective probability associated with electron diffraction. Instead the theory leaves that belief to the individual.

We offer three levels of probabilities: 1) probabilities obtained using the principle of indifference; 2) probabilities obtained from relative frequencies; and 3) probabilities obtained solely from an individual's beliefs. Both the nonextreme subjectivist and the nonextreme frequentist assign probabilities based on this hierarchy. They both say that the probability of an ace of spades is 1/52. It would take many, many repetitions of the experiment for either of them to change that probability. If 15 of 1000 individuals had cancer, they both would say that the probability of cancer is about .015. Moreover, they both would readily change that probability if it were learned that 300 of 10000 individuals had cancer (the frequentist would say that he has learned more about the objective probability, while the subjectivist would say that his belief has changed). They both assign probabilities to statements such as "the Lakers will win the NBA championship." The only real difference between the two is their philosophical justification. The frequentist prefers to postulate the existence of objective probabilities as much as possible, since many probabilities appear to him to have objective realities and he is looking for truth, whereas the subjectivist prefers to view all probabilities as subjective, since many probabilities clearly do not have objective realities and he is primarily concerned with practical applications of the theory.

As may be now be apparent, this author is best classified as a nonextreme frequentist. This will further become evident in the next section, where it is suggested that we stay within the frequency framework wherever possible, and in chapter 10, where we explain the concept of uncertainty in probabilities in terms of the collective. However, these explanations fit into the subjectivistic framework since a collective, for practical purposes, is no different than an infinitely exchangeable sequence.

2.5. PROBABILITY THEORY AND EXPERT SYSTEMS

At the end of chapter 1, we noted that in this book we are concerned with numerically representing uncertainty in experts systems in the case where the uncertainty is in the members of a set of mutually exclusive and exhaustive alternatives. Having briefly reviewed some of the major approaches to probability, we now show that probability theory is the appropriate discipline for our aims.

First we must clarify two approaches to developing expert systems. Recall from the beginning of chapter 1 that an expert system is a program capable of making judgments in a complex area. The exemplar of an expert system is a program which performs medical diagnosis. The fundamental goal is that the expert system makes the best possible judgments. For example, if expert consensus and the results of an autopsy were different, we would want the expert system to agree with the results of the autopsy. Since experts do often make good judgments, one way to develop an expert system is to initially attempt to develop a system which makes the same judgments as the expert. When we do this, we are only concerned that the resultant judgments are the same. That is to say, we are unconcerned with whether the system follows the same process as the expert in reaching those judgments. However, in order to develop systems which reach the same judgments as the expert, many use psychological modeling in an effort to develop systems which follow the same process as the expert. In other words, they attempt to develop systems which "reason" like an expert. In chapter 1, it was argued that backward chaining accomplishes this in the case of categorical knowledge. Another approach to developing expert systems is to be unconcerned with the way the expert reasons; rather our only aim is to create a model which enables us to accurately represent and solve the problem.

If we take the second approach, then the coherency argument of the subjectivists implies that we must combine and update numerical likelihoods, which are assigned to the members of a set of mutually exclusive and exhaustive alternatives, according to the axioms of probability theory and the definition of conditional probability. Any violation of these axioms or this definition would lead to incoherent results. Few would argue with this conclusion.

On the other hand, experts ordinarily make good decisions in the face of uncertainty, and it is reasonable to assume that they do not mentally manipulate numbers according to the axioms of probability theory. If we therefore attempt to create an expert system which simulates the uncertain reasoning of the expert, does it make sense to use probability theory? In subsection 2.5.1 we offer the argument that although humans do not reason numerically, they do reason in terms of dependencies and conditional independencies learned from the frequencies which they experience. If we accept that argument, then we conclude that the use of probability theory is also appropriate when psychological modeling is used in efforts to create systems which reason like an expert.

In subsection 2.5.2 we show that much of the uncertainty in expert systems can be represented by frequency-type probabilities. Moreover, we suggest that when this is the case, it is best to stay within the frequency framework. After presenting these arguments for the use of probability theory in expert systems, we investigate the appropriateness of the disciplines introduced in chapter 1. We end this section with discussions of the meaning of "a priori" probabilities in expert systems, and of the problem in obtaining the probability values needed in expert systems.

2.5.1. Probability Theory and the Modeling of Human Reasoning

An argument for using probability theory even when we are performing psychological modeling is that "although humans hardly remember the exact frequencies of complex events, they learn to think in terms of the dependencies and independencies which are implied by those frequencies" [Pearl, 1989]. For example, if a wife saw her husband purchasing jewelry with his secretary, she might judge the probability of his cheating to be fairly high. However, if it was her birthday and if he had given her a present on almost every previous birthday, her suspicions would be substantially reduced: the more such birthday presents, the lower the suspicion. We see, in this example, human judgments of probabilities (of course, not numerical) derived from experiencing frequencies, and the manipulation of these probabilities according to the qualitative rules of conditional dependence. That is, the proposition that he is cheating is normally perceived to be independent of it being the wife's birthday; however, once the visit to the jeweler is observed, the two become dependent. It is interesting to speculate that, starting in Egypt in 3500 BC, when the first die was manufactured, and before expressions like "the chances are 1 out of 4" were in our vernacular, humans began developing the notion of "equipossible alternatives" in games of chance based on their experiences with the frequencies of occurrence of the alternatives, not based on pure logic. The logic or intuition was then formulated from these experiences.

Thus, although they do not reason numerically, it can be argued that humans do reason according to dependencies and independencies which they have learned from their experience with probabilities (frequencies). If we accept this argument, then the human reasoning process is based on the relationships among probabilities. Therefore when we attempt to create a system which simulates the human reasoning process and use real numbers to represent uncertainty in that system, we should use the probability calculus to manipulate those numbers. If we obtain those numbers solely from an expert, theoretically we could expect the system to make judgments consistent with the expert (this is, of course, an idealization). However, in general, the expert's estimate of the frequencies he has experienced would be fairly crude. If, with the help of statistical data, we could improve on those estimates, possibly we could improve on those judgments.

It clearly would be difficult for a cognitive scientist to test the conjecture that humans reason according to dependencies and independencies learned from the frequencies which they have experienced. However, even if we do not accept that conjecture, the coherency argument of the subjectivists implies that we should use probability theory in expert systems whenever we are numerically representing the uncertainty in the members of a set of mutually exclusive and exhaustive alternatives. That is to say, even if we could use psychological modeling to determine a numerical model which represented some human's uncertain reasoning process and which was not based on probability theory, that model could not lead us to coherent results (in the case where the uncertainty is in the members of a set of mutually exclusive and exhaustive alternatives). Recall that our fundamental goal in expert systems is to obtain the best possible results, not to model human reasoning. An acquaintance of this author believes that he can more easily win in the 4-digit lottery than in the 3-digit lottery because he knows more 4-digit numbers (e.g., his address, his cash card number) and therefore he has an easier time determining a 4-digit number. There would be no advantage in incorporating such reasoning into an expert system.

2.5.2. Relative Frequencies in Expert Systems

Much has been written about the difference between the probability of a statement and the degree of confirmation of a statement. When explaining why they originally did not use a probabilistic approach to deal with uncertainty in MYCIN, Buchanan and Shortliffe [1984] loosely quoted Harré [1970] as follows:

> Harré observes that statistical probability seems to differ syntactically from the sense of problems used in inference problems such as medical diagnosis. He points out that the traditional concept of probability refers to what is likely to turn out to be true (in the future), whereas the other variety of probability examines what has already turned out to be true but cannot be determined directly.

Although these two kinds of problems may be approached on the basis of identical observations, the occurrence or nonoccurrence of future events is subject to the probabilistic analysis of statistics, whereas the verification of a belief, hypothesis, or conjecture concerning a truth in the present requires a process commonly referred to as *confirmation*.

The difference between probability and confirmation is often illustrated with Hempel's [1965] classic paradox of the ravens. If we let

$$H_1 = \text{all ravens are black}$$

$$H_2 = \text{all nonblack things are not ravens,}$$

then clearly H_1 and H_2 are logically equivalent statements. Therefore, if there were conditional probabilities of H_1 and H_2 given some evidence E, those probabilities should be equal. For example, if

$$E = \text{my Cadillac is pink,}$$

then we should have

$$P(H_1 \mid E) = P(H_2 \mid E).$$

The argument in the paradox is that E seems to confirm H_2 but not H_1. That is, when we see a nonblack thing which is not a raven, we might believe more strongly that all nonblack things are not ravens, and if we saw 100,000 such things, we might start believing H_2 very strongly. However, according to the paradox, it does not seem quite right that E or the sight of 100,000 nonblack things should make us start believing that ravens are black. The conclusion then is that the confirmation of H_2 due to E appears to somehow be different from a conditional probability.

It appears that this paradox is resolved once we specify the population from which the 100,000 nonblack things are sampled. If we are sampling from the population of all nonblack objects, then the fact that all 100,000 things are not ravens does confirm H_1, since if H_1 were not true there would be some ravens in the population and therefore we should eventually encounter one of them. On the other hand, if we are purposely sampling from a population of nonblack things which does not include ravens, the sight of 100,000 nonblack things does not confirm H_2 either, since the population can never tell us anything about the color of ravens and therefore our belief that ravens are nonblack remains unchanged by the sightings. Finally, if we are sampling from the population of all objects and all 100,000 things turn out to be nonblack and nonravens, then again the sightings do not confirm H_2 because again we have seen nothing to change our belief that ravens are nonblack (in this case we would have to see ravens to change that belief).

Regardless of one's view on this paradox, there is a difference between a statement such as H_1 and statements which fall within a frequency theory of probability. Any hypothesis in science is the same kind of statement as H_1. In other words, a hypothesis, such as the law of gravity, is a statement which we feel may be true. Evidence tends to confirm or disconfirm our belief in its verity. The statements which fall within the domain of a frequency theory of probability are of a different nature. They too are statements which are either true or false. However, they are statements about a member of a collective (infinitely exchangeable sequence), while scientific hypotheses are not. For example, if we pick a card from a deck, it is either true or not true that the card is the ace of spades, just like it is either true or not true that all ravens are black. The difference, however, is that we can hold the information constant (i.e., put the card back and pick other cards), repeat the experiment, and create instances where the card is the ace of spades and instances where it is not. Therefore the statement "this card is the ace of spades" is a statement about a specific occurrence of picking a card from deck, which is a member of the collective of all picks of a card from a deck. The statement "all ravens are black" is not about a member of a collective. There are no situations, each with the same information, in which the statement is sometimes true and sometimes not true.

Realizing a difference between confirmation and relative frequencies, Carnap [1950] defines two different types of probability:

> (i) Probability$_1$ is the degree of confirmation of a hypothesis h with respect to an evidence statement e; e.g., an observational report. This is a logical semantical concept. A sentence about the concept is based, not on observations of facts, but on logical analysis...

> (ii) Probability$_2$ is the relative frequency (in the long run) of one property of events or things with respect to another. A sentence about this concept is factual, empirical.

Carnap developed a theory of Probability$_1$ for a restricted class of simple languages. Salmon [1966] concludes that the confirmation of scientific hypotheses is best construed within a frequency interpretation of probability. Whether confirmation does indeed fall within the domain of a frequency interpretation of probability theory is a difficult, philosophical question which is beyond the scope of this text. Indeed, the questions, Does it even makes sense to assign numerical likelihoods to confirmation?, and, if we do assign such numbers, What do they mean? do not have easy answers. However, if we do assign such numerical likelihoods, the coherency argument of the subjectivists implies that we are bound to the axioms of probability theory when we manipulate these numbers.

On the other hand, many of the statements in expert systems applications fall within the domain of a frequency theory of probability (Probability$_2$). The statement "this patient has cancer" is a hypothesis which the physician is trying to confirm. However, it is different from a scientific hypothesis in that it is also a statement about a member of a collective, namely the class of all such patients who would ever be considered by this physician (or by this clinic, or in the United States, etc.). Harré's argument that it is a statement about the present rather than about the future is not relevant to whether the statement is about a member of a collective. If we pick a card from a deck but do not look at it, the statement that the card is the ace of spades is about the present. That does not stop us from associating a frequency-type probability with the statement (actually with the collective).

There are advantages to staying with the frequency framework whenever possible. First, we can then either obtain probabilities from a data base or augment the probabilities obtained from an expert with the information in data base. The technique is actually the same in both cases. The only difference is that when there is no expert or any other reason for believing the probabilities should have particular values, we assume uniform prior distributions (i.e., we assume all probability values are equally likely) before augmenting the probabilities with the information in the data base. In chapter 10, we show how to accomplish this augmentation and also how to represent the uncertainty in the probabilities. In some cases probabilities obtained from a data base have proved to be much better than those obtained from an expert. For example, in de Dombel's [1972] successful system for investigating gastrointestinal diseases, a study, based on 304 real patients, revealed that when the probabilities were obtained from statistical data, the system was correct 91.8% of the time, while clinicians were correct only 65–80% of the time. On the other hand, in a study of 600 patients, the system performed significantly less well than clinicians when the probabilities in the system were obtained from experts. Second, if probabilities are given a frequency interpretation, we can eventually substantiate their accuracy if data becomes available. Suppose a physician says that a positive chest X ray confirms lung cancer to degree .4 and we later learn that 70% of a very large number of patients with positive chest X rays have lung cancer. As facetious as this may seem, if the physician does not agree that the sequence is infinitely exchangeable and that his beliefs should be affected by repeated trials (and thereby accepts that the relative frequency should approximately represent his probability), he need not replace the value of .4 by .7.

The medical literature substantiates the fact that many physicians believe that probabilities, according to the frequentist's definition, exist, and that the likelihoods, which they assign, are estimates of these probabilities and are obtained from frequencies which they experience. In his paper cautioning against the blind use of Bayes' theorem, Alvan Feinstein, M.D. [1977] states that

> if such a probability were to be estimated, the clinician would do so by recalling the spectrum of patients with hemoptysis and trying to enumerate the number of people in the spectrum who had lung cancer.

When obtaining likelihoods for the MEDAS expert system, Ben-Bassat, Carlson, et al. [1980] proceeded as follows:

For example, in the urokinase pulmonary embolism trial study [National Co-operative Study, 1974], the publication contains valuable reference to probabilities based on a large series of cases. Using the data accumulated during that study, the conditional probability of abnormal ST and T wave electrocardiographic changes associated with pulmonary thromboembolism is estimated to be 59%... For cases where quantitative data is not available, the clinician subjectively estimates the P_{ij} and $\overline{P}_{ij}$ values, the prior probability and the severity for each order.

In another paper [Ben-Basset, Klove, & Weil, M.D., 1980], Ben-Basset investigates the impact of the deviation of estimated probabilities from "true" (Basset's word) probabilities on the accuracy of the results. We will discuss this impact later in this chapter. The point here is that many physicians believe that there are "true" probabilities which can best be estimated by frequencies, and that the likelihoods, assigned by physicians, are estimates of these probabilities based on their experiences with frequencies.

We conclude then that many of the likelihoods, which would be useful to an expert system, are probabilities, which can be obtained from relative frequencies. Whether we argue that they are objective or are due to the expert's belief in the exchangeability of the sequence is immaterial. In either case we can use a data base to help us determine the probabilities.

2.5.3. The Other Methods for Reasoning Under Uncertainty

We have shown that when representing and manipulating numerical uncertainty in the members of a set of mutually exclusive and exhaustive alternatives, we are bound to the probability calculus. Furthermore, we've shown that if we want to model *coherent* human reasoning numerically in the case of such uncertainty, we should use the probability calculus. We conclude then that probability theory is the appropriate discipline for dealing with much of the uncertainty in expert systems.

Does this imply that the methods introduced in chapter 1 are not appropriate for dealing with uncertainty in expert systems? Not in all cases. However, it is important to clarify precisely where and how they are appropriate. The argument for probability theory has to do with a very specific type of uncertainty, that is, the association of real numbers with the uncertainty in the members of a set of mutually exclusive and exhaustive alternatives. As discussed in chapter 1, fuzzy set theory has to do with a totally different type of uncertainty, namely the uncertainty contained in vague descriptions. Recall that fuzzy set theory is really certain reasoning about partial set membership.

As mentioned in chapter 1, the Dempster–Shafer theory of evidence is an extension of probability theory. We will illustrate this by looking again at the example of the application of the Dempster–Shafer theory given in chapter 1. However, we will look at it in a new light as discussed by Shafer [1986]. Let D be the proposition "the streets are slippery," let E be the proposition "Fred says the streets are slippery," and let rule 1 be "IF E, THEN D." In chapter

1 we associated a bpa of .8 with this rule. We now investigate the relationship between this assigned bpa value and probability theory. Suppose Fred walks into Joe's office and announces that the streets are slippery. Joe hypothesizes that there is a conditional probability of the streets being slippery given that Fred announces that they are. This probability is $P(D \mid E)$. If on 1000 previous days, Fred had announced to Joe that the streets were slippery, then Joe could judge this probability very accurately. Suppose, however, that this situation had never occurred. Then, although Joe feels that there is a probability value associated with the information, he would not know it. Suppose further that Joe and Fred are good friends, and in many other encounters with Fred, Joe had observed that Fred had been careful to say the truth 80% of the time; the other 20% of the time Fred simply answered whatever popped into his head. If we let C be the proposition "Fred is careful," we then have the following probabilities:

$$P(C) = .8 \qquad \text{and} \qquad P(\neg C) = .2.$$

Thus Joe knows the probability values in another probability space. The Dempster–Shafer theory uses the probabilities in this other space to obtain beliefs in events in the space for which Joe does know the probability values. The beliefs are assigned by considering compatible events between the two spaces. For example, given E (i.e., that Fred says the streets are slippery), then D and C can both be true simultaneously. The streets can be slippery and Fred can be being careful. On the other hand, $\neg D$ and C cannot both be true simultaneously (we assume that if Fred is being careful and says that the streets are slippery, then they really are slippery). Therefore, C is compatible with D but not with $\neg D$, and $P(C)$ is assigned as belief in $\{D\}$. Either D or $\neg D$ could simultaneously be true with $\neg C$. Therefore $\neg C$ is compatible with both D and $\neg D$, and $P(\neg C)$ is assigned as belief in $\{D, \neg D\}$. We have then that

$$m_1(\{D\}) = P(C) = .8$$
$$m_1(\{D, \neg D\}) = P(\neg C) = .2$$
$$m_1(\{\neg D\}) = 0,$$

and, as obtained in chapter 1, the belief interval for $\{D\}$ is $(.8, 1)$. Thus, although belief intervals are not probabilities, they are obtained from probabilities. As noted by Shafer [1986],

> All the usual devices of probability are available to the language of belief functions, but in general they are applied in the background...before extending to... the frame of interest.

We see then that the Dempster–Shafer theory extends probability theory to obtain beliefs in propositions in cases where we are unable to obtain the probability values needed to obtain the probabilities of the propositions. In chapters 4 through 9 of this text we assume that we have the probability values necessary to the techniques which we present. In chapter 10 we discuss means of handling our uncertainty in these probability values. Using the methods in chapter 10, we can also represent the above problem directly with probability theory.

One final point about the Dempster–Shafer theory concerns the method for combining beliefs from two separate items of evidence. Suppose Joe's thermometer says that the temperature is 31 degrees Fahrenheit, and Joe knows that ice could not form on the streets at this temperature. If F is the proposition "the thermometer reads 31 degrees," then rule 2 is "IF F, THEN $\neg D$." If we know Joe's thermometer to be absolutely accurate, this would be conclusive evidence that the streets are not slippery, wiping out the evidence of Fred's statement. However, suppose Joe knows that the probability of his thermometer working is .9; that is, if W is the proposition "the thermometer works," Joe knows that $P(W) = .9$. Then, as shown in chapter 1, this evidence would yield a belief interval for D of $(0, .1)$. The question is how we should combine the two separate items of evidence. If we assume that Fred's carefulness and the working of the thermometer are independent, we have

$$P(C \wedge W) = (.8)(.9) = .72 \qquad P(C \wedge \neg W) = (.8)(.1) = .08$$
$$P(\neg C \wedge W) = (.2)(.9) = .18 \qquad P(\neg C \wedge \neg W) = (.2)(.1) = .02.$$

To combine the evidence, and thereby obtain our new beliefs in D and $\neg D$, we determine the compatibility of D and $\neg D$ with the events in this new probability space. This amounts to using the computational technique for combining bpa's illustrated in chapter 1.

In this particular example, it seems reasonable that Fred's carefulness would be independent of the working of the thermometer. In many cases, it is not the case that separate items of evidence are independent. These assumptions of independence will be discussed much more in chapter 4. However, we note here that one of the major recent advances in probabilistic reasoning has been to remove the necessity of making such independence assumptions. Shafer [1986] notes that the combinatoric technique, which we introduced in chapter 1, is not the only way of combining evidence in the Dempster–Shafer theory. We could use a technique which accounts for the dependency between the items of evidence.

Finally, we mention the calculus for manipulating certainty factors. Heckerman [1986] has shown that if this calculus is redefined to remove certain inconsistencies, then the calculus is equivalent to the early probability-based method introduced in chapter 4 and therefore it makes the same assumptions of conditional independencies as that method (the inconsistencies are not in the certainty factor model itself; instead they are between the combination functions in the model and the probabilistic interpretation given to the model). As noted above, one of the major advances of recent research has been to remove the necessity of making such assumptions. The modern probability-based methods, discussed in chapters 5 through 10 of this text, do not make such assumptions. The calculus for combining certainty factors, like the early probability-based method discussed in chapter 4, is of interest primarily for its historical significance.

We are left then with probability theory and its extension, the Dempster–Shafer theory of evidence, as being the most appropriate methods for dealing with numerical uncertainty in the members of a set of mutually exclusive and exhaustive alternatives. The remainder of this text is devoted to the methods which use probability theory directly.

PROBLEM 2.5.3

Consider the Dempster–Shafer problem in which we have the two items of evidence: "Fred says the streets are slippery" and "the thermometer reads 31 degrees." Obtain the beliefs in D and $\neg D$ by considering the compatibility relationships. Notice that no event is compatible with $C \wedge W$.

2.5.4. What Do We Mean by "A Priori" Probabilities in Expert Systems?

Recall from section 2.1, when a probability space is created, we called the probabilities based on the initial information "a priori" probabilities. In an expert system, the initial information would be that in the particular case which is now being considered, all that is known is that it is one of the cases which is ever considered by this particular system. For example, in a medical diagnostic expert system, the a priori probability of the proposition "the patient has lung cancer" is the probability that any patient who is ever considered by this particular system has lung cancer. This may be the probability of a random patient at a particular clinic having lung cancer, the probability of a random individual in the United States or England having lung cancer, or it may be the probability of a random individual in the world having lung cancer. The a priori probability depends on the population for which the system is designed. Similarly, in an expert system for determining the mineral potential of an excavation site, the a priori probability of the proposition "there are massive sulfide deposits" is the probability of there being massive sulfide deposits in the type of land which would ever be considered, by this system, as having sulfide deposits. It may not include swamps, frozen tundra, etc.

A more modern term for a priori probability is prior probability, since, strictly speaking, "a priori" really means independent of experience. Probabilities based on additional information are conditional probabilities. For example, when it is learned that a person is a smoker, there is a new probability of his having lung cancer. This is the conditional probability of a person having lung cancer given that the person is a smoker. When a chest X ray is returned positive, there is a new (conditional) probability of the person having lung cancer given this information. We need a method for determining the new (conditional) probability of the patient having lung cancer given these two items of information. Furthermore, we need a propagation scheme to determine how this information affects the probabilities of other propositions. For example, if patients with lung cancer often have dyspnea, we need to be able to determine the new probability of the patient having dyspnea. The development of these combinatoric techniques and propagation schemes is the focus of much of the remainder of this text.

2.5.5. Where Do We Get the Numbers and How Good Are They?

If we obtain coherent subjective probabilities from an expert, theoretically the system could make judgments consistent with those of the expert. Even though this is somewhat of an idealization, in a case where the probabilities are not related to relative frequencies, we are bound to those probabilities. However, in a case in which the probabilities are related to relative frequencies, we could possibly improve the system if we had probabilities which were better than those obtained solely from the expert. In the last section we saw that de Dombel's system for diagnosing gastrointestinal diseases performed much better when the probabilities were obtained from statistical data. We investigate here the difficulty in obtaining good values of the necessary probabilities. This section only pertains to probabilities which are related to relative frequencies.

We note what probability values are needed by focusing on medical expert systems. In general, a medical expert system has the task of determining the probability of the presence of a disease from the symptoms which are present. If we let

$$D = \text{a particular disease} \quad \text{and} \quad T = \text{a particular symptom,}$$

then, by Bayes' theorem, we have

$$P(D \mid T) = \frac{P(T \mid D)P(D)}{P(T \mid D)P(D) + P(T \mid \neg D)P(\neg D)}.$$

The value actually needed by the expert system is $P(D \mid T)$. All the methods which we will discuss use the definition of conditional probability to obtain $P(D \mid T)$ from $P(T \mid D)$, $P(T \mid \neg D)$, and $P(D)$. We shall now investigate the problems in obtaining those probabilities.

First we must make it clear what we mean by a valid sample point. Suppose that

$$D = \text{patient has cancer}$$

$$T = \text{patient tests positive on xeromammography,}$$

and we wish to determine the $P(T \mid D)$. We would need a sample of patients for whom it can be determined for certain whether they have cancer and whether they test positive. If we do not have an expert supplying probabilities, our value for $P(T \mid D)$ would then be given approximately by

$$P(T \mid D) = \frac{\text{no. of patients who test positive and have cancer}}{\text{no. of patients with cancer}}.$$

TABLE 2.1 Conditional Probabilities for the
Xeromammography Test Obtained from Four
Different Sources

Source	$P(T \mid D)$ (No. of Patients)	$1 - P(T \mid \neg D)$ (No. of Patients)
1	.78 (108/138)	.37 (99/268)
2	.64 (31/48)	.92 (1656/1809)
3	.98 (547/559)	.46 (660/1446)
4	.96 (64/67)	.80 (33/41)

Source: Harris [1981].

In chapter 10 we discuss the exact means of obtaining such probabilities from a data base and a method for representing our uncertainty in the probabilities. In the other chapters, we will treat the probabilities as if they are known for certain. We will see shortly that this may not be as dangerous as it seems. A valid sample point is obtained from a patient who has taken a xeromammography test (the results of which we assume are indubitably positive or negative) and who later had a biopsy which revealed for certain whether cancer was present.

John Harris, M.D. [1981] performed a search through the medical literature for results which compiled data on patients for seven tests, including the xeromammography test. He found four reports for that test; their results are summarized in Table 2.1. We can see that there is considerable variation from report to report even though some of the sample sizes are quite large. Harris found these same large discrepancies in five of the tests investigated, and he found a good correlation in the other two. We might conclude for those other two tests that we are able to obtain good values for the conditional probabilities. However, what values should we use for the remaining five tests? Harris notes that

This difference is hardly trivial. To many clinicians it may well be the difference between further diagnostic studies and no studies. Pooling the published information does not make statistical or clinical sense. Seeing the different values reported, a careful physician may ask which author's patients and techniques most duplicate his own.

In spite of these difficulties, Harris notes that two of the tests show very little variability. He questions whether this is a chance phenomenon or whether some systematic error has been consistently applied by several researchers. Harris concludes that such summary reports are still clearly superior to a semiquantitative or qualitative approach. He recommends that potential authors attempt to standardize their methods and control for sources of bias.

Thus we can get good probabilities (as in the case of two of the tests), but we must take care to do so. What are we to do in the case where we cannot obtain good probabilities? For example, what should we do if the four reports in Table 2.1 are our only source of conditional probabilities for the xeromammography test? A result of Ben-Basset [Ben-Basset, Klove, & Weil, 1980] indicates that even some apparently poor probabilities may be good enough to yield accurate results.

Ben-Basset investigated the effect of the error in the arguments in the above equation for $P(D \mid T)$ on the error in the computed value of $P(D \mid T)$. For example, the likelihood ratio λ is defined by

$$\lambda = \frac{P(T \mid D)}{P(T \mid \neg D)}.$$

Let λ be the "true" value of the likelihood ratio and $\hat{\lambda}$ be our estimate. Suppose

$$P(D) = .5 \qquad \text{and} \qquad \hat{\lambda} = 1.$$

Ben-Basset found that as λ ranged between $.65\hat{\lambda}$ and $1.35\hat{\lambda}$, the "correct" value for $P(D \mid T)$ (computed from Bayes' theorem) only ranged between .39 and .57, whereas the estimated value of $P(D \mid T)$ was .5. Realistically, Ben-Basset did not feel that the deviation of $\hat{\lambda}$ from λ would be worse than the margin he checked. Yet the resultant error in $P(D \mid T)$ was at most .1. He repeated the computations for $\hat{\lambda}$ estimates of 2, 5, and 10, feeling that these values were representative of the entire range of $\hat{\lambda}$ values. The results, which are summarized in Table 2.2, were the same. The resultant error in $P(D \mid T)$ was at most .1. Next Ben-Basset performed the same analysis for $P(D)$ having values of .09, .25, and .83 (again values which he felt were representative of the entire range of values of $P(D)$). The results, which are not presented here, were again the same. The error in $P(D \mid T)$ was at most .1. Finally, holding λ fixed at specific values, Ben-Basset analyzed the effect of the error in the estimate of $P(D)$ on the error in the computed value of $P(D \mid T)$. Again the resultant error was at most .1.

Clearly, the error is most significant when $P(D \mid T)$ is small. However, in this case additional evidence is in order anyway. In the paper Ben-Basset analyzes these results further. He then concludes

> These results indicate that Bayesian models tolerate large deviations in the prior and conditional probabilities. That is, even rough estimates for which qualitative expressions such as "rare," "frequent," and "probable" serve as guidelines may be accurate enough to result in the recommendation of the correct decision.

Ben-Basset's results are not a license to use any old numbers and hope things come out in the wash. Rather we should get the best possible numbers we can. For example, the variation in the estimates of $\hat{\lambda}$ in Table 2.1 is substantially worse than the possibilities considered by Ben-Basset. Therefore we still need further efforts to refine the value of the conditional probabilities for

TABLE 2.2 The Computed Value of $P(D \mid T)$ as a Result of Multiplicative Changes in the Likelihood Ratio λ in the Case Where $P(D) = .5$

λ \ $\hat{\lambda}$	1.00	2.00	5.00	10.00
$.65\hat{\lambda}$	.39	.56	.76	.87
$.90\hat{\lambda}$	.47	.64	.82	.90
$1.00\hat{\lambda}$	.50	.67	.83	.91
$1.10\hat{\lambda}$	.52	.69	.85	.92
$1.35\hat{\lambda}$	.57	.73	.87	.93

Source: Ben-Basset, Klove, and Neil [1980].

the xeromammography test. Ben-Basset's results do show that, in many cases where the indications are that the numbers are not as good as we had hoped, they may still be good enough. In any case, the numbers are a start and they can be refined in the testing stage of the system. Furthermore, by incorporating our uncertainty in the probabilities into the system (as discussed in chapter 10), the system will never claim to be more certain than it actually is.

2.6. THE PRINCIPLE OF MAXIMUM ENTROPY

We return now to the philosophy of probability to discuss one more approach, namely the principle of maximum entropy, because this approach has been applied to expert systems [Lemmer, 1983; Cheeseman, 1984; Shore, 1986; Wen, 1988].

Even though Laplace and his contemporaries believed that the substances in the universe react in a totally deterministic way and that therefore there are no absolute probabilities (see section 2.7), they apparently felt that the information about the substances has *objective* probabilities associated with it. In section 2.1 we noted that the problem in this approach is that the probabilities are based on logic alone [Nagel, 1939]. That is, if we expect the probability to approximate the relative frequency with which the event will occur, there is no proof that this will happen. Furthermore, we noted in that section that Jaynes [1979] feels that this criticism is resolved by removing the criteria involving the convergence of the relative frequency. Jaynes does seem to recognize a need for evidence for the existence of objective probabilities; however, to Jaynes the evidence need not be the approximate convergence of relative frequencies. First we briefly explain Jaynes' extension to the principle of indifference, which is called the principle of maximum entropy. Then we discuss the evidence for the existence of the objective probabilities obtained using this principle. Finally we note some objections to the principle of maximum entropy. Before proceeding, we note that Jaynes can be considered a subjectivist and that he does not really call probabilities, obtained with the principle of maximum entropy, "objective." However, he does state

[Jaynes, 1979]

> But it is just the point of the maximum-entropy principle that it achieves "objectivity" of our inferences, in the sense that we base our predictions only on the information that we do, in fact, have...

and

> The *probabilities*... are an entirely correct description of our *state of knowledge*...

Therefore it seems that Jaynes' approach does consider probabilities objective in the same sense as the logical approaches of Carnap [1950] and Keynes [1948], but not in the physical sense of the frequentists. That is, he feels that there are logically correct probabilities based on the information. However, he goes a step further than the other logical approaches in that in some cases he obtains physical evidence for his probabilities which is not relative frequencies.

Jaynes begins with the principle of indifference, by stating that if the only information is that there are n mutually exclusive and exhaustive events, then the probability of any one of them, relative only to that information, is $1/n$. As noted in section 2.1, according to Jaynes this interpretation of the principle of indifference yields the least presumptive assignment of probabilities. Jaynes argues that probability has to do with a single occurrence, not with repeated trials (notice that this is the opposite of von Mises' point of view). As mentioned previously and as we shall soon show, the evidence for the existence of Jaynes' probabilities is not relative frequencies.

Jaynes then extends the principle of indifference as follows: If there is information about the experiment which implies that the probabilities cannot all be $1/n$, then the least presumptuous assignment of probabilities is the one which comes as close as possible to distributing the probabilities equitably while satisfying the information. Jaynes calls this rule for assigning probabilities the principle of maximum entropy. If we define the entropy H in the experiment by

$$H = -\sum_{i=1}^{n} p_i \ln p_i,$$

where p_i is the probability of the ith alternative, then it can be shown that H is minimized when $p_i = 1$ for some i, and maximized when $p_i = 1/n$ for all i. Thus the principle of maximum entropy states that in the light of information which implies that the probabilities cannot all be $1/n$, the least presumptuous assignment of probabilities is the one which maximizes H.

Jaynes illustrates this principle with his Brandeis dice problem. He supposes a six-sided die for which it is known that, in repeated rolls, the average number of spots is 4.5, not 3.5, as would be expected for a fair die. That is,

$$\sum_{i=1}^{6} i p_i = 4.5.$$

Given this information and nothing else, the problem is to determine the probabilities of the six sides turning up on the next roll. According to the principle of maximum entropy, the solution is the one which maximizes H relative to this mean value constraint. Jaynes solves the problem as a standard variational problem solvable by stationarity using the Lagrange multiplier technique. In this way he determines the probabilities to be

$$\{p_1, p_2, p_3, p_4, p_5, p_6\} = \{.05435, .07877, .11416, .16545, .23977, .34749\}.$$

The evidence for the existence of the probabilities determined using this principle is not relative frequencies, but rather the fact that the principle predicts results in physics which agree with our macroscopic measurements. As noted by Jaynes [1979],

> The price is simply that we must loosen the connections between probability and frequency, by returning to the original viewpoint of Bernoulli and Laplace. The only new feature is that their Principle of Insufficient Reason is now generalized to the Principle of Maximum Entropy. Once this is accepted, the general formalism of statistical mechanics...can be derived in a few lines without wasting a minute on ergodic theory...

> The price we have paid for this simplification is that we cannot interpret the canonical distribution as giving the *frequencies* with which a system goes into the various states. But nobody has ever needed that interpretation anyway. In recognizing that the canonical distribution represents only our state of knowledge when we have certain partial information derived from macroscopic measurements, we are not losing anything we had before.

Thus Jaynes is able to substantiate the existence of probabilities obtained using the principle of maximum entropy with evidence which is not relative frequencies, refuting, in the case of these applications to physics, Nagel's [1939] criticism that the problem with the classical approach is that there is no proof that the probabilities will agree with relative frequencies. Notice, however, that when he extends the principle of maximum entropy to problems such as the Brandeis dice problem, this evidence is no longer present, and we once again need some kind of evidence (perhaps relative frequencies) to establish the existence of objective probabilities. Consider the following example. Suppose we have a three-sided die with 1, 2, and 3 spots inscribed on the respective sides (the construction of a such a die is left to an ingenious engineer), and the only information about the die is that the average number of spots in repeated rolls is known to be 2, the value which would be obtained if the die were fair. For example, possibly the die was purchased at a novelty store and therefore we had no reason to believe that the die was fair or that it had a particular bias. However, a friend later tossed the die many times and found that the mean value was 2. The friend shared the mean value with us, but not the actual results of the individual rolls. Thus we believe almost for certain that the mean value is 2, and this is our only information about the probabilities. Since values of 1/3 for all p_i satisfy this mean value constraint, and since these values yield the absolute maximum entropy, the maximizing entropy solution is easily seen to be $p_i = 1/3$ for all i.

Next we attempt to achieve "objectivity of our inferences" by considering all probability distributions which satisfy the constraints in the problem equally probable. That is, we apply the principle of indifference to the probability values. To that end, if we assume that p_i has some objective value (a propensity) and if we let x_i be a variable standing for the possible values that p_i could have, then the total constraints in this problem are the following (under the assumption that the mean value constraint means that we are essentially certain that the mean value, calculated from the objective probabilities, is equal to 2):

$$0 \leq x_i \leq 1$$

$$x_1 + x_2 + x_3 = 1$$

$$x_1 + 2x_2 + 3x_3 = 2.$$

If we fix x_1 and solve the above two equations for x_2 and x_3, we obtain

$$x_3 = x_1 \quad \text{and} \quad x_2 = 1 - 2x_1.$$

Since $0 \leq x_2 \leq 1$, the only values of x_1 which are possible are ones satisfying the inequality

$$0 \leq 1 - 2x_1 \leq 1.$$

Thus the possible values of x_1 are the ones which are in the interval $[0, .5]$, and it is clear that each of these values can occur in exactly one way; that is, with $x_3 = x_1$ and $x_2 = 1 - 2x_1$. Therefore, based on this application of the principle of indifference, the probability density function for p_1 having value x_1 in the interval $[0, .5]$ is the uniform density function $1/(.5 - 0)$, and the expected value of p_1, based on this distribution, is given by

$$\hat{p}_1 = \int_0^{.5} \frac{x_1}{.5 - 0} dx_1 = .25.$$

In the same way, if we fix x_2, we find that each value of x_2 in the interval $[0, 1]$ can occur in exactly one way and the expected value of p_2, obtained using the principle of indifference, is .5. Finally, fixing x_3, we find that each value of x_3 in the interval $[0, .5]$ can occur in exactly one way and no other values can occur at all, and therefore the expected value of p_3, obtained using the principle of indifference, is .25.

This solution is clearly different from the maximum entropy solution of 1/3 for all p_i. It is also intuitively appealing, since, given that p_2 could be as high as 1 while the other probabilities are bounded above by .5, we may be inclined to bet on a 2. Once the information gives us reason to prefer one alternative over the others, we can no longer claim that the probabilities, based on the

information, are equal. Without some kind of evidence, we are left with two
conflicting intuitively appealing solutions.

Results of Dias and Shimony [1981] shed light on this intuitive conflict. The
event that the mean value is equal to 2 is not an event in the space consisting
of a single roll of the three-sided die. Suppose, however, that this space is
extended to a space which includes this event and, in this extended space, we
compute the conditional probability of a 2 on a given roll given the constraint
that the mean value is r where $0 \leq r \leq 3$. Dias and Shimony show that the
maximum entropy solution, relative to the constraint that the mean value is
r, can be equal to this conditional probability in an extended space only if
the a priori probability (i.e., the probability before knowledge of the mean
value constraint is obtained) is 1 that the mean value will be equal to 2. For
example, consider Jaynes' [1979] following interpretation of the mean value
constraint:

> When a die is tossed, the number of spots up can have any value i in $1 \leq i \leq 6$.
> Suppose a die has been tossed N times and we are told only that the average
> number of spots was not 3.5 as we might expect from an "honest" die but 4.5.
> Given this information *and nothing else*, what probability should we assign to i
> spots on the next toss?

Applying this interpretation to the three-sided die, the maximum entropy solution can be the conditional probability for the $N + 1$st roll given that the
sample average of the first N rolls is equal to r (in a space consisting of
$N + 1$ rolls of the die) only if we are practically certain that r will equal 2
before any rolls are made.

Thus, in order for the maximum entropy solution to be consistent with the
axioms of probability theory and the definition of conditional probability in
an extended space, we must do much more than assume complete ignorance
a priori. Dias and Shimony [1981] note that such additional assumptions are
contrary to Jaynes' intentions:

> This assignment seems like too much knowledge to extract out of a situation
> of presumed ignorance, and indeed it goes directly counter to Jaynes' sensible
> maxim that we should be honest about the extent of our ignorance.

Dias and Shimony [1981] also relate the maximum entropy solution to Carnap's [1952] continuum of inductive methods called "the λ continuum." Carnap has developed a family of probability functions, c_λ, one for each real value
of λ, where $0 \leq \lambda \leq \infty$. They show that precisely one of Carnap's continuum,
namely c_∞, is related to maximum entropy probabilities. A characteristic of
c_∞ is that the same a priori probabilities are assigned to all states. For example, if we rolled the three-sided die 3 times, there are 27 possible states in the
space. The function c_∞ assigns a prior probability of 1/27 to each of them. If

we represent, for example, the outcome that 2 lands on the first roll, 1 on the second, and 3 on the third by the sequence 213, this means that

$$c_\infty(111) = c_\infty(112) = \cdots = c_\infty(333) = 1/27.$$

In the context of the probability functions in the λ continuum, the sample space consists of all possible outcomes of N rolls of the die, the constraint that the mean value is equal to r means that the sample average of the outcomes of the N rolls is equal to the constraining value r, and the event that, say a 2 occurs, is the event that a roll, chosen randomly from the N rolls, turns up a 2. We will denote the event that the sample average equals r in the N rolls by $MV_N = r$ and the event that an i turns up on a roll chosen randomly from the N rolls simply by i. Dias and Shimony show that the maximum entropy solution, relative to the constraint that the mean value is equal to r, is equal to

$$\lim_{N \to \infty} c_\infty(i \mid MV_N = r).$$

Furthermore, they show that the maximum entropy solution does not equal

$$\lim_{N \to \infty} c_\lambda(i \mid MV_N = r)$$

for $\lambda \neq \infty$. Thus only c_∞ can be related to the maximum entropy method. (Dias and Shimony caution that we cannot simply identify c_∞ with maximizing entropy due to the great differences in Carnap's and Jaynes' approaches. A relationship only exists in cases where both methods apply.)

Dias and Shimony use this relationship between c_∞ and maximizing entropy to explain why the maximum entropy method is often successful in statistical mechanics. They note [Dias and Shimony, 1981] that a corollary of the property that all states are assigned equal a priori probabilities by c_∞

> ...is reminiscent of one of the common procedures of statistical mechanics, the use of the microcanonical ensemble, which assigns equal weights to all states in an energy hypersurface. It is well known that once one has a microcanonical ensemble to represent an isolated system, one can straightforwardly derive the canonical distribution to represent a system in contact with a heat bath: one lets the system and the heat bath together constitute an isolated system, characterized by the microcanonical distributions, and then combinatorial considerations lead to the desired conclusion. In this way one comes by the standard route to the canonical distribution, which Jaynes derives by mean of his maximum entropy principle. Does this consideration provide a vindication of Jaynes' method—or even somewhat more than a vindication, for his method arrives at the conclusion more efficiently than do the usual statistical mechanical arguments?
>
> The answer, we believe, is negative. It is only in rather special situations that one can assert the equiprobability of all states of a given class in statistical physics, and when this can be done there is a physical reason, such as ergodicity or an

appropriate symmetry. If one is employing the logical concept of probability, such a reason should be included as part of the evidential proposition.

Suppose next that we represent a state of complete prior ignorance as to the possible outcomes of the roll of the three-sided die by a symmetric Dirichlet distribution in which no prior knowledge is assumed, and we assume that the sequence of rolls is infinitely exchangeable. (In chapter 10 we will show how to accomplish this.) Zabell [1982] has proved that in many cases our prior knowledge must be represented by Dirichlet distributions. (This result is also discussed in chapter 10.) It is a characteristic of this representation that all states are not assigned equal a priori probabilities. Rather if $N = 3$, we have, with this representation, that a priori

$$P(222) = 1/10, \qquad P(122) = 1/30, \qquad P(123) = 1/60.$$

The reason that these probabilities are not equal will become clear in chapter 10, when the method for computing them is discussed. Briefly, note that

$$P(222) = P(2 \mid 22)P(2 \mid 2)P(2) \qquad \text{and} \qquad P(123) = P(3 \mid 12)P(2 \mid 1)P(1).$$

Intuitively, if the first two rolls turn up 2, then we would feel that it is probable that the propensity of the die to land 2 is high. Therefore $P(2 \mid 22)$ is greater than $P(3 \mid 12)$. Recall that we are assuming compete ignorance as to the propensity before any rolls are made.

Within the context of the λ-continuum, this representation is equivalent to taking $\lambda = 3$ in the case of rolling the three-sided die. Dias and Shimony show that, if the mean value constraint is $MV_N = 2$ and again i represents the fact that a roll chosen randomly from the N rolls is equal to i, then

$$c_3(1 \mid MV_N = 2) = .25, \qquad c_3(2 \mid MV_N = 2) = .5, \qquad c_3(3 \mid MV_N = 2) = .25$$

for all even values of N.

Furthermore, if we let P_N represent the probability of a 2 on the $N + 1$st roll given $MV_N = 2$, then, assuming a symmetric Dirichlet distribution with no prior knowledge and infinite exchangeability, it is possible to show [Neapolitan, 1989a], using the methods in chapter 10, that

$$P_N = .5 \quad \text{for all odd } N \qquad \text{and} \qquad \lim_{N \to \infty} P_N = .5 \quad \text{for all } N.$$

We saw in the quote above [Jaynes, 1979] that Jaynes actually meant that the probability obtained from the mean value constraint should be interpreted as the probability of 2 on the $N + 1$st roll given $MV_N = 2$.

Hence the alternative solution, which is obtained by applying the principle of indifference to the probability values subject to the mean value constraint, agrees with the solution obtained by representing prior ignorance by symmetric Dirichlet distributions in which no prior knowledge is assumed.

Neapolitan [1989a] has derived an algorithm for determining the expected probabilities from distributions obtained by applying the principle of indifference to the probability values subject to an arbitrary mean value constraint for the general case of n alternatives. No results of applying this algorithm have agreed with maximum entropy solutions.

Regardless of these objections, the fact remains that in many empirical problems researchers have found maximum entropy solutions useful (note references at the beginning of this section). Seidenfield [1986] observes that

> A pragmatic appeal to successful applications of MAXENT formalism cannot be dismissed lightly. The objections that I raise in this paper are general. Whether (and if so, how) the researchers who apply MAXENT avoid these difficulties remains an open question. Perhaps, by appeal to extra, discipline-specific assumptions they find ways to resolve conflicts within MAXENT theory. A case-by-case examination is called for.

For example, consider again the case of the three-sided die. If one purchased the die at a store which is known to sell fair gambling apparatus, then a priori one would *believe* almost for certain that the die was very close to fair, and thus maximizing entropy relative to a mean value constraint would appear to be a good source of subjective probabilities. On the other hand, in the case where one had no reason to *believe* a priori that the die was fair (the die was purchased at a novelty store), the application of the principle of indifference, given in this section, would appear to be a better source of subjective probabilities.

2.7. ARE THERE ABSOLUTE PROBABILITIES?

We close this chapter with a discussion of absolute probabilities. As mentioned in section 2.1, Laplace and his contemporaries believed in determinism, as is evident from the following statement [Laplace, 1951]:

> We ought then to regard the present state of the universe as the effect of its anterior state and as the cause of the one which is to follow. Given for one instant an intelligence which could comprehend all the forces by which nature is animated and the respective situation of the beings who compose it—an intelligence sufficiently vast to submit these data to analysis—it would embrace in the same formula the movements of the greatest bodies of the universe and those of the lightest atom; for it, nothing would be uncertain and the future, as the past, would be present to its eyes.

Modern definitions of determinism do not embrace the notion of causation. Rather they only say that determinism means that it is possible to predict any future-state description of the universe from a present-state description (this definition is a simplification; for a complete discussion of determinism, the interested reader is referred to Earman [1986]).

In a deterministic universe a chance event is not one which might happen, since all events will happen in a specified manner. Probability therefore only exists relative to partial information. This is clear in the example of drawing a card from a deck. In light of the information as to the top card, all probability values other than 1 and 0 disappear. However, what of the toss of a coin? In this case there is nothing at which to peek. The argument here is that if one knew the position of the coin in the hand, the torque placed on the coin, the atmospheric conditions, etc., one could predict with certainty how the coin would land. It is not that we have sufficiently sophisticated instrumentation to accomplish this, but rather that with such instrumentation one theoretically could accomplish it. This belief is totally consistent with the physics of the 19th century, namely classical physics. However, recall Example 2.10, in which we saw that modern physics, in the case of diffraction, cannot predict the actual location of an observable. It can only give a probability as to the location. Is this an absolute probability which contradicts determinism?

The theory of quantum mechanics implies that if Δx is the uncertainty in the position of a particle and Δp is the uncertainty in the momentum, then

$$\Delta x \Delta p \geq \frac{h}{4\pi},$$

where h is Planck's constant. This statement is the mathematical version of the well-known Heisenberg uncertainty principle. Thus quantum mechanical theory implies that it is not possible to determine the present state (both the position and momentum) of every particle in the universe, and therefore we cannot predict all future events. We offer three possible interpretations to these results:

1. We can maintain that the theory of quantum mechanics implies that there is an observer-independent absence of precise future values of observables. That is, we can embrace an underlying indeterminism. According to this interpretation, the electron "decides" on its path at the instant it goes through the slit. Belinfante [1970] stresses the implications of this intepretation for religion by pointing out that indeterminism is consistent with the belief in God, who makes his own decisions about the universe. We are not capable of predicting these decisions. If the universe were deterministic, there would be nothing for God to do in the universe.

2. We can maintain that the theory of quantum mechanics implies that it is not possible to determine whether there is an observer-independent absence of precise future values for observables. Such an interpretation allows that events could be happening in a specified manner, but that if they are, the ability to conceptualize that manner is beyond our capabilities (manners which we can conceptualize are addressed by hidden variable theories, which are discussed below). In any case, if, by determinism, we mean that it is theoretically possible for *us* to predict the future state of the universe from its present state, then this interpretation is still a belief in indeterminism.

3. We can maintain that the quantum mechanical description of physical reality is incomplete. That is, the calculations in the quantum theory agree very

accurately with the facts of nature, but the theory does not contain a counterpart for every element of physical reality. Einstein et al. [1935] attempted to show that this is the case with the Einstein–Podolsky–Rosen paradox. Their argument proceeds as follows. They assume that a necessary condition for a scientific theory to be complete is that "every element of physical reality must have a counterpart in the physical theory." Furthermore, they assume that the following is a sufficient condition for a physical quantity to be part of physical reality: "If, without in any way disturbing a system, we can predict with certainty (i.e., with probability equal to unity) the value of a physical quantity, then there exists an element of physical reality corresponding to this physical quantity." They proceed to give an example of a particular system, composed of two particles, A and B, which are initially interacting and which separate. They show that, after the separation, it is possible to determine with certainty the momentum of A, without in any way disturbing A, by measuring the momentum of B. Likewise, it is possible to determine the position of A, without in any way disturbing A, by measuring the position of B. They then further assume the locality assumption: If "at the time of measurement...two systems no longer interact, no real change can take place in the second system in consequence of anything that may be done to the first system." Since, at the time when we measure B, A and B are no longer interacting, the measurements at B would therefore have no effect on the values at A. They then conclude that both the position and momentum of A are part of physical reality. Therefore, if quantum mechanical theory were complete, there would be a counterpart in the theory for both the position and momentum of A. That is, the theory would contain a description of both of these quantities and hence they would both be predictable, contrary to the theory.

An immediate criticism of this paradox is that the criteria for physical reality is not sufficiently restrictive. If we require that objects are simultaneously elements of physical reality only if they can be simultaneously measured or predicted, then the paradox breaks down. However, as noted by Einstein et al. [1935], according to this criteria, the reality of the position and momentum of A would depend on the measurement taken on B, a measurement which does not disturb A in any way. They contend that "no reasonable definition of reality could be expected to permit this."

The interested reader is referred to Jammer [1974] for a discussion of other refutations of this paradox. We proceed here with possible conclusions based on an acceptance of the paradox. Einstein et al. conclude the paper with the statement:

> While we have thus shown that the wave function does not provide a complete description of the physical reality, we left open the possibility of whether or not such a description exists. We believe, however, that such a theory is possible.

A complete theory could either extend the existing incomplete theory (i.e., quantum mechanics) or it could be incompatible with the existing theory.

The principal efforts to extend the existing theory were the development of hidden variable theories. Such theories maintain that there are variables, hidden from our sight and beyond our control, which correlate the states of A and B after their separation. Thus we can assume that the positions and momentums of both A and B are determined at the moment they separated, and that if somehow we knew the values of these hidden variables at that moment, we could predict those positions and momentums. A hidden variable interpretation of quantum mechanics has been explicitly constructed [Bohm, 1952]. However, Bell [1964] has mathematically proven that the existence of hidden variables is incompatible with the probabilistic predictions of quantum mechanics. Moreover, experimental results have substantiated that it is not possible to extend quantum mechanics with hidden variable theories.

Consequently, hidden variable theories do not appear very plausible. Although Einstein was sympathetic to these theories, he was not a strong proponent of them. He favored a statistical interpretation of the uncertainty in the location of an electron after diffraction. That is, the probability distribution obtained using the Schrödinger equation does not pertain to the location of a particular electron, but instead to a collective of electrons. This statistical view admits the possibility of hidden variables, but it does not require them. Einstein's real hope was that a future unified field theory would cover quantum phenomena. As noted by Einstein [Heisenberg, 1971], "it is theory which decides what we can observe."

We see then that whether there are absolute probabilities in the universe is a difficult and unanswered question which has concerned philosophers and physicists throughout this century. The goal of this section has only been to point out this difficulty.

APPENDIX 2.A. A DERIVATION OF THE AXIOMS OF PROBABILITY THEORY

The results in this appendix are taken from Neapolitan [1989b]. We shall assume that there is a class of experiments, along with information about the experiments, which can be repeated indefinitely and for which the following axioms hold. That class includes those experiments for which we have evidence that they hold: namely, experiments involving the toss of a coin, the death of a 30-year-old, the location of electrons, the smoking behavior of individuals, etc. That is, this class is the class of all collectives as defined by von Mises. We are actually reobtaining von Mises' results with his definition of a probability as a strict limit replaced by the definition of a probability as a limit in the sense of the weak law of large numbers. We first hypothesize the existence of a probability value. Formally,

Axiom 2.1. Let an experiment be given. Let Ω be a set of mutually exclusive and exhaustive outcomes of the experiment, and let $\mathcal{F}$ be the set of all subsets of Ω (technically, if Ω is the set of real numbers or any cartesian product of sets of reals, then the members of $\mathcal{F}$ must be Borel sets). Then, based on certain information, there is associated with each $E \in \mathcal{F}$ a real number $P(E)$, called the probability of E based on this information.

As mentioned in section 2.2, in many cases our experiences can best be described by hypothesizing the existence of such probabilities and by assuming that the relative frequency of occurrence of an event converges to its probability in the sense of the weak law of large numbers.

Axiom 2.2. For every experiment, information, and set of mutually exclusive and exhaustive outcomes Ω:

1. The $P(\Omega)$ is always equal to the same value. We shall call that value C.
2. The $P(\text{NULL})$ is always equal to the same value. We shall call that value Z.
3. C and Z are not equal.

This axiom hypothesizes that regardless of the experiment, a certain event always has the same probability, an event which cannot occur always has the same probability, and the probability of the certain event and the event which cannot occur are not equal.

Axiom 2.3. If E_1 and E_2 are two events such that $E_1 \subseteq E_2$, then

$$|P(E_2) - C| \le |P(E_1) - C|.$$

This axiom states that if the occurrence of a given event E_1, implies the occurrence of another event E_2, then the probability of E_2 must be closer to certainty than that of E_1.

Axiom 2.4. Let an experiment and information be given, let Ω be a set of mutually exclusive and exhaustive outcomes of the experiment, let $\mathcal{F}$ be the set of all subsets of Ω, and suppose each $E \in \mathcal{F}$ has a probability $P(E)$, as hypothesized in Axiom 2.1. Then for every $n \ge 1$ the experiment of repeating the original experiment n times is also in the class of experiments which have associated probability values. That is, if we let Ω^n be the set of n-tuples of Ω, then clearly Ω^n is the set of mutually exclusive and exhaustive outcomes of the experiment of repeating the original experiment n times. If by $\mathcal{F}^n$ we mean the set of all subsets of Ω^n, then each event $E^n \in \mathcal{F}^n$ also has a probability $P^n(E^n)$.

This axiom says that if each event associated with an experiment has a probability, then each event associated with n repetitions of the experiment also has a probability. We have called this probability P^n to make it clear that it is a probability associated with the experiment of repeating the original experiment n times.

Definition 2.8. Let an experiment and information be given, let Ω be a set of mutually exclusive and exhaustive outcomes of the experiment, let $\mathcal{F}$ be the set of all subsets of Ω, and suppose each $E \in \mathcal{F}$ has a probability $P(E)$, as hypothesized in Axiom 2.1. A random variable R on Ω is a function from Ω to the reals. If J is a subset of the reals, $(R \in J)$ means the event which is the subset of all sample points in Ω which are mapped into J by R (again, technically J must be a Borel subset).

Definition 2.9. Let an experiment and information be given, let Ω, $\mathcal{F}$, Ω^n, and $\mathcal{F}^n$ be as in Axiom 2.4, and let $E \in \mathcal{F}$. Then $S^n(E)$ means the random variable on Ω^n which assigns to each n-tuple in Ω^n the count of the number of elements of E in the n-tuple.

Example 2.13. If J is the interval $[0, \delta)$, then the event $(S^n(E) \in J)$ can be written as $(S^n(E) < \delta))$. This is the event which is the subset of all those n-tuples in Ω^n such that the number of members of E in the n-tuple is less than δ. It is clearly the event that E occurs less than δ times in n repetitions of the experiment.

Example 2.14. A real valued function of a random variable is also a random variable. For example,

$$\left| P(E) - \frac{S^n(E)}{n} \right|$$

is a random variable of Ω^n. We have that the event

$$\left(\left| P(E) - \frac{S^n(E)}{n} \right| < \delta \right)$$

is the event which is the subset of n-tuples in Ω^n such that the ratio of the count, of the number of members of E in the n-tuple, to n is within δ of $P(E)$. This is the event that, in n repetitions of an experiment, the relative frequency of occurrence of E will be within δ of the $P(E)$.

Lemma 2.2. Let an experiment and information be given, let Ω be a set of mutually exclusive and exhaustive outcomes of the experiment, let $\mathcal{F}$ be the set of all subsets of Ω, and let R be a random variable on Ω which assigns a constant K to every member of Ω. Then the event $(R < \delta)$ is equal to Ω if $K < \delta$ and is equal to NULL if $K \geq \delta$.

Proof. The proof is obvious. $\square$

Axiom 2.6'. Let an experiment and information be given and let Ω, $\mathcal{F}$, Ω^n, and $\mathcal{F}^n$ be as in Axiom 2.4. Then, for every $E \in \mathcal{F}$ and every ϵ, $\delta > 0$, there exists an $N(\epsilon,\delta)$ such that for $n > N(\epsilon,\delta)$

$$\left| P^n\left(\left| P(E) - \frac{S^n(E)}{n} \right| < \delta \right) - C \right| < \epsilon.$$

This axiom is the statement of the weak law of large numbers. Notice that the axiom is called Axiom 2.6'. This axiom is actually a special case of Axiom 2.6. It is more illustrative to introduce this and another special case (Axiom 2.6") first. The general case (Axiom 2.6) will be given near the end of this appendix. Axiom 2.5 has to do with conditional probability and will be stated later in this appendix.

Next we prove several theorems and lemmas (proofs of the axioms of probability theory are called theorems, while other results are called lemmas):

Theorem 2.5. The $P(\Omega)$ is 1; that is, the value of C is 1.

Proof. Since $C = P(\Omega)$, we have by Axiom 2.6 that for every ϵ, $\delta > 0$, there exists an $N(\epsilon,\delta)$ such that for $n > N(\epsilon,\delta)$

$$\left| P^n\left(\left| C - \frac{S^n(\Omega)}{n} \right| < \delta \right) - C \right| < \epsilon.$$

Since every n-tuple in Ω^n is comprised only of members of Ω, the random variable $S^n(\Omega)$ is equal to n for every n-tuple in Ω^n. We therefore have for $n > N(\epsilon,\delta)$ that

$$|P^n(|C - 1| < \delta) - C| < \epsilon.$$

Notice that $|C - 1|$ is a constant random variable on Ω^n for any n. That is, all n-tuples map into $|C - 1|$. For a given $\delta > 0$, suppose there exists no n such that the event

$$(|C - 1| < \delta)$$

is Ω^n. Then, by Lemma 2.2, $(|C - 1| < \delta)$ is equal to NULL for every n. We therefore have for $n > N(\epsilon,\delta)$ that

$$|P^n(\text{NULL}) - C| < \epsilon.$$

However, by Axiom 2.2 this implies that for $n > N(\epsilon,\delta)$

$$|Z - C| < \epsilon.$$

Since this inequality is true for an arbitrary ϵ, this implies that $Z = C$, which contradicts Axiom 2.2. Therefore for every δ there exists some n such that the event

$$(|C - 1| < \delta)$$

is equal to Ω^n.
   Now if

$$|C - 1| = \delta' > 0$$

then, by Lemma 2.2, the event

$$(|C - 1| < \delta')$$

is equal to NULL for every n. This contradiction completes the proof. $\square$

Henceforth we will replace C by 1 in references to $P(\Omega)$.

Lemma 2.3. The $P(\text{NULL})$ is 0; that is, the value of Z is 0.

Proof. The proof is analogous to that in Theorem 2.5 and is left as an exercise. $\square$

Theorem 2.6. For any event E, $P(E) \geq 0$.

Proof. Suppose $P(E) < 0$. Then the random variable

$$\left| P(E) - \frac{S^n(E)}{n} \right| = -P(E) + \frac{S^n(E)}{n}$$

and Axiom 2.6 implies that for any $\epsilon > 0$ and $\delta > 0$ there exists an $N(\epsilon, \delta)$ such that for $n > N(\epsilon, \delta)$

$$\left| P^n \left(\frac{S^n(E)}{n} < P(E) + \delta \right) - 1 \right| < \epsilon.$$

Taking $\delta = |P(E)/2|$, we have

$$\left| P^n \left(\frac{S^n(E)}{n} < \frac{P(E)}{2} \right) - 1 \right| < \epsilon.$$

However, since $P(E) < 0$ and, for any n-tuple in Ω^n, $S^n(E) \geq 0$, the event

$$\left(\frac{S^n(E)}{n} < \frac{P(E)}{2} \right)$$

is equal to NULL by Lemma 2.2, and therefore, by Lemma 2.3, has probability 0 for any n. Therefore taking $\epsilon = 1/2$, we have the contradiction that $1 < 1/2$. $\square$

Lemma 2.4. For any event E, $P(E) \leq 1$.

Proof. Suppose $P(E) > 1$. Then the random variable

$$\left| P(E) - \frac{S^n(E)}{n} \right| = P(E) - \frac{S^n(E)}{n},$$

and Axiom 2.6 implies that for any ϵ, $\delta > 0$, there exists an $N(\epsilon, \delta)$ such that for $n > N(\epsilon, \delta)$

$$\left| P^n \left(P(E) < \frac{S^n(E)}{n} + \delta \right) - 1 \right| < \epsilon.$$

Taking $\delta = P(E) - 1$, we have

$$\left| P^n \left(1 < \frac{S^n(E)}{n} \right) - 1 \right| < \epsilon.$$

However, since the random variable $S^n(E) \leq n$ for any n-tuple, the event

$$\left(1 < \frac{S^n(E)}{n} \right)$$

is equal to NULL and therefore by Lemma 2.3 has probability 0 for any n. Taking $\epsilon = 1/2$, we have the contradiction that $1 < 1/2$. $\square$

Lemma 2.5. If E_1 and E_2 are two events such that $E_1 \subseteq E_2$, then

$$P(E_2) \geq P(E_1).$$

Proof. The proof follows directly from Lemma 2.4, Axiom 3, and Theorem 2.5. $\square$

We now state the second special case of Axiom 2.6. It is stated in the simpler limit format (i.e., with no reference to ϵ) because we will have no further occasion to refer specifically to ϵ.

Axiom 2.6''. Let an experiment and information be given and let Ω, $\mathcal{F}$, Ω^n, and $\mathcal{F}^n$ be as defined in Axiom 2.4. If $E_1, E_2, \ldots$, and E_k are a finite set of events in $\mathcal{F}$, then given an arbitrary $\delta > 0$,

$$\lim_{n \to \infty} P^n \left(\left(\left| P(E_1) - \frac{S^n(E_1)}{n} \right| < \delta \right) \wedge \cdots \wedge \left(\left| P(E_k) - \frac{S^n(E_k)}{n} \right| < \delta \right) \right) = 1.$$

This axiom states that we can be almost certain of the simultaneous convergence of the relative frequencies of occurrences of a finite set of events to their respective probabilities. The validity of this assumption is again based on our experience with repeating experiments and observing manifestations of the weak law of large numbers. Axiom 2.6' is a special case of this axiom in which there is only one event. The final version of this axiom is even more general.

Before proving additivity, we give another example:

Example 2.15. A real valued function of several random variables is also a random variable. For example, if E_1 and E_2 are in $\mathcal{F}$, then

$$\left| P(E_1 \vee E_2) - \frac{S^n(E_1 \vee E_2)}{n} + \frac{S^n(E_1)}{n} - P(E_1) + \frac{S^n(E_2)}{n} - P(E_2) \right|$$

is a random variable on Ω^n. We have that

$$\left(\left| P(E_1 \vee E_2) - \frac{S^n(E_1 \vee E_2)}{n} + \frac{S^n(E_1)}{n} - P(E_1) + \frac{S^n(E_2)}{n} - P(E_2) \right| < \delta \right)$$

is the event that is the subset of n-tuples of Ω^n such that if the number of occurrences of E_1, E_2, and $E_1 \vee E_2$ in the n-tuple replace $S^n(E_1)$, $S^n(E_2)$, and $S^n(E_1 \vee E_2)$, respectively, in the expression of the random variable, then the absolute value of the expression is less than δ.

Theorem 2.7. If E_1 and E_2 are mutually exclusive events, that is, $E_1 \wedge E_2 = $ NULL, then

$$P(E_1 \vee E_2) = P(E_1) + P(E_2).$$

Proof. Since $E_1 \wedge E_2 = $ NULL, the random variable (on Ω^n)

$$\left| P(E_1 \vee E_2) - \frac{S^n(E_1 \vee E_2)}{n} + \frac{S^n(E_1)}{n} - P(E_1) + \frac{S^n(E_2)}{n} - P(E_2) \right|$$

is equal to the constant random variable

$$|P(E_1 \vee E_2) - P(E_1) - P(E_2)|.$$

Thus for any n,

$$P^n(|P(E_1 \vee E_2) - P(E_1) - P(E_2)| < \delta)$$

is equal to

$$P^n \left(\left| P(E_1 \vee E_2) - \frac{S^n(E_1 \vee E_2)}{n} + \frac{S^n(E_1)}{n} - P(E_1) + \frac{S^n(E_2)}{n} - P(E_2) \right| < \delta \right).$$

However, by Lemma 2.5 this last probability is

$$\geq P^n \left(\left| P(E_1 \vee E_2) - \frac{S^n(E_1 \vee E_2)}{n} \right| + \left| P(E_1) - \frac{S^n(E_1)}{n} \right| \right.$$
$$\left. + \left| P(E_2) - \frac{S^n(E_2)}{n} \right| < \delta \right),$$

which again by Lemma 2.5 is

$$\geq P^n \left(\left(\left| P(E_1 \vee E_2) - \frac{S^n(E_1 \vee E_2)}{n} \right| < \frac{\delta}{3} \right) \wedge \left(\left| P(E_1) - \frac{S^n(E_1)}{n} \right| < \frac{\delta}{3} \right) \right.$$
$$\left. \wedge \left(\left| P(E_2) - \frac{S^n(E_2)}{n} \right| < \frac{\delta}{3} \right) \right).$$

Due to Axiom 2.6, the limit as $n \to \infty$ of this last probability is equal to 1. We therefore have for any $\delta > 0$ that

$$\lim_{n \to \infty} P^n(|P(E_1 \vee E_2) - P(E_1) - P(E_2)| < \delta) \geq 1.$$

By Lemma 2.4 this limit must therefore equal 1, and thus, by the same argument as that used in the proof of Theorem 2.5, for every $\delta > 0$ there is some n such that the event

$$(|P(E_1 \vee E_2) - P(E_1) - P(E_2)| < \delta)$$

is equal to Ω^n. The proof is now completed in the exact same fashion as the conclusion of the proof of Theorem 2.5.   □

We have now derived the axioms in the definition of a probability space. We need still to derive the definition of conditional probability. First we need an axiom concerning conditional probability:

Axiom 2.5. Let an experiment and information be given and let Ω be a set of mutually exclusive and exhaustive outcomes of the experiment. Let $\mathcal{F}$ be the set of all subsets of Ω, and suppose each $E \in \mathcal{F}$ has a probability $P(E)$, as hypothesized in Axiom 2.1. Then, if $E' \in \mathcal{F}$ and $P(E') > 0$, for every event $E \in \mathcal{F}$ there is a conditional probability, denoted $P(E \mid E')$, of the event E occurring given that E' has occurred.

This axiom says that conditional probabilities exist. Their existence must be taken as axiomatic, just like the existence of any probability. Our final version of Axiom 2.6 says that relative frequencies of occurrences also converge to conditional probabilities in accordance with the weak law of large numbers.

Axiom 2.6. Let an experiment and information be given and let Ω, $\mathcal{F}$, Ω^n, and $\mathcal{F}^n$ be as defined in Axiom 2.4. If we define the random variable (on Ω^n)

$$S^n(E \mid E') = \begin{cases} \dfrac{S^n(E \wedge E')}{S^n(E')} & \text{if} \quad S_n(E') > 0 \\[2mm] 0 & \text{if} \quad S_n(E') = 0, \end{cases}$$

then $S^n(E \mid E')$ approaches $P(E \mid E')$ in the sense of the weak law of large numbers as described in Axiom 2.6'. Furthermore, each probability in Axiom 2.6'' may be either a conditional or an unconditional probability. That is, an event of the form

$$(|P(E \mid E') - S^n(E \mid E')| < \delta)$$

may appear in the intersection in that axiom.

$S^n(E \mid E')$ is the fraction of times that E occurs of the total number of times E' occurs in repeated trials. The reason that this fraction should approach $P(E \mid E')$ is again our experience in observing manifestations of the weak law of large numbers. This final version of Axiom 2.6 is our assumption of the weak law of large numbers.

We can now derive the definition of conditional probability:

Theorem 2.8. If $P(E') > 0$, then for any event $E \in \mathcal{F}$ we have that

$$P(E \mid E') = \frac{P(E \wedge E')}{P(E')}.$$

Proof. Let $\delta > 0$ be given. We will show that

$$\lim_{n \to \infty} P^n \left(\left| P(E \mid E') - \frac{P(E \wedge E')}{P(E')} \right| < \delta \right) = 1. \qquad (2.1)$$

The probability expression in (2.1) is equal to

$$P^n \left(\left| P(E \mid E') - S^n(E \mid E') + S^n(E \mid E') - \frac{P(E \wedge E')}{P(E')} \right| < \delta \right),$$

which by Lemma 2.5 is

$$\geq P^n \left(\left(|P(E \mid E') - S^n(E \mid E')| < \frac{\delta}{2} \right) \wedge \left(\left| S^n(E \mid E') - \frac{P(E \wedge E')}{P(E')} \right| < \frac{\delta}{2} \right) \right).$$

$$(2.2)$$

The event on the left in (2.2) is an event which is allowed in the intersection in Axiom 2.6. The event on the right in (2.2) is equal to

$$\left(\left| S^n(E \mid E') - \frac{S^n(E \wedge E')}{nP(E')} + \frac{S^n(E \wedge E')}{nP(E')} - \frac{P(E \wedge E')}{P(E')} \right| < \frac{\delta}{2} \right),$$

which contains as a subset the event

$$\left(\left| S^n(E \mid E') - \frac{S^n(E \wedge E')}{nP(E')} \right| < \frac{\delta}{4} \right) \wedge \left(\left| \frac{S^n(E \wedge E')}{nP(E')} - \frac{P(E \wedge E')}{P(E')} \right| < \frac{\delta}{4} \right).$$

$$(2.3)$$

The event on the right in the intersection in (2.3) is equal to

$$\left(\left| \frac{S^n(E \wedge E')}{n} - P(E \wedge E') \right| < P(E') \frac{\delta}{4} \right),$$

which is an event which is allowed in the intersection in Axiom 2.6. We will proceed backward to show that the set on the left in (2.3) contains as a subset an event which is allowed in the intersection in Axiom 2.6. Clearly, the event

$$\left(\left| P(E') - \frac{S^n(E')}{n} \right| < P(E') \frac{\delta}{4} \right) \qquad (2.4)$$

is such an event. Any n-tuple in the event in (2.4) for which $S^n(E') = 0$ is clearly in the event on the left in (2.3). For any n-tuple in the event in (2.4) for which $S^n(E') \neq 0$ we have, after rearranging terms in (2.4), that

$$|S^n(E')| \left| \frac{1}{S^n(E')} - \frac{1}{nP(E')} \right| < \frac{\delta}{4}$$

and therefore

$$|S^n(E \wedge E')| \left| \frac{1}{S^n(E')} - \frac{1}{nP(E')} \right| < \frac{\delta}{4}$$

or

$$\left| \frac{S^n(E \wedge E')}{S^n(E')} - \frac{S^n(E \wedge E')}{nP(E')} \right| < \frac{\delta}{4},$$

which is the same as

$$\left| S^n(E \mid E') - \frac{S^n(E \wedge E')}{nP(E')} \right| < \frac{\delta}{4}.$$

Therefore the event in (2.4) is a subset of the event on the left in (2.3). Thus by Axioms 2.6 and Lemma 2.5, we have that

$$\lim_{n \to \infty} P^n \left(\left| P(E \mid E') - \frac{P(E \wedge E')}{P(E')} \right| < \delta \right) \geq 1$$

and therefore that limit is equal to 1 by Lemma 2.4. Thus, by the same argument used in the proof of Theorem 2.5, for every $\delta > 0$ there is some n such that the event

$$\left(\left| P(E \mid E') - \frac{P(E \wedge E')}{P(E')} \right| < \delta \right)$$

is equal to Ω^n. The proof is now completed due to the same argument at the end of Theorem 2.5. □

Finally, note that we have made a stronger assumption than the axioms of probability theory. That is to say, the axioms of probability theory could hold without having convergence in the sense of the weak law of large numbers. However, along with the assumption that separate trials of the same experiment are probabilistically independent, it is possible to deduce our axioms from the axioms of probability theory (it is not possible to deduce definite convergence as assumed by von Mises). On the other hand, the assumption of both the axioms of probability theory and the independence of separate trials is stronger than our assumption. In other words, we cannot conclude that separate trials of the same experiment are independent from our assumptions. Indeed, it is a simple exercise to create an example which shows that they need not be independent.

PROBLEMS APPENDIX 2.A

1. Using the method in the proof of Theorem 2.5, prove Lemma 2.3.

2. Create an example which illustrates that it is possible to have convergence in the sense of the weak law of large numbers without separate trials of the experiment being probabilistically independent. Hint: Assume that if heads turns up on one toss of a coin, then tails will definitely turn up on the next toss, and if tails turn up, then heads will definitely turn up on the next toss.

CHAPTER 3

GRAPH THEORETIC CONSIDERATIONS

The early rule-based techniques for using probability theory and networks to reason under uncertainty, which are discussed in chapter 4, require little graph theory. However, as we shall see, these techniques have severe limitations. The techniques based on causal networks, which are discussed in chapters 5 through 10, do not possess these limitations. However, the removal of these limitations requires the use of more sophisticated mathematics, in particular graph theory. The necessary background in graph theory is reviewed in this chapter. The topics are not those which one would ordinarily find in the first chapter of a book on graph theory. Rather they are the ones necessary to the development of the methods in the chapters on causal networks. A complete introduction to graph theory can be found in [Bondy and Murty, 1982], while a discussion of perfect graphs is in [Golumbic, 1980].

3.1. BASIC DEFINITIONS AND NOTATION

3.1.1. Functions and Relations

Definition 3.1. A binary relation R on a set X is a set of ordered pairs of elements (x, x') where $x \in X$ and $x' \in X$.

If $(x, x') \in R$, then x' is said to be a relative of x. Notice that this does not imply that x is a relative of x'. For each $x \in X$, the set of all relatives of x is denoted by $\text{Rel}(x)$. We have

$$(x, x') \in R \qquad \text{if and only if} \quad x' \in \text{Rel}(x).$$

Example 3.1. Let $X = \{a,b,c,d,e,f\}$. A possible binary relation R on X is

$$\{(a,b),(a,c),(a,d),(c,a),(c,d),(d,d),(e,f)\}.$$

The set of relatives of a, Rel(a), equals $\{b,c,d\}$.

Definition 3.2. A binary relation R is reflexive if $x \in \text{Rel}(x)$ for all $x \in X$.

Example 3.2. Let $X = \{a,b,c,d\}$. If we define

$$R = \{(a,a),(a,c),(b,b),(c,a),(c,c),(d,d)\},$$

then R is reflexive.

Definition 3.3. A binary relation R is irreflexive if $x \notin \text{Rel}(x)$ for all $x \in X$.

Example 3.3. Let $X = \{a,b,c,d\}$. If we define

$$R = \{(a,b),(a,c),(b,c),(c,a),(d,c)\},$$

then R is irreflexive.

Notice that the relation in Example 3.1 is neither reflexive nor irreflexive.

Definition 3.4. A binary relation R is symmetric if for all $x \in X$

$$x' \in \text{Rel}(x) \qquad \text{implies} \qquad x \in \text{Rel}(x').$$

Example 3.4. Let $X = \{a,b,c,d\}$. If we define

$$R = \{(a,b),(b,a),(c,d),(d,c),(d,e),(e,d)\},$$

then R is symmetric.

Definition 3.5. A binary relation R is antisymmetric if for all $x \in X$

$$x' \in \text{Rel}(x) \qquad \text{implies} \qquad x \notin \text{Rel}(x').$$

Example 3.5. Let $X = \{a,b,c,d\}$. If we define

$$R = \{(a,b),(c,d),(c,e),(e,d)\},$$

then R is antisymmetric.

Notice that the relation is Example 3.1 is neither symmetric nor antisymmetric.

3.1.2. Graphs

Definition 3.6. A directed graph (or digraph) G consists of a finite set V of vertices or nodes and an irreflexive binary relation E on V. The graph G is denoted as (V,E).

The relation is called the adjacency relation. If w is a relative of v (i.e., $(v,w) \in E$), then w is adjacent to v and there is said to be an arc or edge from v to w. The set of all vertices which are adjacent to v is denoted as Adj(v). Clearly,

$$(v,w) \in E \qquad \text{if and only if} \quad w \in \text{Adj}(v).$$

The irreflexivity of the relation Adj implies that for all $v \in V$

$$(v,v) \notin E.$$

In some texts it is not required that the relation be irreflexive. That is, (v,v) could be in E for some v. The definition which includes irreflexivity is more appropriate for the applications in this text.

Digraphs are usually represented by diagrams such as those in Figure 3.1. The vertices are represented by circles, while the arcs are represented by arrows. We adopt a convention here which will be used throughout the text. The small letters u, v, w, x, y, and z will be used to denote variables which have value in V, while the capital letters A, B, C, and so forth will be used to denote the actual values in V. We see then that in Figure 3.1(a),

$$V = \{A,B,C,D,E\}$$

and

$$E = \{(A,B),(A,D),(A,C),(E,C)\}.$$

Definition 3.7. An undirected graph $G = (V,E)$ is a graph in which the adjacency relation is symmetric. That is,

$$(v,w) \in E \qquad \text{implies} \quad (w,v) \in E.$$

Since in an undirected graph there is an arc from v to w only if there is an arc from w to v, the arcs in this case are represented in diagrams simply by lines connecting two vertices. Figure 3.2 contains examples of undirected graphs. In the case of undirected graphs, if $(v,w) \in E$, it is simply said that v and w are adjacent.

3.1.3. Chains and Paths

Definition 3.8. Let $G = (V,E)$ be a graph (directed or undirected). A sequence of vertices $[v_0, v_1, \ldots, v_m]$ is a chain of length m in G between v_0 and

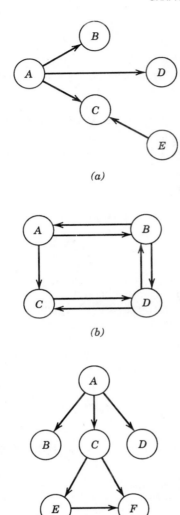

(a)

(b)

(c)

FIGURE 3.1 Examples of digraphs.

v_m if

$$(v_{i-1}, v_i) \in E \qquad \text{or} \qquad (v_i, v_{i-1}) \in E \qquad \text{for} \quad i = 1, 2, \dots, m.$$

Definition 3.9. Let $G = (V, E)$ be a graph (directed or undirected). A sequence of vertices $[v_0, v_1, \dots, v_m]$ is a path of length m in G from v_0 to v_m if

$$(v_{i-1}, v_i) \in E \qquad \text{for} \quad i = 1, 2, \dots, m.$$

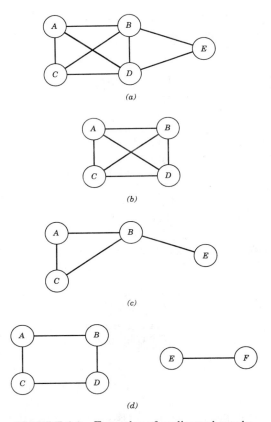

FIGURE 3.2 Examples of undirected graphs.

Notice that in the case of an undirected graph the concepts of a chain and a path are identical. Furthermore, in this case, there is a path from v_0 to v_m only if there is a path from v_m to v_0. Therefore, for an undirected graph, it is simply said that there is a path between v_0 and v_m.

In the graph in Figure 3.1(c) there are paths from A to all the other vertices, there are two paths from C to F, and there are paths from C and E to F. These are the only paths in that graph. In the graphs in Figure 3.2(a), (b), and (c) there are paths between every vertex and every other vertex.

Definition 3.10. A path $[v_0, v_1, \ldots, v_m]$ is a simple path if $v_i \neq v_j$ for $i \neq j$ for $1 \leq i < j \leq m$.

The path $[A, C, F]$ in the graph in Figure 3.1(c) is simple, whereas the path $[A, B, D, C, D]$ in the graph in Figure 3.1(b) is not. In the graph in Figure 3.2(a), the path $[A, B, D, E]$ is simple, while the path $[A, B, D, C, B, E]$ is not.

Definition 3.11. A graph $G = (V, E)$ (directed or undirected) is connected if for every two distinct vertices $v \in V$ and $w \in V$ there is a chain between v and w.

Definition 3.12. A graph $G = (V, E)$ (directed or undirected) is strongly connected if for every two distinct vertices $v \in V$ and $w \in V$ there is a path from v to w.

Notice that if any graph (directed or undirected) is strongly connected, then it is connected, and that if an undirected graph is connected, then it is strongly connected. Therefore in the case of undirected graphs, it is just said that the graph is connected.

Only the graph in Figure 3.1(b) is strongly connected, while the graphs in Figure 3.2(a), (b), and (c) are all connected.

Definition 3.13. A cycle of length m is a path $[v_0, v_1, \ldots, v_{m-1}, v_0]$ from a vertex v_0 to v_0. The cycle is a simple cycle if $[v_0, v_1, \ldots, v_{m-1}]$ is a simple path.

In the graph in Figure 3.1(b), there is a cycle of length 4 from every vertex to itself. Note that in the graph in Figure 3.1(c) $[C, E, F, C]$ is not a cycle.

The cycle $[A, C, D, B, A]$ in the graph in Figure 3.1(b) is simple. In the graph in Figure 3.2(a), the cycle $[A, B, C, A]$ is simple, while the cycle $[A, B, D, E, B, C, A]$ is not.

Definition 3.14. In an undirected graph, a simple path or a simple cycle possesses a chord if there is an arc between two nonconsecutive vertices of the path or cycle. If a simple path or a simple cycle does not possess a chord, it is said to be chordless.

In the graph in Figure 3.2(b) the cycle $[A, B, D, C, A]$ possesses the chords (A, D) and (B, C), while in the graph in Figure 3.2(d) the cycle $[A, B, D, C, A]$ possesses no chords.

3.1.4. Complete Sets and Cliques

Definition 3.15. An undirected graph is complete if every pair of distinct vertices is adjacent.

The graph in Figure 3.2(b) is an example of a complete graph.

Definition 3.16. Let $G = (V, E)$ be a graph (directed or undirected). Given a subset $W \subseteq V$, the subgraph induced by W is denoted G_W. This subgraph's edges are defined by

$$E_W = \{(v, w) \text{ such that } (v, w) \in E \text{ and } v \in W \text{ and } w \in W\}.$$

The graph in Figure 3.2(c) is the subgraph induced by $\{A,B,C,E\}$ from the graph in Figure 3.2(a).

Definition 3.17. In an undirected graph $G = (V,E)$, a subset $W \subseteq V$ is called a complete set of G if it induces a complete subgraph.

Definition 3.18. In an undirected graph $G = (V,E)$, a subset $W \subseteq V$ is called a clique of G if W is a complete set and W is maximal. (That is, there is no complete set which properly contains W as a subset.)

The cliques of the graph in Figure 3.2(a) are $\{A,B,C,D\}$ and $\{B,D,E\}$. In Figure 3.2(b) the only clique is the set V itself. In Figure 3.2(c) the cliques are $\{A,B,C\}$ and $\{B,E\}$. In some texts cliques are defined as complete sets. That is, cliques need not be maximal. The above definition is more appropriate for the applications in this text.

Definition 3.19. Let X be an arbitrary set and I the set of integers. A collection $\{X_i$ such that $i \in I\}$ of subsets of X is called a cover of X if their union equals X.

Notice that the set of cliques of an undirected graph $G = (V,E)$ always covers V.

Definition 3.20. Let $G = (V,E)$ be an undirected graph in which V contains n vertices and $\alpha = [v_1, v_2, \ldots, v_n]$ be a total ordering of all the vertices in V. If v is labeled before w, it is said that $v <_\alpha w$.

Definition 3.21. Let $G = (V,E)$ be an undirected graph in which V contains n vertices. An ordering $\alpha = [v_1, v_2, \ldots, v_n]$ of all the vertices in V is called perfect if for every i

$$\text{Adj}(v_i) \cap \{v_1, v_2, \ldots, v_{i-1}\}$$

is a complete set of G.

The vertices in the graph in Figure 3.3 have been numbered by a perfect ordering. For example,

$$\text{Adj}(v_5) \cap \{v_1, v_2, v_3, v_4\} = \{v_3, v_4, v_6, v_7\} \cap \{v_1, v_2, v_3, v_4\}$$
$$= \{v_3, v_4\}$$

is a complete set of G.

Definition 3.22. Let $G = (V,E)$ be an undirected graph containing p cliques. An ordering $[C_1, C_2, \ldots, C_p]$ of the cliques has the running intersection property if for every $j > 1$ there exists an $i < j$ such that

$$C_j \cap (C_1 \cup C_2 \cup \cdots \cup C_{j-1}) \subseteq C_i.$$

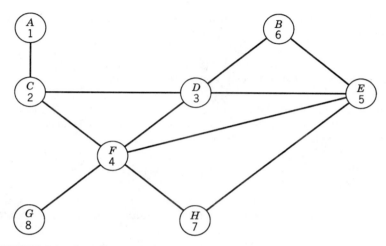

FIGURE 3.3   A perfect ordering obtained using maximum cardinality search.

The cliques in the graph in Figure 3.3 are the following:

$$C_1 = \{A, C\} \qquad C_2 = \{C, D, F\} \qquad C_3 = \{D, E, F\}$$
$$C_4 = \{B, D, E\} \qquad C_5 = \{F, E, H\} \qquad C_6 = \{F, G\}.$$

This ordering has the running intersection property, as can be seen by direct verification. For example,

$$C_5 \cap (C_1 \cup C_2 \cup C_3 \cup C_4) = \{E, F\} \subseteq C_3.$$

Any clique C_i with $i < j$ which contains $C_j \cap (C_1 \cup C_2 \cup \cdots \cup C_{j-1})$ is called a parent of C_j. It is possible for a clique to have more than one parent. In the example above, C_2, C_3, and C_5 are all parents of C_6.

It is no coincidence that it was possible to obtain an ordering of the cliques which has the running intersection property for the graph in Figure 3.3. The following lemmas and theorem prove that it is always possible to obtain such a clique ordering from a perfect ordering of the vertices.

Lemma 3.1. Let $G = (V, E)$ be an undirected graph in which V contains n vertices and let $\alpha = [v_1, v_2, \dots, v_n]$ be a perfect ordering of the vertices in V. Let $[C_1, C_2, \dots, C_p]$ be an ordering of the cliques according to their highest labeled vertex. Then, for each j, the highest labeled vertex in C_j is not in $C_1 \cup C_2 \cup \cdots \cup C_{j-1}$.

Proof. Let v_k be the highest labeled vertex in C_j. Then for $i \le j$,

$$C_i \subseteq \{v_1, v_2, \dots, v_k\}. \tag{3.1}$$

If this were not the case, there would be some $i \le j$ such that there is a vertex $w \in C_i$ and $w \notin \{v_1, v_2, \dots, v_k\}$. This implies, however, that w is a higher labeled vertex than v_k, and therefore the highest labeled vertex in C_i has a label greater than the highest labeled vertex in C_j. Therefore C_i would have to have been labeled after C_j and $i > j$, a contradiction.

The lemma will now be proved by showing that if, for some $i \leq j$, $v_k \in C_i$, then i must equal j. To this end, if $v_k \in C_i$, then, since C_i is a complete set,

$$C_i - \{v_k\} \subseteq \mathrm{Adj}(v_k).$$

Along with (3.1), this implies that

$$C_i - \{v_k\} \subseteq \mathrm{Adj}(v_k) \cap \{v_1, v_2, \ldots, v_{k-1}\}.$$

Similarly,

$$C_j - \{v_k\} \subseteq \mathrm{Adj}(v_k) \cap \{v_1, v_2, \ldots, v_{k-1}\}.$$

Since the ordering is assumed to be perfect, $\mathrm{Adj}(v_k) \cap \{v_1, v_2, \ldots, v_{k-1}\}$ is a complete set. Therefore, if $w \in C_i - \{v_k\}$ and $u \in C_j - \{v_k\}$ with $w \neq u$, then $(w, u) \in E$. Furthermore, since $v_k \in C_i$, $(v_k, w) \in E$ for $w \in C_i - \{v_k\}$. We have then that $(C_i - \{v_k\}) \cup C_j$ is a complete set. However, since $v_k \in C_j$,

$$(C_i - \{v_k\}) \cup C_j = C_i \cup C_j$$

and therefore $C_i \cup C_j$ is a complete set. Since C_i and C_j are both maximal, this implies that $C_i = C_j$ and $i = j$, which completes the proof. $\square$

Lemma 3.2. Let $G = (V, E)$ be an undirected graph with p cliques. If $[C_1, C_2, \ldots, C_p]$ are the cliques of G, then, for $m \leq p$, $[C_1, C_2, \ldots, C_m]$ are the cliques of the subgraph G_W induced by $W = C_1 \cup C_2 \cup \cdots \cup C_m$.

Proof. Clearly, C_i is a complete set of G_W for $i \leq m$ and their union is W. Suppose for some $i \leq m$, C_i is not maximal in G_W. Then there exists a vertex $w \in W$ such that $w \notin C_i$ and, for every $u \in C_i$, $(u, w) \in E$. However, then $C_i \cup \{w\}$ is a complete set of G, contradicting the fact that C_i is maximal in G. $\square$

Next it is shown that an ordering of the the cliques with the running intersection property can be obtained from a perfect ordering of the vertices.

Theorem 3.1. If the vertices of an undirected graph $G = (V, E)$ are numbered with a perfect ordering α, and the cliques are ordered according to their highest labeled vertex, then the clique ordering has the running intersection property.

Proof. Assume p is the number of cliques and $[C_1, C_2, \ldots, C_p]$ is the ordering of the cliques obtained from the perfect ordering α of the vertices. Let $j \leq p$ be given and v_k be the highest labeled node in C_j. Then, for $w \in C_j$ and $w \neq v_k$, $(v_k, w) \in E$. Therefore

$$C_j \subseteq \mathrm{Adj}(v_k) \cup \{v_k\}$$

and

$$C_j \cap (C_1 \cup C_2 \cup \cdots \cup C_{j-1}) \subseteq (\mathrm{Adj}(v_k) \cup \{v_k\}) \cap (C_1 \cup C_2 \cup \cdots \cup C_{j-1}).$$

Since, due to Lemma 3.1, v_k is not in $C_1 \cup C_2 \cup \cdots \cup C_{j-1}$, we then have that

$$C_j \cap (C_1 \cup C_2 \cup \cdots \cup C_{j-1}) \subseteq \mathrm{Adj}(v_k) \cap (C_1 \cup C_2 \cup \cdots \cup C_{j-1}). \qquad (3.2)$$

As shown in the proof of Lemma 3.1,

$$C_1 \cup C_2 \cup \cdots \cup C_j \subseteq \{v_1, v_2, \ldots, v_k\}.$$

Therefore, again since $v_k \notin C_1 \cup C_2 \cup \cdots \cup C_{j-1}$, we have

$$C_1 \cup C_2 \cup \cdots \cup C_{j-1} \subseteq \{v_1, v_2, \ldots, v_{k-1}\}.$$

Together with (3.2), this implies that

$$C_j \cap (C_1 \cup C_2 \cup \cdots \cup C_{j-1}) \subseteq \mathrm{Adj}(v_k) \cap \{v_1, v_2, \ldots, v_{k-1}\}. \qquad (3.3)$$

The proof will be completed by showing that the set on the left in (3.3) is a subset of C_i for some $i < j$. To that end, let

$$W = C_1 \cup C_2 \cup \cdots \cup C_{j-1}.$$

By Lemma 3.2, $[C_1, C_2, \ldots, C_{j-1}]$ are the cliques of G_W. Now the set on the right in (3.3) is a complete set of G due to the ordering of the vertices being perfect. Therefore the set on the left in (3.3) is also a complete set of G. However, since the set on the left is a subset of W, this implies that it is a complete set of G_W. Therefore the set on the left is a subset of one of the cliques of W. That is, it is a subset of some C_i for $i \leq j$, which completes the proof. $\square$

3.2. TRIANGULATED GRAPHS

Triangulated graphs play a central role in the technique for propagating probabilities in trees of cliques [Spiegelhalter, 1986b, 1987; Lauritzen & Spiegelhalter, 1988]. This technique is developed in chapter 7. All the graph theoretic concepts necessary to the technique are covered in this section.

3.2.1. Basic Concepts

Definition 3.23. An undirected graph is called triangulated if every simple cycle of length strictly greater than 3 possesses a chord.

Some examples of triangulated graphs are shown in Figure 3.4. Although the graph in Figure 3.4(c) appears to be composed of triangles, it is not

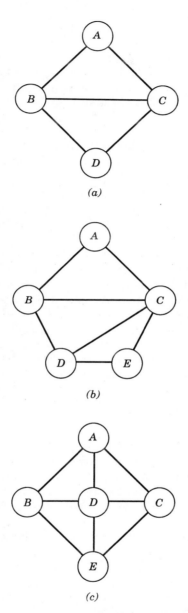

FIGURE 3.4   (a) and (b) are examples of triangulated graphs, while (c) is not.

triangulated because the simple cycle $[A, B, E, C, A]$ does not possess a chord.

In the technique for propagating probabilities in trees of cliques, it will be necessary to obtain an ordering of the cliques, of an arbitrary undirected graph, which has the running intersection property. Theorem 3.1 proved that a perfect ordering of the vertices yields such an ordering of the cliques. We will now

1. Develop a method for adding arcs to an arbitrary undirected graph so that the new graph is triangulated. This process is called triangulating the graph.
2. Develop a method for ordering the vertices of an undirected graph, which always yields a perfect ordering of the vertices if the graph is triangulated. The method is called maximum cardinality search.

In the technique for propagating probabilities in trees of cliques, these steps will be applied to obtain an ordering of the cliques which has the running intersection property. The method in step 2 is developed in subsection 3.2.2. In subsection 3.2.3, we will develop a method for triangulating a graph. Furthermore, in that subsection, we will see that we can obtain our desired results (i.e., a perfect ordering of the vertices) by performing maximum cardinality search first and then triangulating the graph. In the exercises (first problem in Problems) you are asked to prove that only triangulated graphs admit an ordering of the cliques which has the running intersection property.

3.2.2. Maximum Cardinality Search

Definition 3.24. Maximum Cardinality Search. Let $G = (V, E)$ be an undirected graph. An ordering of the vertices in V according to maximum cardinality search is obtained by assigning 1 to an arbitrary vertex. For the next vertex to number, select the vertex adjacent to the largest number of previously numbered vertices, breaking ties arbitrarily.

The ordering obtained in Figure 3.3 has been obtained using maximum cardinality search.

In an unpublished work, Tarjan [1976] proves that an undirected graph is triangulated if and only if maximum cardinality search yields a perfect ordering of the vertices. In the exercises (second problem in Problems), you are asked to prove that if a graph admits a perfect ordering, then it must be triangulated. We will show that if a graph is triangulated then maximum cardinality search yields a perfect ordering. To prove that result, we need the following lemmas. The first is taken from Tarjan and Yannakakis [1984].

Lemma 3.3. Let $G = (V, E)$ be an undirected graph and α an ordering of the vertices obtained using maximum cardinality search. Then α has the following

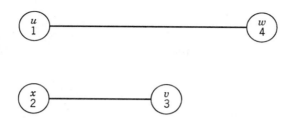

FIGURE 3.5 The arcs among u, v, w, and x illustrating property (P) in Lemma 3.3. A possible ordering is depicted.

property, which we call property (P):

PROPERTY (P): If $(u,w) \in E$ and $(u,v) \notin E$ and

$$u <_\alpha v <_\alpha w,$$

then there is a vertex $x \in V$ such that

$$x <_\alpha v$$

and $(v,x) \in E$, but $(w,x) \notin E$.

Proof. Since $v <_\alpha w$ when v is numbered, it must be adjacent to at least as many numbered vertices as w. Thus, since w but not v is adjacent to u, v is adjacent to some other numbered vertex x not adjacent to w. Since this holds for all such u, v, and w, ordering α has property (P). $\square$

The arcs among u, v, w, and x illustrating property (P) are depicted in Figure 3.5. The ordering shown is only a possible one. In general the four vertices need not be numbered consecutively, and x could be numbered before u.

Lemma 3.4. Let $G = (V,E)$ be a triangulated graph and α an ordering of the vertices obtained using maximum cardinality search. Then there are no two distinct vertices $u \in V$ and $v \in V$ which have the following property, which we call property (Q):

PROPERTY (Q): $(u,v) \notin E$ and there is a chordless path of length ≥ 2 between u and v such that every vertex on the path, excluding u and v, is labeled after both u and v.

Proof. The lemma will be proved by showing that if there are two such vertices, u and v, then there must be two more vertices, u' and v', also having property (Q), where either $u' <_\alpha u$ and $v' = v$ or $v' <_\alpha v$ and $u' = u$. An induction argument will then be used to conclude that there must be an infinite number of vertices in the graph.

Suppose then that u and v are two vertices having property (Q). Without loss of generality, assume $u <_\alpha v$. Let ρ be the chordless path between u and

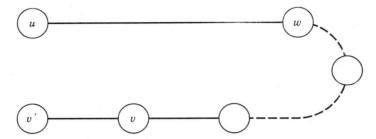

FIGURE 3.6   The vertices from Lemma 3.4, with vertices depicted to the right having higher labels in α. The simple path between u and v is the one in property (Q).

v in property (Q), and let w be the vertex on ρ which is adjacent to u. Then $(u,w) \in E$. Furthermore, $(u,v) \notin E$ and $v <_\alpha w$ due to property (Q). Therefore u, v, and w satisfy property (P) and by Lemma 3.3 there exists a $v' <_\alpha v$ such that $(v',v) \in E$, but $(v',w) \notin E$. These vertices are depicted in Figure 3.6, with vertices depicted to the right having higher labels in order α. There may be an arc between v' and one of the vertices on ρ. However, since $(v',w) \notin E$ and ρ is a chordless path with length ≥ 2, we are guaranteed that there is a chordless path with length ≥ 3 between u and v'. Furthermore, all the vertices on this path are labeled after both u and v'. Therefore, since the graph is triangulated, $(u,v') \notin E$, and u and v' have property (Q).

Now let W be the set of all vertices which have property (Q) with some other vertex in V. Since V is finite, there is some vertex $y \in W$ which has the minimum label (with respect to α) over all vertices in W. There is at least one vertex $x_1 \in W$ such that x_1 and y have property (Q). By way of induction, let x_n be a vertex having property (Q) with y. Due to the argument above, there must be vertices x' and y' also having property (Q) with $x' <_\alpha x$ and $y' = y$ or $y' <_\alpha y$ and $x' = x$. However, since y is minimal in W (with respect to α), we must have that $x' <_\alpha x$ and $y' = y$. If we label x' as x_{n+1}, we then have a new vertex x_{n+1} such that x_{n+1} and y have property (Q) and $x_{n+1} <_\alpha x_n$. Thus we can create an infinite sequence of vertices $[x_1, x_2, \ldots, x_n, x_{n+1}, \ldots]$ such that

$$\cdots <_\alpha x_{n+1} <_\alpha x_n <_\alpha \cdots <_\alpha x_1.$$

However, since the graph is finite, this is a contradiction. $\square$

We can now prove that in the case of triangulated graphs, maximum cardinality search yields a perfect ordering of the vertices.

Theorem 3.2. Let $G = (V, E)$ be a triangulated graph. If α is an ordering of the vertices in V according to maximum cardinality search, then α is a perfect ordering.

Proof. If V contains exactly one vertex, the proof is trivial. Assume as an induction hypothesis that the theorem is true for all triangulated graphs with

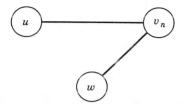

FIGURE 3.7 The vertices in Theorem 3.2, with vertices depicted on right having higher label in ordering α.

fewer than n vertices. Let G be a triangulated graph containing n vertices and $\alpha = [v_1, v_2, \ldots, v_n]$ be an ordering of the vertices according to maximum cardinality search. Then $\alpha' = [v_1, v_2, \ldots, v_{n-1}]$ is an ordering, according to maximum cardinality search, of the vertices of the subgraph G_W induced by $W = \{v_1, v_2, \ldots, v_{n-1}\}$. Due to the induction assumption,

$$\text{Adj}(v_i) \cap \{v_1, v_2, \ldots, v_{i-1}\}$$

is a complete set of G_W and therefore a complete set of G for $i \leq n - 1$. Therefore we need only show that

$$\text{Adj}(v_n) \cap \{v_1, v_2, \ldots, v_{n-1}\}$$

is a complete set of G. Suppose it is not. Then there exists vertices u and w such that

$$(u, v_n) \in E, \qquad (w, v_n) \in E, \qquad \text{and} \qquad (u, w) \notin E.$$

These vertices are depicted in Figure 3.7. We see that the vertices u and w have property (Q), which implies, due to Lemma 3.4, that the graph is not triangulated, a contradiction. $\square$

We therefore have established a method, namely maximum cardinality search, for creating a perfect ordering of the vertices of a triangulated graph. Before developing a method for triangulating an arbitrary undirected graph, we give an algorithm for maximum cardinality search.

This algorithm is taken from Tarjan and Yannakakis [1984]. We proceeed as follows. Maintain an array of sets of vertices, set[i] for $0 \leq i \leq n - 1$. Store in set[i] all unnumbered vertices adjacent to exactly i numbered vertices. Initially, set[0] contains all the vertices. Maintain the largest index j such that set[j] is nonempty. To carry out a step in the search, remove a vertex v from set[j] and number it. For each unnumbered vertex w adjacent to v, move w from the set containing it, say set[i], to set[$i + 1$]. Next add one to j. Then while set[j] is empty, repeatedly decrement j. If each set is represented by a doubly linked list of vertices and for each vertex an index is maintained of the set containing it, Tarjan and Yannakakis show that this search runs in $O(n + e)$ time, where n is the number of vertices and e the number of arcs.

Pseudocode for the algorithm described above follows. For any unnumbered vertex v, size$[v]$ is the number of numbered vertices adjacent to v. If v is a numbered vertex, size$[v]$ is set to -1.

MAXIMUM CARDINALITY SEARCH

```
Constant V = set of all vertices in graph;
         n = number of vertices in V;
         E = set of all vertices in graph;

Var i, j: integer;
    v, w: vertex;
     set: array[0, n − 1] of subset of vertices;
    size: array[V] of integer;
       α: array[V] of integer;
    α⁻¹: array[1...n] of vertex;

Begin
    for i := 0 to (n − 1) do
        set[i] := ∅;
    for v ∈ V do
        begin
          size[v] := 0;
          add v to set[0]
        end;
    i := 1; j := 0;

    while i ≤ n do
        begin
          v := delete any from set[j];
          α[v] := i; α⁻¹[i] := v; {assign v the number i}
          size[v] := −1;
          for w ∈ V − {v} do
              if (v, w) ∈ E and size(w) ≥ 0 then
                begin
                    delete w from set[size[w]];
                    size[w] := size[w] + 1;
                    add w to set[size[w]]
                end;
          i := i + 1;
          j := j + 1;
          while j ≥ 0 and set[j] = ∅ do
              j := j − 1
        end
end.
```

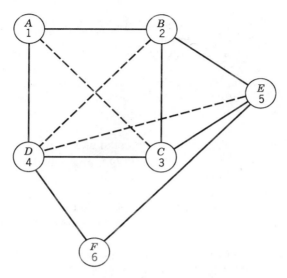

FIGURE 3.8 A graph ordered by maximum cardinality search and the corresponding elimination graph. Original arcs are solid; fill-in arcs are dashed.

3.2.3. Filling in an Undirected Graph

Next we develop a method for adding arcs to an arbitrary undirected graph so that the new graph is triangulated. This procedure is called triangulating the graph. The following results, also taken mostly from Tarjan and Yannakakis [1984], accomplish this:

Definition 3.25. Let $G = (V, E)$ be an undirected graph. Let α be a total ordering of the vertices in V. Then

    $F(\alpha) = \{(v, w)$ such that $(v, w) \notin E$ and there is a path between v and w containing only v, w, and vertices ordered after both v and w. That is, if x is a vertex on the path other than v or w, then $x >_\alpha v$ and $x >_\alpha w.\}$

    $F(\alpha)$ is called the fill-in of G with respect to α.

Definition 3.26. Let $G = (V, E)$ be an undirected graph. The elimination graph $G(\alpha)$ of G with respect to α is defined by

$$G(\alpha) = (V, E \cup F(\alpha)).$$

Definition 3.27. If $F(\alpha) = \emptyset$, α is called a zero fill-in of G.

The graph in Figure 3.8 has been ordered by maximum cardinality search, and the resultant fill-in has been added to the graph.

We will now develop a method for triangulating an arbitrary undirected graph $G = (V, E)$ by 1) proving that any ordering α of the vertices in V is a zero fill-in order of $G(\alpha)$; 2) proving that a graph which admits a zero fill-in order is triangulated and therefore $G(\alpha)$ is triangulated; and 3) developing an algorithm for generating $G(\alpha)$ from G and α. We can then triangulate G by creating an arbitrary order α of the vertices in V and generating $G(\alpha)$. The following lemmas and theorems obtain these results:

Lemma 3.5. Let $G = (V, E)$ be an undirected graph and α a complete ordering of the vertices in V. Then α is zero fill-in if and only if α has the following property, which we shall call property (R):

PROPERTY (R): For all distinct vertices u, v, and w, if $(u, v) \in E$, $(u, w) \in E$, and u is ordered (with respect to α) after both v and w, then $(v, w) \in E$.

Proof. Suppose α is a zero fill-in ordering. Let u, v, and w be vertices satisfying the conditions in property (R). There is then a path between v and w containing only v, w, and vertices labeled after both v and w (namely u). Therefore either $(v, w) \in E$ or $(v, w) \in F(\alpha)$. However, since $F(\alpha) = \emptyset$, this implies $(v, w) \in E$. Thus α has property (R).

Next suppose α has property (R). Let ρ be a path of length l, where $l \geq 2$, between v and w containing only v, w, and vertices labeled after both v and w. Let z be the vertex with highest label on ρ, and x and y the vertices on ρ adjacent to z. $(x, y) \in E$ due to property (R), and therefore there is a path of length $l - 1$ between v and w. By induction, there is a path of length 1 between v and w. That is, $(v, w) \in E$, and therefore α is zero fill-in. □

Lemma 3.6. Let $G = (V, E)$ be an undirected graph and α a total ordering of the vertices in V. Then α is a zero fill-in order for $G(\alpha) = (V, E \cup F(\alpha))$, the elimination graph of G relative to α.

Proof. Suppose $(u, v) \in E \cup F(\alpha)$, $(u, w) \in E \cup F(\alpha)$, and u is ordered (with respect to α) after both v and w. Then either $(u, w) \in E$ or there is a path in G between u and w containing only u, w, and vertices labeled after both u and w. Similarly, either $(u, v) \in E$ or there is a path in G between u and v containing only u, v, and vertices labeled after both u and v. Therefore there is a path in G between v and w containing only v, w, and vertices labeled after both v and w. Thus $(v, w) \in E \cup F(\alpha)$ and α is a zero fill-in for $G(\alpha)$ due to Lemma 3.5. □

Lemma 3.7. If $G = (V, E)$ is an undirected graph which admits a zero fill-in order, then G is triangulated.

Proof. Let α be a zero fill-in order of G, ρ a chordless cycle of length greater than 3 in G, u the vertex on ρ with highest label (relative to α), and v and

w the vertices on ρ adjacent to u. Then $(u,v) \in E$, $(u,w) \in E$, and u is ordered after both v and w. Hence, by Lemma 3.5, $(v,w) \in E$ and G is triangulated. □

Theorem 3.3. Let $G = (V,E)$ be an undirected graph and α a complete ordering of the vertices in V. Then $G(\alpha) = (V, E \cup F(\alpha))$, the elimination graph of G relative to α, is triangulated.

Proof. By Lemma 3.6 $G(\alpha)$ admits a zero fill-in, and therefore by Lemma 3.7, $G(\alpha)$ is triangulated. □

We therefore see that to triangulate an arbitrary undirected graph we need only order the vertices in an arbitrary manner and compute the fill-in for that order. The elimination graph relative to that order will always be triangulated. However, if the graph is already triangulated, we do not want to add any arcs. The following theorem shows that if instead of ordering the vertices arbitrarily, we order them according to maximum cardinality search, the fill-in is zero.

Theorem 3.4. Let $G = (V,E)$ be a triangulated graph and α an ordering of the vertices in G according to maximum cardinality search. Then α is a zero fill-in order.

Proof. Let u, v, and w be vertices in G satisfying the condition in property (R) as stated in Lemma 3.5. If $(v,w) \notin E$, then v and w have property (Q) as stated in Lemma 3.4, and therefore, by that lemma, either G would not be triangulated, or α would not be according to maximum cardinality search. Therefore $(v,w) \in E$, and by Lemma 3.5 α is zero fill-in. □

We still need an algorithm for computing the fill-in for an undirected graph. The following results, again taken from Tarjan and Yannakakis [1984], accomplish this.

Lemma 3.8. Let $G = (V,E)$ be an undirected graph, α a total ordering of the vertices in V, and $G(\alpha) = (V, E \cup F(\alpha))$ be the elimination graph of G relative to α. Let $v \in V$ and $w \in V$. Then $(v,w) \in E \cup F(\alpha)$ if and only if either $(v,w) \in E$ or there is a vertex u such that (u,v) and (u,w) are both in $E \cup F(\alpha)$ and u is ordered after both v and w.

Proof. Suppose $(v,w) \in E \cup F(\alpha)$. If $(v,w) \notin E$, then $(v,w) \in F(\alpha)$ and there is a path in G between v and w containing only v and w and vertices ordered after both v and w by the definition of a fill-in. Let u be the vertex on this path of smallest label, excluding v and w. If $(u,w) \notin E$, there is then a path between u and w containing only u, w, and vertices ordered after both u and w. Therefore either $(u,w) \in E$ or $(u,w) \in F(\alpha)$, which implies $(u,w) \in E \cup F(\alpha)$. Similarly, $(u,v) \in E \cup F(\alpha)$.

Conversely, suppose either $(v,w) \in E$ or there is a vertex u such that (u,v) and (u,w) are both in $E \cup F(\alpha)$ and u is ordered after both v and w. If $(v,w) \in E$, then $(v,w) \in E \cup F(\alpha)$. If $(v,w) \notin E$, then there is a path in $G(\alpha)$ between v and w containing only v and w and vertices labeled after both v and w (namely the path $[v,u,w]$). Therefore (v,w) is either in the fill-in of $G(\alpha)$ relative to α or is in $F(\alpha)$. However, by Lemma 3.6 the fill-in of $G(\alpha)$ relative to α is equal to $\emptyset$. Therefore $(v,w) \in F(\alpha)$, which implies $(v,w) \in E \cup F(\alpha)$. $\square$

Definition 3.28. Let $G = (V,E)$ be an undirected graph, α a total ordering of the vertices in v, and $G(\alpha) = (V, E \cup F(\alpha))$ the elimination graph of G relative to α. For any $v \in V$, $f(v)$, the follower of v, is the vertex of largest number (relative to α) which is both adjacent to v in $G(\alpha)$ and has a number smaller than v. For $i \geq 0$,

$$f^0(v) = v \qquad \text{and} \qquad f^{i+1}(v) = f(f^i(v)).$$

Notice that $f^1(v)$ is $f(v)$. Figure 3.9 contains graphs with the followers of vertices labeled. Note that a vertex need not have a follower. For example, the vertex with lowest label never has a follower. In Figure 3.9(b) two vertices have no follower.

Lemma 3.9. Let $G = (V,E)$ be an undirected graph with n vertices, α a total ordering of the vertices in V, and $G(\alpha) = (V, E \cup F(\alpha))$ be the elimination graph of G relative to α. If $(v,w) \in E \cup F(\alpha)$ with $v >_\alpha w$, then $f^i(v) = w$ for some $i \geq 1$.

Proof. The proof is by induction from 1 to n. It is trivially true for v_1, the lowest labeled vertex. Let v and w be arbitrary vertices satisfying the conditions of the lemma. Since w is adjacent to v in $G(\alpha)$ and has a label lower than v, we know $f^1(v)$ exists. Suppose $f^1(v) \neq w$. Then $v >_\alpha f^1(v) >_\alpha w$ by the definition of f^1. By Lemma 3.8, $(f^1(v),w) \in E \cup F(\alpha)$. Suppose by way of induction that the lemma is true for all vertices labeled before v. Then $f^i(f(v)) = w$ for some $i \geq 1$, and therefore $f^{i+1}(v) = w$. $\square$

Theorem 3.5. Let $G = (V,E)$ be an undirected graph, α a complete ordering of the vertices in V, and $G(\alpha) = (V, E \cup F(\alpha))$ the elimination graph of G relative to α. Suppose x and w are two vertices in V with $x >_\alpha w$. Then $(x,w) \in E \cup F(\alpha)$ if and only if there is a vertex v such that $(v,w) \in E$ and $f^i(v) = x$ for some $i \geq 0$.

Proof. Let x and w be two vertices with $x >_\alpha w$. Suppose there is a vertex v such that $(v,w) \in E$ and $f^i(v) = x$ for some $i \geq 0$. Then $f^k(v) \geq_\alpha x$ for $k \leq i$ by the definition of f. Therefore there is a path in $G(\alpha)$ between x and w containing only x and w and vertices labeled after both x and w. Since α is a zero fill-in order for $G(\alpha)$ by Lemma 3.6, $(x,w) \in E \cup F(\alpha)$ by the definition of fill-in.

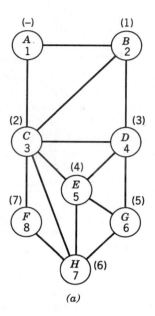

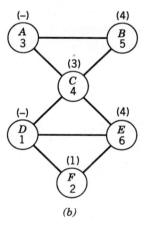

FIGURE 3.9 (a) contains a triangulated graph ordered by maximum cardinality search; (b) contains a triangulated graph ordered arbitrarily. In both (a) and (b), followers of vertices are labeled in parenthesis.

Conversely, suppose $x >_\alpha w$ and $(x, w) \in E \cup F(\alpha)$. We use induction to prove that there is a vertex v such that $(v, w) \in E$ and $f^i(v) = x$ for some $i \geq 0$. For a given vertex w, let z be the highest labeled vertex such that $(z, w) \in E \cup F(\alpha)$ and $z >_\alpha w$ for a vertex w. Then by the definition of a fill-in $(z, w) \notin F(\alpha)$, which implies $(z, w) \in E$. Therefore the theorem holds with $v = z$ and $i = 0$. Next let x be any other vertex such that $(x, w) \in E \cup F(\alpha)$ and $x >_\alpha w$. Assume by the way of induction that the theorem is true for all vertices with higher labels than x. If $(x, w) \in E$, then $v = x$ and $i = 0$ satisfy the theorem. Otherwise, by Lemma 3.8 there is a vertex u such that (u, x) and (u, w) are both in $E \cup F(\alpha)$ and $u >_\alpha x$. By the induction assumption, there is a vertex v such that $(v, w) \in E$ and $f^i(v) = u$ for some $i \geq 0$. By Lemma 3.9, $f^j(u) = x$ for some $j \geq 0$. Therefore v and $i + j$ satisfy the theorem. $\square$

We use Theorem 3.5 to create a fast algorithm for computing a fill-in as follows. We process the vertices in order from the vertex numbered n to the vertex numbered 1. When processing vertex w, we determine the set $A(w)$ of all vertices such $(x, w) \in E \cup F(\alpha)$ and $x >_\alpha w$. We also determine $f(x)$ for all vertices x such that $f(x)$ is w. In this way if $f(x) >_\alpha w$, its value will be determined when the followers of w are processed and hence its value is available when w is processed. To accomplish this, when processing vertex w, we initialize $A(w)$ to contain all vertices v such that $(v, w) \in E$ and $v >_\alpha w$. Then we repeat the following step until there are no more vertices satisfying the condition in the step: Select a vertex $x \in A(w)$ such that $f(x)$ has been determined (i.e., $f(x) >_\alpha w$ and $f(x) \notin A(w)$), and add $f(x)$ to $A(w)$. After constructing $A(w)$, set $f(x)$ to w for all vertices $x \in A(w)$ for which $f(x)$ has not yet been determined.

Tarjan and Yannakakis [1984] have implemented this algorithm to run in $O(n + e')$ time, where n is the number of vertices and e' is the number of arcs in $G(\alpha)$. The implementation is as follows: We maintain two arrays, f and index, and process from i equal n to 1. When vertex x with $\alpha(x) = i$ is processed, we initialize $f(x)$ to be x and index(x) to be i. The first time we process a lower numbered vertex w adjacent to x, we set index$(x) = \alpha(w)$. In this way, index(x) is always the minimal label of the vertices in the set

$$\{x\} \cup \{w \text{ such that } (x, w) \in E \text{ and } w \text{ has been processed}\}.$$

To process a vertex w, we repeat the following step for each vertex v such that $(v, w) \in E$ and $v >_\alpha w$:

General Step:

Initialize x to v. While index$(x) > \alpha(v)$, set index$(x) = \alpha(w)$ and add (x, w) to $F(\alpha)$. Set x equal to $f(x)$ and repeat. When index$(x) = \alpha(x)$, if $f(x) = x$ set $f(x) = w$.

The following is pseudocode for this implementation of the algorithm:

FILL-IN COMPUTATION

Constant V = set of all vertices in graph;
$\qquad n$ = number of vertices in V;
$\qquad E$ = set of all edges in graph;
$\qquad \alpha$ = array $[V]$ of integer; {contains the ordering}
$\qquad \alpha^{-1}$ = array $[1...n]$ of vertex; {of the vertices}

$\quad$Var i: integer;
v, w, x: vertex;
$\qquad f$: array $[V]$ of vertex;
$\quad$index: array $[V]$ of integer;
$E \cup F(\alpha)$: set of edges;

Begin
$\quad$for $i := n$ to 1 do
$\qquad$begin
$\qquad\quad w := \alpha^{-1}(i); f(w) := w; \text{index}(w) := i;$
$\qquad\quad$for $v \in V$ do
$\qquad\qquad$if $(v, w) \in E$ and $\alpha(v) > i$ then
$\qquad\qquad\quad$begin
$\qquad\qquad\qquad x := v;$
$\qquad\qquad\qquad$while $\text{index}(x) > i$ do
$\qquad\qquad\qquad\quad$begin
$\qquad\qquad\qquad\qquad \text{index}(x) := i;$
$\qquad\qquad\qquad\qquad$add (x, w) to $E \cup F(\alpha);$
$\qquad\qquad\qquad\qquad x := f(x)$
$\qquad\qquad\qquad$end;
$\qquad\qquad\qquad$if $f(x) = x$
$\qquad\qquad\qquad\quad$then $f(x) := w$
$\qquad\qquad$end
$\quad$end
end.

We have one more important theorem in this section:

Theorem 3.6. Let $G = (V, E)$ be a undirected graph, α an arbitrary ordering of the vertices in V, and $G(\alpha)$ the elimination graph relative to α. Then α is a perfect ordering of the vertices in V relative to $G(\alpha)$.

Proof. The theorem follows easily from Lemma 3.6. The details of the proof are left as an exercise. $\quad \square$

Corollary to Theorem 3.2 Let $G = (V, E)$ be a triangulated undirected graph and let α be an ordering of the vertices in V according to maximum cardinality search. Then α is a perfect ordering.

Proof. By Theorem 3.6, α is a perfect ordering of the vertices in V relative to $G(\alpha)$, and by Theorem 3.4, α is a zero fill-in order and therefore $G(\alpha) = G$. $\square$

Theorem 3.2 is therefore a consequence of the results in this section. However, it was illustrative to obtain that result directly in section 3.2.2. Theorem 3.6 implies that, after triangulating the graph, there is no need to perform a second maximum cardinality search in order to obtain a perfect ordering of the vertices in V relative to $G(\alpha)$. We see then that to obtain an ordering of the cliques with the running intersection property, as we will need to do in chapter 7, we do not even need Theorem 3.2. Rather our reason for using maximum cardinality search is so that the fill-in is zero if the graph is already triangulated.

Finally, we summarize the procedure which yields an ordering of the cliques with the running intersection property. That procedure is as follows:

1. Order the vertices according to maximum cardinality search.
2. Use the fill-in computation algorithm to triangulate the graph. If the graph was already triangulated, no new arcs will be added due to Theorem 3.4. Due to Theorem 3.6, the original ordering of the vertices is a perfect ordering of the triangulated graph.
3. Determine the cliques of the triangulated graph.
4. Order the cliques according to their highest labeled vertices to obtain an ordering of the cliques with the running intersection property.

Note that this procedure only yields a possible triangulation and clique ordering. A different maximum cardinality search and corresponding fill-in can add different arcs and form different cliques. Both the number of added arcs and the maximal clique size depend on the particular maximum cardinality search. The problem of finding the triangulation which adds the minimum number of arcs has been shown to be NP-Hard [Yannakakis, 1981]. NP-Hard problems are discussed more in subsection 7.7.2. For a complete introduction to NP-Hard problems the reader is referred to [Garey and Johnson, 1979] and [Balcazar, Diaz, and Gabarro, 1988].

PROBLEMS 3.2

1. Prove that if a graph admits an ordering of its cliques with the running intersection property then the graph is triangulated.

2. Prove that if a graph admits a perfect ordering of its vertices then the graph is triangulated. Hint: Consider the highest labeled vertex on any simple cycle length ≥ 4. (Also see if this result can be deduced from Theorem 3.1, Theorem 3.2, and Problem 3.2.1.)

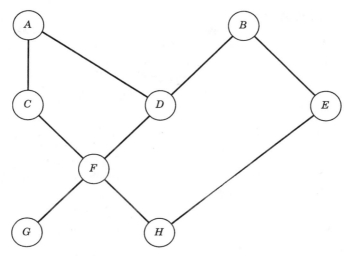

FIGURE 3.10 An undirected graph.

3. Prove that an order is zero fill-in if and only if it is perfect (Theorem 3.6 then follows immediately). Note that this problem, Problem 3.2.1, Theorem 3.1, and Theorem 3.2 imply that the following conditions are equivalent for an undirected graph G:

   (a) G is triangulated.
   (b) G admits a perfect ordering of its vertices.
   (c) G admits a zero fill-in ordering of its vertices.
   (d) G admits an ordering of its cliques which has the running intersection property.
   (e) Maximum cardinality search yields a perfect ordering of the vertices of G.

4. Perform the following steps on the graph in Figure 3.10:

   (a) Order the vertices by walking through the maximum cardinality search algorithm.
   (b) Triangulate the graph by walking through the fill-in computation Algorithm. The original ordering should be a perfect ordering for the triangulated graph. Check that this is the case.
   (c) Determine the cliques of the triangulated graph.
   (d) Order the cliques according to their highest labeled vertices. This ordering should have the running intersection property. Check that this is the case.

COMPUTER PROBLEMS 3.2

1. Using your favorite programming language, write a program which performs maximum cardinality search on an undirected graph $G = (V, E)$ to produce an ordering α of the vertices in V. The inputs to the program should be the vertices in V and arcs in E; the output should be the vertices in V and their labels according to α.

2. Let $G = (V, E)$ be an undirected graph and α an ordering of the vertices in V. Write a program which determines the elimination graph $G(\alpha) = (V, E \cup F(\alpha))$ of G relative to α, thereby triangulating G. The inputs to the program should be the vertices in V and their labels in α, and the arcs in E; the output should be the arcs in $F(\alpha)$.

3. Write a program which determines the cliques of a triangulated graph $G = (V, E)$. The inputs to the program should be the vertices in V and arcs in E; the output should be the set of cliques of G. Note that in general this problem is also NP-Hard. However, Golumbic [1980] has derived an algorithm for the special case of triangulated graphs, which runs in $O(n + e)$ time, where n is the number of vertices and e is the number of arcs.

4. Let $G = (V, E)$ be an undirected graph and α a total ordering of the vertices in V. Write a program which orders the cliques of G according to their highest labeled vertex in α. Note that you should use the program written in Computer Problem 3 to determine the cliques. The inputs should be the vertices in V and their labels in α, and the arcs in E; the output should be the set of cliques and their labels.

5. Use the programs written in Computer Problems 1–4 to write a program which takes as input an undirected graph $G = (V, E)$, triangulates the graph, and orders the cliques of the triangulated graph so that the ordering has the running intersection property. The inputs should be the vertices in V and the arcs in E; the outputs should be the additional arcs in the triangulated graph, and the cliques of the triangulated graph and their labels. Test the program on the graph in Figure 3.10.

3.3. DIRECTED ACYCLIC GRAPHS

Directed acyclic graphs (DAGs) play a central role in causal networks, which are the main focus of this text and are discussed in chapters 5 through 10. The necessary graph theoretic concepts are covered in this section.

3.3.1. Basic Concepts

Definition 3.29. Let $G = (V, E)$ be a graph (directed or undirected). Then G is acyclic if G contains no cycles.

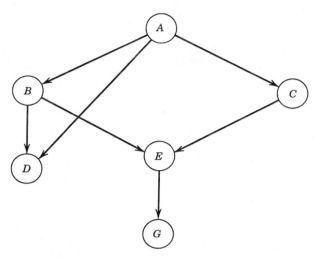

FIGURE 3.11   A directed acyclic graph (DAG).

Definition 3.30. A graph $G = (V, E)$ is a directed acyclic graph (DAG) if G is both directed and acyclic.

The graph in Figure 3.11 is a DAG.

Definition 3.31. Let $G = (V, E)$ be a digraph, and u and v vertices in V. Then u is a parent (predecessor) of v and v is a child (successor) of u if $(u, v) \in E$; u is an ancestor of v and v is a descendent of u if there is a path from u to v. A vertex with no parents is called a root.

In the graph in Figure 3.11, B and C are parents of E; B is an ancestor of D, E, and G; G is a descendent of every vertex except D.

Although Definition 3.31 is made for digraphs in general, its use is found mostly in DAGs. In DAGs it is a simple exercise to show that no vertex is a descendent or ancestor of itself. Additionally, we have the following definition and theorem, which will be needed when we discuss causal networks.

Definition 3.32. Let $G = (V, E)$ be a digraph with n vertices. Then $\beta = [v_1, v_2, \ldots, v_n]$ is an ancestral ordering of the vertices in V if for every $v \in V$ all the ancestors of v are ordered before v.

Example 3.6. $[A, B, C, D, E, G]$, $[A, B, C, E, G, D]$, and $[A, C, B, D, E, G]$ are all ancestral orderings of the vertices in the DAG in Figure 3.11.

Theorem 3.7. Let $G = (V, E)$ be a digraph. Then an ancestral ordering of the vertices in V exists if and only if G is a DAG.

Proof. The proof is left as an exercise.   □

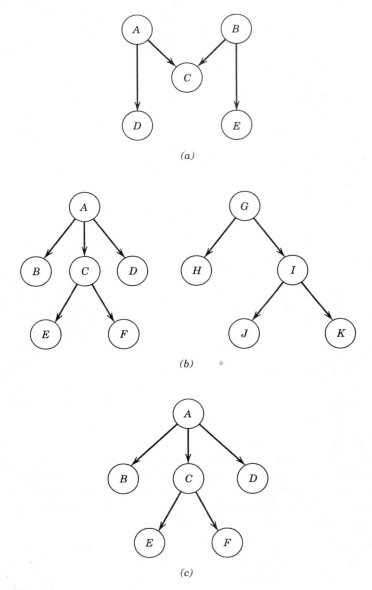

FIGURE 3.12   The DAG in (a) is singly connected, the one in (b) is a forest, and the one in (c) is a tree.

Theorem 3.8. Let $G = (V,E)$ be a DAG and $v \in V$. Then it is always possible to obtain an ancestral ordering of the vertices in V such that only the descendents of v are labeled after v.

Proof. Let X be the set of all the descendents of v, $U = X \cup \{v\}$, and $W = V - U$. Then trivially G_W is also a DAG, and therefore by Theorem 3.7 it is possible to create an ancestral ordering α_1 of the vertices in W. Similarly, G_U is a DAG in which v is the only root, and it is possible to create an ancestral ordering α_2 of the vertices in U. Clearly, v has label 1 in this order. If we then order the vertices in V by first ordering all those in W according to α_1, next ordering v, and finally ordering the remaining vertices in U according to their positions in α_2, we will have the desired ordering of the vertices in V. $\square$

Example 3.7. $[A,B,C,D,E,G]$ is an ancestral ordering of the vertices in the DAG in Figure 3.11 in which only descendents of E are labeled after E. $[A,B,D,C,E,F]$ is an ancestral ordering in which only descendents of C are labeled after C.

Definition 3.33. A DAG $= (V,E)$ is singly connected if for every $u \in V$ and $v \in V$ there is at most one chain between u and v.

Definition 3.34. A DAG $= (V,E)$ is called a forest if every vertex $v \in V$ has at most one parent.

It is a simple exercise to show that a forest is singly connected.

Definition 3.35. A forest is called a tree if exactly one vertex has no parent. That vertex is called the root of the tree.

Figure 3.12 contains examples of a singly connected network, a forest, and a tree.

PROBLEMS 3.3

1. Prove Theorem 3.7. Hint: To show that if G is a DAG then an ancestral ordering of the vertices in V exists first show that there must be at least one root, that is, a vertex with no parents. This can be done by using an inductive argument to show that if this were not the case, there would be an infinite number of vertices. Next, create an order by arbitrarily picking a root and labeling it 1, then label one vertex at a time, always picking as the next vertex to label one whose ancestors, if any, are already labeled.

2. For every vertex v in the DAG in Figure 3.11 find an ancestral ordering of the vertices in which only the descendents of v are labeled after v.

3. Prove that if a DAG is a forest then it is singly connected.

CHAPTER 4

RULE-BASED SYSTEMS VERSUS CAUSAL (BELIEF) NETWORKS

Since rule-based systems were successful in the representation of categorical knowledge, it was natural to attempt to use them to represent uncertain knowledge. In section 4.1, we describe how probability theory has been used to handle uncertainty in rule-based systems. Our goal is not to cover the rule-based approach in detail, but rather to illuminate the many problems which are encountered when we attempt to represent and update uncertainty within the rule-based framework. In section 4.2, we show that, in many cases, the rule-based framework is not even appropriate for representing uncertain relationships. Finally, in section 4.3, we show that it is not plausible that the human uncertain reasoning process could be modeled within a rule-based structure. Along the way, we informally introduce causal networks and show that they have none of these shortcomings. Causal networks are formally defined and described in detail in the next chapter.

4.1. PROBABILITY PROPAGATION IN RULE-BASED EXPERT SYSTEMS

In this section we discuss the representation of uncertainty in rule-based expert systems by probability values, and the propagation of those probability values. By the propagation of probability values we mean the computation of the change in the probabilities of certain assertions given that the probabilities of other assertions have changed (due to information supplied by the user about a specific case). We shall see that in order to propagate these values, it will be necessary to make ad hoc adjustments and assumptions of conditional independence which may not be justified.

4.1.1. The Direction of Implication in a Rule

In a rule which represents a pure logical relationship, there is only one choice for the direction of implication in a rule. However, in a rule which represents uncertainty, the direction of implication can go in either direction. For example, we could have the following two rules:

IF	the patient is a smoker		IF	the patient has lung cancer
THEN	the patient has lung cancer.		THEN	the patient is a smoker.

The rules are not meant to represent a causal relationship between the assertions, but rather that the premise is evidence for the conclusion. If a person smokes, that is evidence that he may have lung cancer, and if a person has lung cancer, that is evidence that he may be a smoker. Thus, in rule-based systems which reason under uncertainty, we are free to choose the direction of implication in a rule. In practice, the direction is placed from the evidence, which we are able to gather either directly or from the use of other rules, to the desired conclusion. For example, in a diagnostic medical expert system, we want to determine whether the patient has lung cancer. We can easily determine whether the patient smokes, and this is evidence for the presence of lung cancer. Therefore we would use the first rule above.

4.1.2. Inference Networks

A set of rules, which does not contain cyclic rules, determine a directed acyclic graph (DAG). We will illustrate this with a simplified subset of the rules in the PROSPECTOR expert system (Duda, Hart, Nilsson, & Sutherland, 1978). PROSPECTOR is a system which helps geologists evaluate the mineral potential of exploration sites. The subset of rules, which is in Table 4.1, can be represented by the DAG in Figure 4.1. The capital letter next to a proposition in the table is used to represent that proposition in the DAG. Each rule is represented by an arc from the antecedent to the conclusion. The values stored at each vertex and each arc in Figure 4.1 will be explained in subsection 4.1.4. Such a DAG is called an inference network, which is formally defined as follows:

Definition 4.1. An inference network is a directed acyclic graph (DAG) in which the vertex set is a set of propositions.

The "roots" of an inference network contain propositions about which the user is expected to supply information, while the "leaves" contain the propositions of interest (hypotheses). For example, in the inference network in Figure 4.1, the user is expected to determine with some accuracy whether bleached rocks are present, whereas the geologist is interested in whether there are massive sulfide deposits.

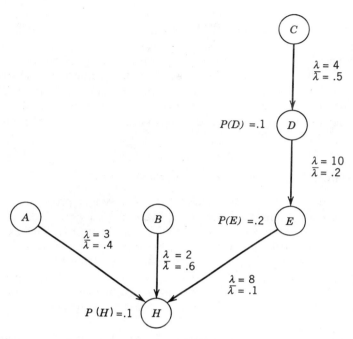

FIGURE 4.1 The inference network which represents the rules in Table 4.1. λ and $\bar{\lambda}$ values are stored on the arcs, and a priori probabilities are stored at each vertex, except for roots.

TABLE 4.1 A Simplified Subset of the Rules in the PROSPECTOR Expert System

1.	IF	barite is overlying sulfide (A)
	THEN	there are massive sulfide deposits (H).
2.	IF	galena, sphalerite, or chalcopyrite fill cracks in rhyolite or dacite (B)
	THEN	there are massive sulfide deposits (H).
3.	IF	there are bleached rocks (C)
	THEN	there is a reduction process (D).
4.	IF	there is a reduction process (D)
	THEN	there are clay minerals (E).
5.	IF	there are clay minerals (E)
	THEN	there are massive sulfide deposits (H).

Source: Duda, Hart, and Sutherland (1978).

In general, in an inference network, the verity of the assertions cannot be determined for certain. Furthermore, the truth of a premise is indicative of the truth of a conclusion, but not absolutely confirmative. For example, the user might determine that there is a good chance that the rocks are bleached, and the presence of bleached rocks might usually indicate that there is a reduction process. In subsection 4.1.4 we show how to propagate probabilities from the leaves to the roots in an inference network. First, in subsection 4.1.3, we investigate the case in which there are more than one premise in the antecedent of a rule.

4.1.3. Multiple Premises in an Antecedent

In inference networks we treat each vertex as a single proposition. Yet, in many rule-based systems, the antecedent is the conjunction of several propositions (recall the sample rule from MYCIN in chapter 1). The user would enter individual probabilities for each of the separate premises. We therefore need to determine the probability of the conjunction from the probabilities of the separate premises. In general, there is no way to do this. However, the following theorem gives a lower bound for the probability of the conjunction of premises in terms of the probabilities of separate premises.

Theorem 4.1. If $(\Omega, \mathcal{F}, P)$ is a probability space, then for every integer n and every set of events $\{E_1, E_2, \ldots, E_n\}$ we have that

$$P(E_1 \wedge \cdots \wedge E_n) \geq M_n$$

where

$$M_n = \max\left(0, \sum_{i=1}^{n} p_i - n + 1\right) \quad \text{and} \quad p_i = P(E_i).$$

Furthermore, this inequality cannot be improved upon for any value of n. That is, for any n, there exists a probability space $(\Omega', \mathcal{F}', P')$ and set of events $\{E_1', E_2', \ldots, E_n'\}$ such that $p_i = P'(E_i')$ and

$$M_n = P'(E_i' \wedge \cdots \wedge E_n').$$

Therefore no sharper lower bound for $P(E_1 \wedge \cdots \wedge E_n)$ can be deduced from the values of p_i alone.

Proof. It is left as an exercise to use induction to show that M_n is a lower bound. The proof that there is no sharper lower bound is quite lengthy and can be found in Neapolitan [1986]. □

Example 4.1. Suppose that $P(E_1) = .9$, $P(E_2) = .8$, and $P(E_3) = .7$. Then by Theorem 4.1,

$$P(E_1 \wedge E_2 \wedge E_3) \geq \max(0, .9 + .8 + .7 - 3 + 1) = .4.$$

The lower bound obtained from Theorem 4.1 becomes meaningless unless all the individual probabilities are fairly high. In some cases it may be reasonable to assume that if A is one of the premises in a rule and W is the conjunction of any of the other premises, then $P(A \mid W) \geq P(A)$. For example, in the sample rule from MYCIN in chapter 1, it may be reasonable to assume that the information that the organism grows in pairs and in chains should not lower the probability that it grows in clumps. The following theorem, taken from Neapolitan et al. [1987], obtains a better lower bound when this assumption is valid:

Theorem 4.2. Let $(\Omega, \mathcal{F}, P)$ be a probability space and $\{E_1, E_2, \ldots, E_n\}$ be a set of events such that for $1 \leq i \leq n$ and for every W which is the intersection of the events in any subset of $\{E_1, E_2, \ldots, E_n\}$, we have that

$$P(W) > 0 \quad \text{and} \quad P(E_i \mid W) \geq P(E_i).$$

Then

$$P(E_1 \wedge \cdots \wedge E_n) \geq P(E_1)P(E_2)\ldots P(E_n).$$

Proof. Since for all such W the $P(W)$ is positive, we obtain the following "chain rule" from the definition of conditional probability:

$$P(E_1 \wedge \cdots \wedge E_n) = P(E_1 \mid E_2 \wedge \cdots \wedge E_n)P(E_2 \mid E_3 \wedge \cdots \wedge E_n)\ldots P(E_{n-1} \mid E_n)P(E_n)$$

$$\geq P(E_1)P(E_2)\ldots P(E_{n-1})P(E_n).$$

The last inequality is by the assumptions for this theorem, and proves the theorem. $\square$

The following theorem shows that Theorem 4.2 always yields a greater lower bound than Theorem 4.1:

Theorem 4.3. Let n values, $p_1, p_2, \ldots, p_n$, be given such that for $1 \leq i \leq n$, we have that $0 \leq p_i \leq 1$. Then

$$\prod_{i=1}^{n} p_i \geq \sum_{i=1}^{n} p_i - n + 1.$$

Proof. The proof is by induction and is left as an exercise. $\square$

PROBLEMS 4.1.3

1. Prove that M_n is a lower bound in Theorem 4.1.

2. Prove Theorem 4.3.

4.1.4. The Method of Odds Likelihood Ratios

The best-known method for propagating probabilities in an inference network is the method of odds likelihood ratios developed by Duda, Hart, and Nilsson [1976] for the PROSPECTOR expert system. We briefly discuss this method to elucidate the problems in using rule-based expert systems to reason under uncertainty.

Updating in a Single Rule. We assume that for a given inference network we are able to determine the a priori probabilities of the propositions at each vertex, except for the roots, and, if A is the premise in a rule and H is the conclusion, $P(A \mid H)$ and $P(A \mid \neg H)$. In addition, we assume that all these probabilities are positive. We wish to determine a method for computing $P(H \mid A)$ and $P(H \mid \neg A)$ from these values. First we need the following definition:

Definition 4.2. Let $(\Omega, \mathcal{F}, P)$ be a probability space and $E \in \mathcal{F}$. Then the odds $O(E)$ are defined by

$$O(E) = \frac{P(E)}{1 - P(E)}.$$

If we have a conditional probability $P(E \mid F)$, the conditional odds $O(E \mid F)$ are defined analogously. Notice that

$$P(E) = \frac{O(E)}{1 + O(E)}.$$

Definition 4.3. If A is the premise in a rule and H is the conclusion, the likelihood ratio λ of A and the likelihood ratio $\overline{\lambda}$ of $\neg A$ are defined by

$$\lambda = \frac{P(A \mid H)}{P(A \mid \neg H)} \quad \text{and} \quad \overline{\lambda} = \frac{P(\neg A \mid H)}{P(\neg A \mid \neg H)}.$$

Theorem 4.4. If A is the premise in a rule and H is the conclusion, then

$$O(H \mid A) = \lambda O(H) \quad \text{and} \quad O(H \mid \neg A) = \overline{\lambda} O(H). \qquad (4.1)$$

Proof. The proof follows easily from the definition of conditional probability. $\square$

We have assumed that we know $P(A \mid H)$ and $P(A \mid \neg H)$. Since

$$P(\neg A \mid H) = 1 - P(A \mid H) \quad \text{and} \quad P(\neg A \mid \neg H) = 1 - P(A \mid \neg H),$$

we can compute λ and $\overline{\lambda}$. These are the values which are actually stored at each rule. Therefore we can use formula (4.1) to compute the new probability of H when either A or $\neg A$ becomes known for certain. In the development of the PROSPECTOR system, the experts were actually asked to estimate the values of λ and $\overline{\lambda}$, rather than $P(A \mid H)$ and $P(A \mid \neg H)$. Theorem 4.4 implies that if $\lambda = 1$, the occurrence of A has no effect on H, if $\lambda > 1$, A is evidence for H, and, if $\lambda < 1$, A is evidence against H. $\overline{\lambda}$ has a similar relationship to $\neg A$. It may be thought that the expert is free to estimate λ and $\overline{\lambda}$ individually based solely on his judgments. The following theorem shows that this is not the case:

Theorem 4.5. With λ and $\overline{\lambda}$ defined as in Definition 4.3, we have that

$$\overline{\lambda} = \frac{1 - \lambda P(A \mid \neg H)}{1 - P(A \mid \neg H)}.$$

Proof. The proof is left as an exercise. $\square$

Theorem 4.5 implies that $\lambda = 1$ if and only if $\overline{\lambda} = 1$ and $\lambda > 1$ if and only if $\overline{\lambda} < 1$. This contradicts what many people believe. They feel that A can be evidence for H, while A's absence is not evidence against H. That is, they feel that λ can be greater than 1 while $\overline{\lambda}$ is equal to 1. The following rule illustrates this erroneous line of reasoning:

IF   a husband dines with a strange lady

THEN the husband is cheating.

If a man dined with a strange lady, his wife would become suspicious that he was cheating. However, if she had reason to believe that her husband was cheating based on other evidence (an example of other evidence would be the repeated occurrence of the other party hanging up when she answered the phone), one might feel that the fact that he had never dined with a strange lady should hardly change her suspicions. Therefore the fact that he definitely did not dine with a strange lady should have no bearing on conclusions drawn from other evidence. Based on such reasoning, we would assign a value greater than 1 to λ and the value of 1 to $\overline{\lambda}$ in the above rule. However, in actuality, each time the wife eliminates one possible item of evidence, the wife is indeed bringing down the likelihood that her husband is cheating. If she initially became suspicious due to incidents involving the phone, and subsequently she eliminated every other possible item of evidence for cheating, she would conclude that it is unlikely that he is cheating.

Probability Propagation and Uncertain Evidence. In Figure 4.1, the λ and $\overline{\lambda}$ values are stored at each arc and a priori probabilities are stored at each vertex, except for roots. Suppose C became known for certain. We need to compute the impact of this evidence on the probabilities of the other vertices in the network. The probabilities of all of them will change. However, in rule-based expert systems we are ordinarily only interested in the probabilities of leaves. Therefore we need only devise a method for propagating this evidence down to H. First we compute the new probability of D:

$$O(D \mid C) = \lambda O(D) = 4\frac{.1}{1-.1} = .444$$

$$P(D \mid C) = \frac{.444}{1+.444} = .307.$$

Next we must propagate to compute $P(E \mid C)$. We know how to compute the probability of E if D became known for certain. However, D is not known for certain. The following theorem enables us to obtain a propagation formula:

Theorem 4.6. Let $(\Omega, \mathcal{F}, P)$ be a probability space and E_1, E_2, and E_3 events in $\mathcal{F}$. Then

$$P(E_1 \wedge E_2 \mid E_3) = P(E_1 \mid E_2 \wedge E_3)P(E_2 \mid E_3)$$

if all a priori probabilities and conditional probabilities are positive.

Proof. The proof is left as an exercise. $\square$

By the above theorem, we have that

$$P(E \mid C) = P(E \wedge D \mid C) + P(E \wedge \neg D \mid C)$$
$$= P(E \mid D \wedge C)P(D \mid C) + P(E \mid \neg D \wedge C)P(\neg D \mid C).$$

We next assume that

$$P(E \mid D \wedge C) = P(E \mid D) \qquad \text{and} \qquad P(E \mid \neg D \wedge C) = P(E \mid \neg D).$$

That is, E and C are independent in the space that D has occurred and in the space that $\neg D$ has occurred. This assumption is clearly not valid in networks in which there are multiple paths from the evidence to the conclusion. If C is also evidence for E through another path, this assumption would not be valid. Using the above assumption, we obtain that

$$P(E \mid C) = P(E \mid D)P(D \mid C) + P(E \mid \neg D)P(\neg D \mid C). \qquad (4.2)$$

Formula (4.2) is our operative formula for propagating probabilities. We compute the needed probabilities in formula (4.2) using the method described in the previous subsection. Once we have $P(E \mid C)$, formula (4.2) can be used to compute $P(H \mid C)$ from $P(E \mid C)$ and $P(\neg E \mid C)$. It is left as an exercise to actually perform these calculations.

We also use formula (4.2) when the user is not certain as to whether a proposition at a root is true or false. For example, suppose the user says that $P(B) = .8$. This means, if we call B' all the evidence which is available to the user, that $P(B \mid B') = .8$. Formula (4.2) can then be used to compute $P(H \mid B')$.

An Inconsistency in Probability Propagation in Inference Networks.

Suppose that, based on his evidence, the user enters a value for $P(C \mid C')$ (C'

is the user's evidence for C) which, after an application of formula (4.2), results in $P(D \mid C')$ being equal to .1, the same value as the a priori probability of D. If we use the propagation technique outlined in the previous subsection, we would obtain that $P(E \mid C')$ equals .114. However, since C' is only evidence for E through D, we should have that if $P(D \mid C') = P(D)$, then $P(E \mid C')$ is equal to $P(E)$ (i.e., .2). We can see the reason for this problem by rewriting the propagation equation as follows:

$$P(E \mid C') = P(E \mid D)P(D \mid C') + P(E \mid \neg D)P(\neg D \mid C')$$

$$= P(E \mid D)P(D \mid C') + P(E \mid \neg D)(1 - P(D \mid C'))$$

$$= P(E \mid \neg D) + (P(E \mid D) - P(E \mid \neg D))P(D \mid C'). \qquad (4.3)$$

This formula is the equation of a straight line which maps values of $P(D \mid C')$ to $P(E \mid C')$. The line is graphed in Figure 4.2. $P(D)$ should map into $P(E)$; however, instead another point, which we shall call $P_c(D)$, maps into $P(E)$. A resultant problem is that if $P(D \mid C')$ is between $P(D)$ and $P_c(D)$, then the evidence will decrease the probability of E, when in fact, it should increase the probability of E. The problem is a result of overspecification. It is necessary to specify the values of $P(E \mid D)$, $P(E \mid \neg D)$ (or equivalently, λ and $\bar{\lambda}$), and $P(E)$ so that we can update the probability of E using equation (4.1). By formula (4.3) these values uniquely determine the consistent value of the probability of D. That is, they determine $P_c(D)$. However, it is also necessary to specify the $P(D)$ so we can update the probability of D using equation (4.1). Unfortunately, if all these values are obtained from a human, the values of $P_c(D)$ and $P(D)$ would rarely be equal. We could try to relax the specifications by changing the value of $P(D)$ to $P_c(D)$ to achieve consistency. However, since D and E are embedded in a network, this would call for propagating the relaxation through the entire network. At best, we would end up specifying only one a priori probability. In practice, consistency is obtained by making ad hoc function adjustments to equation (4.3) like the one illustrated in Figure 4.3. The point $P(D)$ is forced to map into the point $P(E)$.

Multiple Evidence.   For the inference network in Figure 4.1, we have that

$$O(H \mid A) = 3(.111) = .333 \qquad \text{and} \qquad O(H \mid B) = 2(.111) = .222.$$

If both A and B become known for certain, we need to be able to combine this evidence to compute the $P(H \mid A \wedge B)$. The following theorem enables us to accomplish this:

Theorem 4.7. Suppose, in an inference network, for $1 \leq i \leq n$ we have arcs from E_i to H. Denote the λ and $\bar{\lambda}$ values on the arc from E_i by λ_i and $\bar{\lambda}_i$, respectively. If for $1 \leq i \leq n$, the E_i's are conditionally independent in the space that H occurs and the space that H does not occur, then

$$O(H \mid E_1 \wedge \cdots \wedge E_n) = \lambda_1 \times \cdots \times \lambda_n O(H)$$

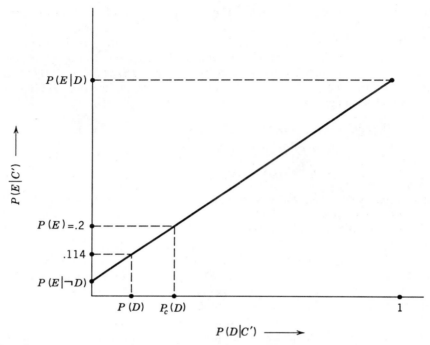

FIGURE 4.2 The graph of the line which maps values of $P(D \mid C')$ to values of $P(E \mid C')$.

and

$$O(H \mid \neg E_1 \wedge \cdots \wedge \neg E_n) = \overline{\lambda}_1 \times \cdots \times \overline{\lambda}_n O(H).$$

Proof. We obtain the first formula. The second one is obtained analogously. By the definitions of odds and conditional probability, we have that

$$
\begin{aligned}
O(H \mid E_1 \wedge \cdots \wedge E_n) &= \frac{P(H \mid E_1 \wedge \cdots \wedge E_n)}{P(\neg H \mid E_1 \wedge \cdots \wedge E_n)} \\
&= \frac{P(E_1 \wedge \cdots \wedge E_n \mid H) P(H) / P(E_1 \wedge \cdots \wedge E_n)}{P(E_1 \wedge \cdots \wedge E_n \mid \neg H) P(\neg H) / P(E_1 \wedge \cdots \wedge E_n)} \\
&= \frac{P(E_1 \wedge \cdots \wedge E_n) \mid H)}{P(E_1 \wedge \cdots \wedge E_n \mid \neg H)} O(H).
\end{aligned}
$$

The proof now follows from the independence assumptions. □

The independence assumptions in Theorem 4.7 have become notorious in probabilistic reasoning. We discuss them in the following section. We will now show how to use this theorem to combine evidence. By Theorem 4.7 we have

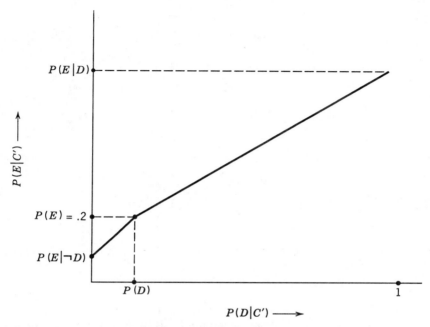

FIGURE 4.3 The graph of the line which maps values of $P(D \mid C')$ to $P(E \mid C')$ with an ad hoc function adjustment.

that:

$$O(H \mid A \wedge B) = \lambda_A \lambda_B O(H)$$

$$= (3)(2)(.111) = .666$$

and

$$P(H \mid A \wedge B) = \frac{.666}{1 + .666} = .400.$$

Similarly, we could compute $P(H \mid \neg A \wedge \neg B)$. However, we must also consider the case where A and B do not become known for certain. For example, suppose the user enters values of $P(A \mid A')$ and $P(B \mid B')$ which, using the propagation formula from the previous subsection, results in the following values:

$$P(H \mid A') = .2 \quad \text{and} \quad P(H \mid B') = .15.$$

We need to combine this evidence to determine the $P(H \mid A' \wedge B')$. The following definition and theorem enable us to accomplish this:

Definition 4.4. Suppose, in an inference network, there is an arc from A to H, and A' is all the evidence for A. Then we define the effective likelihood ratio λ' on that arc as follows:

$$\lambda' = \frac{O(H \mid A')}{O(H)}.$$

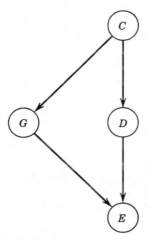

FIGURE 4.4   A nonsingly connected inference network in which there is more than one path from C to E.

Theorem 4.8. Suppose, in an inference network, for $1 \leq i \leq n$, there is an arc from E_i to H, and E_i' is all the evidence for E_i. If all E_i''s are independent in the space that H occurs and in the space that H does not occur, then

$$O(H \mid E_1' \wedge \cdots \wedge E_n') = \lambda_1' \times \cdots \times \lambda_n' O(H), \qquad (4.4)$$

where λ_i' is the effective likelihood ratio on the arc from E_i to H.

Proof. The proof is analogous to that for Theorem 4.7. $\square$

As mentioned previously, we discuss these independence assumptions in the next subsection. We note here that they could never hold in a network in which there are multiple paths from the evidence to the conclusion. Consider the network in Figure 4.4. If it becomes known for certain that C is true, then the evidence, G', for G, and the evidence D', for D, are both equal to C. The proposition C is certainly not independent of itself.

Theorem 4.8 enables us to easily combine evidence. If $P(H \mid B') = .15$, then

$$O(H \mid B') = \frac{.15}{1 - .15} = .176$$

$$O(H) = \frac{.1}{1 - .1} = .111$$

$$\lambda_B' = \frac{O(H \mid B')}{O(H)} = \frac{.176}{.111} = 1.59.$$

Similarly, if $P(H \mid A') = .2$, then λ'_A equals 2.25. Therefore

$$O(H \mid A' \wedge B') = (2.25)(1.59)(.111) = .397.$$

$$P(H \mid A' \wedge B') = \frac{.397}{1 + .397} = .284.$$

PROBLEMS 4.1.4

1. Prove Theorems 4.5 and 4.6.

2. Compute $P(H \mid C)$ for the inference network in Figure 4.1 using formula (4.2).

4.1.5. Concluding Remarks

While describing the propagation of probabilities in rule-based systems, we have alluded to problems in the propagation scheme. These problems are summarized in this section. As we discuss each problem, we note that it either disappears or is minimized in causal networks.

Problems in Obtaining the Necessary Probability Values. We need to be able to ascertain the a priori probabilities of all propositions, except for those stored at the roots, in the network and the conditional probabilities of evidence given conclusions. Ordinarily, it is an easier task to obtain the values of the conditional probabilities than those of the a priori probabilities. For example, the probability of a symptom given a disease is, for the most part, a fixed value regardless of the population. On the other hand, the a priori probability of a disease is very population specific, and therefore more difficult to ascertain. We will see in chapter 5 that it is only necessary to obtain the a priori probabilities of propositions which are stored at the roots in a causal network. Moreover, in many cases in causal networks, the roots contain propositions which are hypotheses. It is often easier to obtain the a priori probability of a hypothesis than that of evidence. For example, we may have data on the a priori probability of a disease because that is a value in which we are interested. On the other hand, the a priori probability of a symptom is of little interest.

Problems With the Assumptions of Conditional Independence. It was necessary to assume that items of evidence are conditionally independent in the space that a hypothesis occurs and in the space that it does not. Much has been written about these assumptions. In particular, Charniak [1983] notes that symptoms are certainly not independent in the space at large, but in many cases they are independent given that a disease has occurred. For example, vomiting and diarrhea occur together so often that $P(\text{vomiting} \mid \text{diarrhea})$ is much greater than $P(\text{diarrhea})$. However, once it is known that the patient has stomach flu, then both symptoms are usually present, and knowledge of one would have little bearing on the probability of the other. That is, $P(\text{vomiting} \mid$

diarrhea $\wedge$ stomach flu) is about equal to $P(\text{vomiting} \mid \text{stomach flu})$. Charniak offers justifications for the use of these conditional independence assumptions. However, he is arguing for a method for determining the probabilities of direct causes given their effects, as discussed in section 7.7 of this text. In a network, items of evidence are often separated from a hypothesis by many intermediate propositions and rules. There is little reason to believe that the independence assumptions are valid. In particular, we have shown that they would never be valid in a network in which there are multiple paths from the evidence to the conclusion.

It may be thought that, since for the most part we are interested in the new (conditional) probabilities of the leaves and since evidence comes in at the roots, we could just store the conditional probability of each leaf given all combinations of values of the roots. Then we could forget about propagating entirely, thereby eliminating these propagation problems. First, this would only work if it became known for certain whether each root was or was not true. Second, since the leaves and the roots are not directly related, these conditional probabilities would be harder to obtain. Last, and most significantly, we would need an astronomical amount of data even for a relatively small network. If there were only ten roots, we would need to store 2^{10} values for each leaf.

A more reasonable solution would be to find a propagation scheme which makes no probabilistic assumptions about the propositions in the network. The following example, taken from Neapolitan [1987], shows that this is not possible. Suppose that we have three propositions, A, B, and H, and we need a method for computing $P(H \mid A \wedge B)$ from the $P(H \mid A)$ and $P(H \mid B)$, or, if this were not possible, from the probabilities of one- and two-member combinations of A, B, and H. That is, perhaps we could settle for a method for computing $P(H \mid A \wedge B)$ from $P(A)$, $P(B)$, $P(H)$, $P(H \mid A)$, $P(A \mid H)$, $P(H \mid \neg A)$, $P(A \mid \neg H)$, $P(A \wedge H)$, $P(A \wedge \neg H)$, $P(A \vee H)$, $P(A \vee \neg H)$, $P(\neg A \vee H)$, $P(B \mid A)$, $P(A \mid B)$, etc. If we could compute $P(H \mid A \wedge B)$ from some combination of these quantities, perhaps we could devise a propagation scheme which yielded the necessary quantities. The two probability distributions depicted by the Venn diagrams in Figure 4.5 show that this is not possible. Each x represents a point of equal probability (1/4). Ω is the set of all the points. A proposition such as A is represented by the set of all points in A. It is left as an exercise to show that all the probabilities of one- and two-member combinations of A, B, and H are the same for these two probability distributions. However,

$$P_1(H \mid A \wedge B) = 0 \quad \text{and} \quad P_2(H \mid A \wedge B) = 1.$$

Hence a method for computing the desired conditional probabilities from the probabilities of one- and two-member combinations would have to yield a value of 0 and a value of 1 simultaneously. This example shows that it is not even possible to obtain an informative upper or lower bound for the $P(H \mid A \wedge B)$ from the probabilities of one- and two-member combinations.

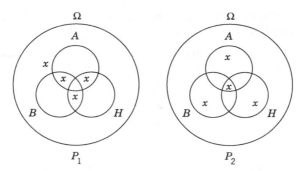

FIGURE 4.5 Two probability distributions in which the probabilities of all one- and two-member combinations of the propositions A, B, and H are equal.

Thus rule-based systems are doomed to making some assumptions about the probability space. We will see in chapter 5 that causal networks also make assumptions. However, they are not arbitrarily made just to make things work; rather they are the assumptions of conditional independencies which we feel actually do hold for the propositions in the network.

The Problem With Ad Hoc Function Adjustments. In order to achieve consistency it was necessary to make ad hoc function adjustments. When we make such adjustments we often offer intuitive justifications for them, and test their validity by observing how well the system ends up working. Although such a practice is often necessary, it is clear that it should be avoided whenever possible. It is easier to test and evaluate a system if we use axioms which are based on sound philosophical and/or empirical grounds, and only make valid deductions from those axioms. The methods for propagating probabilities in causal networks only assume the axioms of probability theory and the definition of conditional probability. We have already seen that these axioms and this definition have a sound basis.

In closing, we mention that the work of Duda, Hart, Nilsson, and Sutherland, although now dated, was of importance in the development of a discipline for probabilistic reasoning. Together with the early work on certainty factors, their results began a serious effort to propagate probabilities from proposition to proposition through a network. This problem is ignored in most probability and statistics texts, which to this day are totally devoid of diagrams. The next major step in this effort was the development of techniques for the propagation of probabilities in causal networks.

PROBLEM 4.1.5

Show that the probabilities of all one- and two-member combinations of the propositions in Figure 4.5 are the same for P_1 and P_2.

4.2. THE INAPPROPRIATENESS OF RULE-BASED SYSTEMS FOR REPRESENTING UNCERTAINTY

Many of the results in this section are obtained from Heckerman and Horvitz [1987].

4.2.1. The Problem in Representing Mutually Exclusive and Exhaustive Alternatives

Consider first the classic case of balls in urns. Specifically, suppose we have three urns, one containing two black balls, one containing two white balls, and one containing one white ball and one black ball, as depicted in Figure 4.6. All the urns appear identical. We choose one of the urns and select one ball from the chosen urn. Our goal is to write rules which can be used to determine the probabilities of our urn being each of the three possible urns based on the color of the ball selected. If we define

H_1 = our urn contains 1 white ball and 1 black ball

H_2 = our urn contains two white balls

H_3 = our urn contains two black balls

B = the chosen ball is black,

we then have three rules:

1) IF B, THEN H_1 2) IF B, THEN H_2 3) IF B, THEN H_3.

We do not need a proposition, "the chosen ball is white," and rules involving that proposition, since that proposition is equal to $\neg B$. The inference network which represents this rule base is depicted in Figure 4.7. The λ value for rule 1 is given by

$$\lambda = \frac{P(B \mid H_1)}{P(B \mid \neg H_1)}$$

$$= \frac{P(B \mid H_1)}{P(B \mid H_2 \vee H_3)}$$

$$= \frac{P(B \mid H_1)}{P(B \mid H_2)P(H_2 \mid H_2 \vee H_3) + P(B \mid H_3)P(H_3 \mid H_2 \vee H_3)}$$

$$= \frac{.5}{(0)(.5) + 1(.5)} = 1.$$

It is left as an exercise to compute the other λ values, the $\overline{\lambda}$ values, and the a priori probabilities. They are also in Figure 4.7. Given this inference network, we could then update the probabilities of H_1, H_2, and H_3, when the ball is drawn, using the technique in section 4.1.

FIGURE 4.6   There are three urns, each with two balls, and each ball can be one of two possible colors.

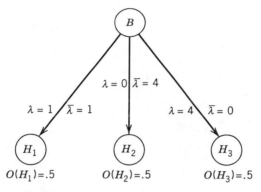

FIGURE 4.7   The inference network which represents the case where one ball is drawn from the chosen urn, and there are two possible colors of the ball.

The above inference network is simple enough. However, suppose next that we choose two balls, with replacement, from our urn. There are now three mutually exclusive and exhaustive alternatives for the possible evidence instead of just two. A severe limitation of inference networks is that we can only store a single proposition at each vertex. Thus, when we have a set of more than two mutually exclusive and exhaustive alternatives, we must represent each alternative as a separate vertex in the network. In this case we need vertices for the following propositions:

$$BB = \text{both balls are black}$$

$$BW = \text{one ball is black and the other is white}$$

$$WW = \text{both balls are white.}$$

The inference network which represents this case is depicted in Figure 4.8. Notice that it is substantially more complex than the network in Figure 4.7; it

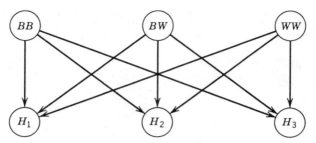

FIGURE 4.8 The inference network which represents the case where two balls are drawn, with replacements from the chosen urn, and there are two possible colors of the balls.

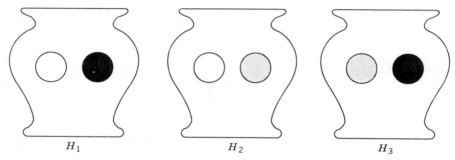

FIGURE 4.9 There are three urns, each with two balls, and each ball can be one of three possible colors.

requires nine arcs instead of three. It is left as an exercise to compute the λ and $\overline{\lambda}$ values for this network.

Suppose now that there are three possible colors for the balls, as depicted in Figure 4.9, and we choose one ball from our urn. Again, since there are three alternatives for the color of the ball, we need vertices for the following propositions:

$$B = \text{the chosen ball is black}$$

$$W = \text{the chosen ball is white}$$

$$G = \text{the chosen ball is gray.}$$

The inference network which represents this case is shown in Figure 4.10. Again it is substantially more complex than the network in Figure 4.7, requiring nine arcs instead of three. It is again left as an exercise to compute the λ and $\overline{\lambda}$ values for this network.

Finally, suppose that we draw two balls, with replacement, from an urn which has been chosen from the urns in Figure 4.9. In this case there are six mutually exclusive and exhaustive alternatives for the possible evidence. The

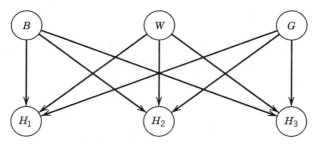

FIGURE 4.10 The inference network which represents the case where one ball is drawn from the chosen urn and there are three possible colors for the balls.

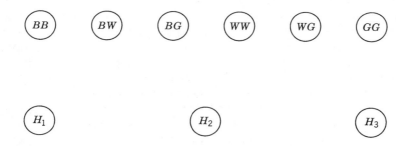

FIGURE 4.11 The inference network (without the arcs) which represents the case where two balls are drawn, with replacement, and there are three possible colors for the balls. There is an arc from every vertex on the top to every vertex on the bottom.

resultant inference network is depicted in Figure 4.11. There is an arc from every vertex on the top to every vertex on the bottom; they are omitted to avoid a spaghetti appearance. This network requires 18 arcs. If we draw three balls, we will need 30 arcs. We see then that the network must become quite complex in order to represent a very simple situation.

We will formally define causal networks and discuss them in detail in chapter 5. Briefly, one property of a causal network is that it is a DAG in which each vertex can contain a set of mutually exclusive and exhaustive alternatives. Therefore a major advantage of causal networks is that all of the alternatives can be stored at one vertex. A causal network which represents the case where we have two different colored balls and one draw is depicted in Figure 4.12. There are only two vertices in the DAG; each of the vertices is a "propositional variable" instead of a simple proposition. The term "propositional variable" will also be formally defined in chapter 5. Briefly, it is simply a variable whose value can be one of a set of mutually exclusive and exhaustive alternatives. Thus H can take one of these values:

h_1 = our urn contains one white ball and one black ball

h_2 = our urn contains two white balls

h_3 = our urn contains two black balls.

C can take one of these values:

c_1 = the chosen ball is black

c_2 = the chosen ball is white.

We will see in the next chapter that the direction of the arc can be arbitrarily chosen to go in either direction. However, as will be illustrated in the next section, it often goes from a perceived cause to a perceived effect, thus the name "causal network." In a causal network, it is necessary to store the a priori probability of every propositional value of each root and the conditional probability of every propositional value of each nonroot given the propositional values of its parents. These values have been computed and are also in Figure 4.12. From these values it is possible to determine the new (conditional) probability of each propositional value of every vertex given that any other vertex has been instantiated for one of its values. For example, we could compute $P(h_1 \mid c_1)$ or $P(c_1 \mid h_1)$ for the causal network in Figure 4.12. We will see how to do this in the following three chapters (obviously, it is an easy task in the simple network in Figure 4.12). The point here is the simplicity with which we have represented the relationships.

If we draw two balls, with replacement, from our urn, we need only add another vertex to represent the second ball drawn. This case is represented by the causal network in Figure 4.13, where D is a propositional variable whose possible values are

d_1 = the second ball drawn is black

d_2 = the second ball drawn is white.

Not only is this network much simpler than the one in Figure 4.8, it is also more expressive. In Figure 4.13 we could instantiate C for one of its values, D for one of its values, or both C and D, each for one of their values. Thereby we could determine the probability of each value of H given the result of one draw or the result of two draws. Using the inference network in Figure 4.8, we could only determine the probabilities based on two draws.

If there are three possible colors for the balls, we need only add another possible value to the variables C and D. The network in Figure 4.14 represents the case where one ball is drawn and our urn has been chosen from the set of urns in Figure 4.9. Again, if two balls are drawn, with replacement, we need only add another arc D. It is left as an exercise to draw this network. Not only is the resultant network simpler than the one in Figure 4.11, it is also

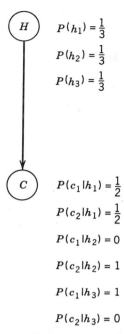

$$P(h_1) = \frac{1}{3}$$

$$P(h_2) = \frac{1}{3}$$

$$P(h_3) = \frac{1}{3}$$

$$P(c_1|h_1) = \frac{1}{2}$$

$$P(c_2|h_1) = \frac{1}{2}$$

$$P(c_1|h_2) = 0$$

$$P(c_2|h_2) = 1$$

$$P(c_1|h_3) = 1$$

$$P(c_2|h_3) = 0$$

FIGURE 4.12   A causal network which represents the case where one ball is drawn from the chosen urn and there are two possible colors for the balls.

more expressive. That is, we can determine the probabilities based on one draw from this network. In fact, we can even determine the probabilities of the values of the second draw from the result of the first draw. If more balls are drawn, with replacement, we need only add one arc for each draw.

PROBLEMS 4.2.1

1. Compute the remaining λ and $\overline{\lambda}$ values for the inference networks in Figures 4.7, 4.8, and 4.10.

2. Create the causal network which represents the case where two balls are drawn, with replacement, from an urn which is chosen from the set of urns in Figure 4.9.

4.2.2. The Problem in Representing Multiple Causes

We will illustrate the problem in representing multiple causes in an inference network by looking at a problem which will be analyzed in detail in chapters 5 and 6. Suppose that in the past few years, Mr. Holmes has noticed that frequently earthquakes have caused his burglar alarm to sound. At present, he is sitting in his office, and his wife calls, informing him that the burglar alarm

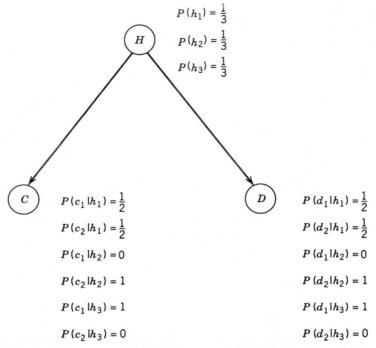

FIGURE 4.13   The causal network which represents the case where two balls are drawn, with replacement, and there are two possible colors of the balls.

has sounded. He then rushes home, assuming that there is a good chance that his residence has been burglarized. On the way home, however, he hears on the radio that there is an earthquake. He now relaxes substantially, assuming that there is a good chance that the earthquake triggered his alarm.

It is quite difficult to represent this situation with an inference network. Let

A = Mr. Holmes' burglar alarm sounds

B = Mr. Holmes' residence is burglarized

C = there is an earthquake.

Since A is evidence for B and C is evidence against B, our initial inclination might be to represent the situation by the inference network in Figure 4.15. When Mr. Holmes learns that the alarm sounded, we instantiate A, thereby increasing the probability of B. Then, when he learns of an earthquake, we instantiate C, thereby decreasing the probability of B. Notice first that the independence assumptions in the odds likelihood ratio method are violated in this network. That is, A and C are not independent given B. Unless a burglar always definitely triggers the alarm, the probability of A given B and C is greater than the probability of A given only B. Second, consider the situ-

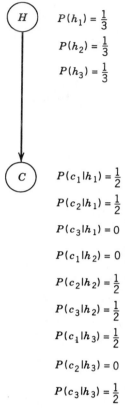

$$P(h_1) = \tfrac{1}{3}$$
$$P(h_2) = \tfrac{1}{3}$$
$$P(h_3) = \tfrac{1}{3}$$

$$P(c_1|h_1) = \tfrac{1}{2}$$
$$P(c_2|h_1) = \tfrac{1}{2}$$
$$P(c_3|h_1) = 0$$
$$P(c_1|h_2) = 0$$
$$P(c_2|h_2) = \tfrac{1}{2}$$
$$P(c_3|h_2) = \tfrac{1}{2}$$
$$P(c_1|h_3) = \tfrac{1}{2}$$
$$P(c_2|h_3) = 0$$
$$P(c_3|h_3) = \tfrac{1}{2}$$

FIGURE 4.14   The causal network which represents the case where one ball is drawn and there are three possible colors for the balls.

ation in which Mr. Holmes hears there is an earthquake, but has not heard that the alarm sounded. Using this network, he would have to conclude that the news of the earthquake would reduce the probability of his having been burglarized. If he suspected that he had been burglarized based on some other grounds (perhaps someone was spotted lurking around his house), news of an earthquake should not relax him. In reality, C only decreases B when it is known that A has occurred. This problem is actually a result of the violation of the independence assumptions. Thus we must consider all combinations of A and C and represent the situation by the inference network in Figure 4.16. This network can express the situation correctly when we know for certain both whether A has occurred and whether C has occurred. But what of the case where we know A but have no idea as to the occurrence of C? We would need a separate arc from A to B. However, the influence of this arc would have to be negated when we learned C, not combined by the method used

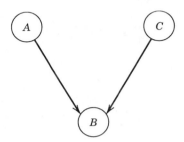

FIGURE 4.15 A first attempt to create an inference network which represents the uncertainty in Mr. Holmes' residence being burglarized.

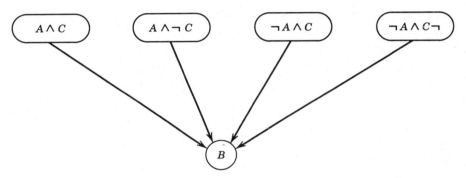

FIGURE 4.16 A second attempt to create an inference network which represents the uncertainty in Mr. Holmes' residence being burglarized.

in the previous chapter. That is, the influence of this arc would have to be replaced by the influence of the arc from $A \wedge C$.

Suppose next that Mr. Holmes' neighbor calls Mr. Holmes during his trip home (assuming Mr. Holmes has a car phone) and informs him that he spotted a person lurking around the house. If we define the proposition L as follows:

$$L = \text{a person is spotted lurking around the house,}$$

then we would need the inference network in Figure 4.17 to represent the situation. Besides being unnecessarily complex, again we have no vertex and arc representing the situation where all that is known is that A has occurred.

Thus we see that it is not really possible to represent this problem in a rule-based framework. Moreover, when we attempt to do so, the representation becomes complex, and the underlying relationships become obscured. Next we represent the problem by a causal network. Let A, B, and C be propositional

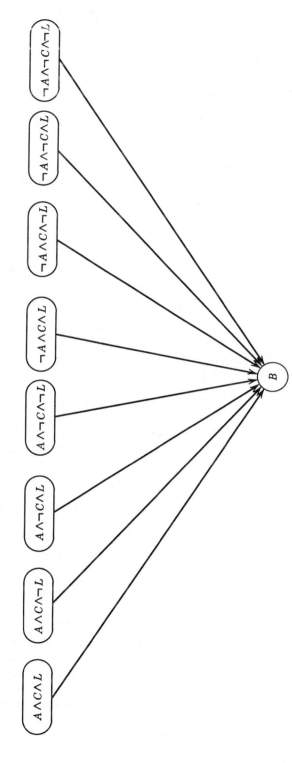

FIGURE 4.17 An attempt to create an inference network which represents the uncertainty in Mr. Holmes' residence being burglarized, which includes information about a lurker.

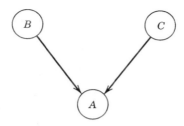

FIGURE 4.18 A causal network which representss the uncertainty in Mr. Holmes' residence being burglarized.

variables and

$$a_1 = \text{Mr. Holmes' burglar alarm sounds}$$

$$a_2 = \text{Mr. Holmes' burglar alarm does not sound}$$

$$b_1 = \text{Mr. Holmes' residence is burglarized}$$

$$b_2 = \text{Mr. Holmes' residence is not burglarized}$$

$$c_1 = \text{there is an earthquake}$$

$$c_2 = \text{there is not an earthquake.}$$

In causal network, the arcs are often from perceived causes to their effects. Since both an earthquake and a burglary can cause the alarm to sound and since there is no causal relationship between earthquakes and burglaries (except in the case of looting after a catastrophe), we represent the situation by the causal network in Figure 4.18. Recall that in causal networks we need to be able to ascertain the a priori probabilities of propositional values of the roots and the conditional probabilities of propositional values of each nonroot given the propositional values of its parents. We therefore need to ascertain

$$P(b_1) \qquad\qquad P(c_1)$$
$$P(a_1 \mid b_1 \wedge c_1) \qquad P(a_1 \mid b_1 \wedge c_2)$$
$$P(a_1 \mid b_2 \wedge c_1) \qquad P(a_1 \mid b_2 \wedge c_2).$$

(Since each variable only has two alternatives, we need only ascertain the probability of one of them; the probability of the other is then equal to 1 minus the ascertained value.) Once we have done this, we can use the network to compute the change in the probability of a propositional value of any vertex based on the instantiations of any set of vertices. Hence we could determine the probability of being burglarized given only that the alarm has sounded; when it is also learned that there was an earthquake, we could determine the new probability of being burglarized based on this additional evidence. The actual propagation methods are covered in chapters 6 and 7. The point here is the simplicity and naturalness of the representation. All we needed to do was identify the obvious causal relationships among the variables to establish the network.

Consider the variable L which represents the observation of a person lurking around the house. The only causal relationship this variable has to the others in the network is that B could cause L. That is, the occurrence of a burglary could cause someone to be observed lurking. Thus we need only add create an arc from B to L.

PROBLEMS 4.2.2

1. Create the causal network which represents the uncertainty in Mr. Holmes' residence being burlgarized with a vertex added for the observation of a lurker. Identify the new conditional probabilities which need to be ascertained.

2. Determine the relationship between the number of conditional probabilities which need to be ascertained at a given vertex in a causal network and the number of parents of that vertex.

4.3. THE MODELING OF HUMAN UNCERTAIN REASONING

As mentioned in chapter 2, many use psychological modeling in an effort to create systems which reason in the same way as humans. It can be argued that the rule-based framework, together with the forward and backward chaining inference engines, model human certain reasoning at some level. For example, it is not unreasonable to postulate that a botanist would store in his personal data bank an item of knowledge such as

IF   stem is green
THEN type is herb.

Furthermore, if the botanist were determining the family of a particular plant, it is plausible that he would proceed in a fashion similar to the decision tree in Figure 1.2, which is the equivalent of backward chaining. That is, the botanist would first ask questions which determined the type, then ones that determined the class, and finally ones that determined the family.

The situation is substantially different in the case of uncertain reasoning. Let us look again at the inference network in Figure 4.17. Aside from being complex, that network only allows us to determine probabilities in the direction of B. For example, we could not determine the probability of the alarm sounding given that Mr. Holmes' residence had been burglarized, or the probability of an earthquake given that the alarm had sounded, or the probability of the alarm sounding given that someone had been observed lurking around the house. Yet these are all judgments Mr. Holmes would be capable of making. To enable the network to be capable of making these determinations we could add rules for every possible situation. However, such a representation apparently has nothing to do with human reasoning. It seems unreasonable to postulate that the human mind contains millions of rules, each of which individually represents a complex probabilistic relationship. For example, it does not seem that Mr. Holmes' mind would contain an item of knowledge such as

$$\text{IF } A \wedge C \wedge L, \text{ THEN } B.$$

together with a likelihood of B given the combination of assertions in the antecedent of the rule. In addition, it is equally as implausible that we reach our uncertain judgments by combining the certainties implied by such rules. A more reasonable conjecture is that we identify local (nonnumeric) probabilistic relationships between individual propositions, and that the change in the certainty in one proposition changes our certainty in a related one, which in turns changes our certainty in propositions related to that one. For example, it seems plausible that the knowledge possessed by Mr. Holmes would include the following: a burglary would usually cause his alarm to sound; a burglary would sometimes cause a lurker to be observed around his house; and earthquakes often cause his alarm to sound. Therefore, when Mr. Holmes learns that there was an earthquake, it is not unreasonable to hypothesize that he would reason that the earthquake could have caused his alarm to sound, and that this deduction in turn would cause him to reason that there is less of a likelihood that his home has been burglarized. If Mr. Holmes' knowledge is truly in the form of such localized probabilistic dependencies between propositions, then a causal network represents the structure of his uncertain knowledge at some level. This conjecture—that causal networks represent the structure of human uncertain knowledge—will be discussed more in section 5.6. In chapter 6, we will develop a propagation method in causal networks, which, according to its creator, models human uncertain reasoning by tracing the arcs in the network. The point here is that although the conjecture that causal networks represent the structure of human uncertain knowledge is certainly disputable, this conjecture is far more plausible than the conjecture that rule-based systems could represent that structure.

CHAPTER 5

CAUSAL (BELIEF) NETWORKS

Causal networks, which were briefly introduced in the last chapter, are also called Bayesian networks, belief networks, and, when they are augmented with decision vertices (to be discussed in chapter 9), influence diagrams. There is disagreement among researchers as to which of these terms is most appropriate. This author is not satisfied with any of them and will suggest a new name at the end of this chapter. Throughout this text, however, we will call them causal networks. They are not new, but rather have a history in decision analysis [Miller et al., 1976; Howard & Matheson, 1984]. However, recently causal networks have been used in applications other than ones involving traditional decision analysis; that is, they have been used solely to perform probabilistic reasoning. The techniques which accomplish this are discussed in chapters 6 through 8. In chapter 9, we show how to apply these techniques to traditional problems in decision analysis. Before discussing these techniques and applications in the following chapters, we explain the concept of a causal network and derive many important properties of causal networks in this chapter. However, before we can discuss causal networks starting in section 5.3, we must first clarify the concepts of propositional variables, and joint and marginal probability distributions in the following two sections.

5.1. PROPOSITIONAL VARIABLES

Definition 5.1. Let a probability space $(\Omega, \mathcal{F}, P)$ be given. A propositional variable A on the probability space is a function from Ω to a subset $\mathcal{A}$ of $\mathcal{F}$ containing mutually exclusive and exhaustive events. A is called a finite propositional variable if its range $\mathcal{A}$ contains a finite number of events.

As a convention we will denote propositional variables by capital letters (A, B, C, etc). In our applications the subset of $\mathcal{F}$ will usually contain a finite number of events (or at least a countable number). When this is so, its members will be represented by the lowercase letter corresponding to the capital letter used to denote the variable. For example, if A is a propositional variable, the set of its possible values will be represented by

$$\mathcal{A} = \{a_i \text{ such that } i \in I, \text{ the integers}\}.$$

We saw in chapter 2 that events can be specified by propositions rather than by sets. We will ordinarily specify the events which are the values of propositional variables by propositions which describe the events (hence the name propositional variable).

Example 5.1. Pick a card at random from an ordinary poker deck. Then $\Omega = \{x \text{ such that } x \text{ is one of the 52 different cards}\}$. Let

$$a_1 = \text{a club is picked}$$

$$a_2 = \text{a heart is picked}$$

$$a_3 = \text{a diamond is picked}$$

$$a_4 = \text{a spade is picked}$$

$$\mathcal{A} = \{a_1, a_2, a_3, a_4\}.$$

Then the propositional variable A maps any club to a_1, any heart to a_2, etc.

Example 5.2. [Pearl, 1986a]. Suppose there are three mutually exclusive and exhaustive suspects in a murder trial. Label the suspects person 1, person 2, and person 3. Let

$$a_i = \text{person } i \text{ is the killer}$$

$$b_i = \text{person } i \text{ last held the murder weapon}$$

$$\Omega = \{a_i \wedge b_j \text{ for } i = 1,2,3, \ j = 1,2,3\}$$

$$\mathcal{F} = \text{set of all subsets of } \Omega$$

$$\mathcal{A} = \{a_1, a_2, a_3\}$$

$$\mathcal{B} = \{b_1, b_2, b_3\}.$$

Note that a_i and b_i are defined to be propositions and Ω is defined to be the set of all logical conjunctions of a_i and b_i for $1 \leq i$, $j \leq 3$. If we define a probability space on Ω, then the propositional variable A maps $(a_i \wedge b_j)$ for $j = 1,2,3$ to the following member of $\mathcal{F}$:

$$\{a_i \wedge b_1, \ a_i \wedge b_2, \ a_i \wedge b_3\}.$$

This event is the event that person i is the killer and is more easily represented simply by the proposition a_i. The propositional variable B maps $(a_i \wedge b_j)$ for $i = 1,2,3$ to the event

$$\{a_1 \wedge b_j,\ a_2 \wedge b_j,\ a_3 \wedge b_j\},$$

which is represented by the proposition b_j. Notice that the intersection of the event $\{a_i \wedge b_1,\ a_i \wedge b_2,\ a_i \wedge b_3\}$ with the event $\{a_1 \wedge b_j,\ a_2 \wedge b_j,\ a_3 \wedge b_j\}$ is the event $\{a_i \wedge b_j\}$ which contains the single proposition $a_i \wedge b_j$. Hence we have consistency in the use of the logic symbol $\wedge$ to represent the intersection of events. As mentioned in chapter 2, probability theory can be expressed entirely within the conceptual framework of propositions, conjunctions, and disjunctions. The intersection of two sets becomes the logical "and" of two propositions, while the union of two sets becomes the logical "or" of two propositions.

For a propositional variable A the event $A = a_i$ is defined to be the event a_i. Therefore $P(A = a_i)$ can be written simply as $P(a_i)$, the conditional expression $P(A = a_i \mid B = b_j)$ can be written as $P(a_i \mid b_j)$, and $P(A = a_i \wedge B = b_j)$ can be written as $P(a_i \wedge b_j)$. Since the events in $\mathcal{A}$ are mutually exclusive and exhaustive, the sum of the probabilities of all the events in $\mathcal{A}$ is always 1. For example, in Example 5.1,

$$P(a_1) + P(a_2) + P(a_3) + P(a_4) = 1/4 + 1/4 + 1/4 + 1/4 = 1.$$

When we want to denote the probability that a propositional variable A takes one of its possible values, we will only write $P(A)$. For example, $P(A \mid B) = P(A)$ means that $P(a_i \mid b_j) = P(a_i)$ for all i,j; that is, the probability of A taking any of its values is not changed if it is known that B has taken one of its values.

5.2. JOINT AND MARGINAL PROBABILITY DISTRIBUTIONS

Before defining joint and marginal distributions, we adopt a standard convention. When taking the intersection of events from the ranges of several propositional variables, instead of using the logic symbol $\wedge$ or the intersection symbol $\cap$ we will use a comma. Thus, if A, B, and C are propositional variables, the event

$$(a_i, b_j, c_k) \qquad \text{means} \qquad a_i \wedge b_j \wedge c_k.$$

When we are referring to the variables rather than specific values, the event

$$(A, B, C) \qquad \text{means} \qquad A \wedge B \wedge C.$$

The parentheses are used to avoid confusion with sets, which are enclosed in braces. For example $\{A, B, C\}$ is the set containing the three propositional variables A, B, and C, while $\{(A, B, C)\}$ is the set containing all intersections of possible values of the propositional variables A, B, and C. When there is no danger of confusion, we will only write

$$A, B, C \quad \text{to mean} \quad A \wedge B \wedge C.$$

If $\aleph = \{X_1, X_2, \ldots, X_n\}$ is a finite set of propositional variables, the event

$$\aleph \quad \text{means} \quad X_1 \wedge X_2 \wedge \cdots \wedge X_n.$$

Example 5.3. [Pearl, 1986a]. Let

$$a_1 = \text{Mr. Holmes' burglar alarm sounds}$$

$$a_2 = \text{Mr. Holmes' burglar alarm does not sound}$$

$$\mathcal{A} = \{a_1, a_2\}$$

$$b_1 = \text{Mr. Holmes' residence is burglarized}$$

$$b_2 = \text{Mr. Holmes' residence is not burglarized}$$

$$\mathcal{B} = \{b_1, b_2\}$$

$$c_1 = \text{there is an earthquake}$$

$$c_2 = \text{there is not an earthquake}$$

$$\mathcal{C} = \{c_1, c_2\}$$

$$\Omega = \{a_i, b_j, c_k \text{ such that } 1 \leq i, j, \ k \leq 2\}$$

$$\mathcal{F} = \text{set of all subset of } \Omega$$

and suppose we have a probability space defined on Ω. Then the propositional variable A maps (a_i, b_j, c_k) to a_i for $1 \leq i, j, \ k \leq 2$, the propositional variable B maps (a_i, b_j, c_k) to b_j for $1 \leq i, j, \ k \leq 2$, and the propositional variable C maps (a_i, b_j, c_k) to c_k for $1 \leq i, j, \ k \leq 2$.

Henceforth we will usually drop the reference to Ω. When we refer to a set of propositional variables $\{X_1, X_2, \ldots, X_n\}$, it is understood that they are all defined on the same probability space. We will only be interested in the joint probability distribution of these variables, which is defined as follows:

Definition 5.2. Let $\{X_1, X_2, \ldots, X_n\}$ be a set of n finite propositional variables defined on the same probability space $(\Omega, \mathcal{F}, P)$. If we define

$$\Omega' = \{(X_1, X_2, \ldots, X_n)\}$$

$$\mathcal{F}' = \text{set of all subsets of } \Omega'$$

$$P'(X_1, X_2, \ldots, X_n) = P(X_1, X_2, \ldots, X_n),$$

then clearly, since Ω' is a set of mutually exclusive and exhaustive events, P' defines a probability measure on $\mathcal{F}'$. The distribution $(\Omega', \mathcal{F}', P')$ is called a joint probability distribution of $\{X_1, X_2, \ldots, X_n\}$.

Therefore it is immaterial as to whether Ω is equal to $\{(X_1, X_2, \ldots, X_n)\}$ or is a more fundamental set. As long as we are not interested in the more fundamental set, we can just take Ω to be $\{(X_1, X_2, \ldots, X_n)\}$ as was done in Examples 5.2 and 5.3, and we can call the joint probability measure P.

Example 5.4. Continuing Example 5.3, we can define a joint probability distribution of $\{A, B, C\}$ as follows:

$$P(a_1, b_1, c_1) = .0000099 \qquad P(a_1, b_1, c_2) = .008991$$

$$P(a_1, b_2, c_1) = .000495 \qquad P(a_1, b_2, c_2) = .0098901$$

$$P(a_2, b_1, c_1) = .0000001 \qquad P(a_2, b_1, c_2) = .000999$$

$$P(a_2, b_2, c_1) = .000495 \qquad P(a_2, b_2, c_2) = .9791199.$$

Since the events in $\{(A, B, C)\}$ are mutually exclusive and exhaustive, when we specify the values in the joint distribution, we must specify values which sum to 1.

Notice that for each i, the probability of X_i can be computed from the joint distribution. When this is done, the result is a probability distribution on $\{(X_i)\}$. This distribution is called a marginal distribution, which is formally defined as follows:

Definition 5.3. Let $\{X_1, X_2, \ldots, X_n\}$ be a set of n finite propositional variables defined on the same probability space, and let $(\Omega, \mathcal{F}, P)$ be their joint probability distribution. Then clearly,

$$P(X_i) = \sum_{X_j \neq X_i} P(X_1, X_2, \ldots, X_n).$$

where summing over $X_j \neq X_i$ means all the other propositional variables are running through all their possible values while X_i is being held constant. If we define

$$\Omega' = \{(X_i)\}$$

$$\mathcal{F}' = \text{set of all subsets of } \Omega'$$

$$P'(X_i) = P(X_i),$$

then clearly P' defines a probability measure on $\mathcal{F}'$. This probability distribution is called the marginal probability distribution of $\{X_i\}$ relative to the joint probability distribution.

The distribution is called marginal because of the similarity to the process of adding across a row or column. The definition and the summation formula also hold when computing the joint distribution of a subset of $\{X_1, X_2, \ldots, X_n\}$

from the joint distribution of $\{X_1, X_2, \ldots, X_n\}$. For example,

$$P(X_1, X_2) = \sum_{X_j \neq X_1, X_2} P(X_1, X_2, \ldots, X_n).$$

Example 5.5. Continuing Example 5.4, we compute the marginal distribution of $\{A\}$ relative to the joint distribution of $\{A, B, C\}$:

$$P'(A) = \sum_{B,C} P(A, B, C)$$

$$P'(a_1) = \sum_{B,C} P(a_1, B, C)$$

$$= P(a_1, b_1, c_1) + P(a_1, b_1, c_2) + P(a_1, b_2, c_1) + P(a_1, b_2, c_2)$$

$$= .0000099 + .008991 + .000495 + .0098901 = .019386.$$

The computation of $P(a_2)$ is left as an exercise. Notice that $P'(a_1)$ and $P(a_1)$ are equal. This is clearly the case for any event on which both the joint distribution and the marginal distribution are defined.

   We will see more examples of these computations in many of the examples in the coming sections.

PROBLEMS 5.2

1. Perform the remaining computations in Example 5.5.

2. Compute the marginal distribution of $\{A, B\}$ relative to the joint distribution of $\{A, B, C\}$ in Example 5.4.

5.3. THE MATHEMATICAL FOUNDATIONS OF CAUSAL NETWORKS

When referring to a finite set of propositional variables, we have thus far listed the variables as ordered as a matter of notational convenience. That is, we denoted a finite set as $\{X_1, X_2, \ldots, X_n\}$. Note that the ordering has nothing to do with the definition of joint and marginal probability distributions, but it does make the notation clearer. In what follows, the propositional variables will be vertices in a DAG. It will occasionally be necessary to produce an ancestral ordering of these vertices as defined in Definition 3.32. Therefore, to avoid confusion with the cases where an ordering becomes necessary, we will no longer denote a finite set of propositional variables as ordered. We can now define a causal network.

Definition 5.4. Let V be a finite set of finite propositional variables defined on the same probability space, let $(\Omega, \mathcal{F}, P)$ be their joint probability distribution,

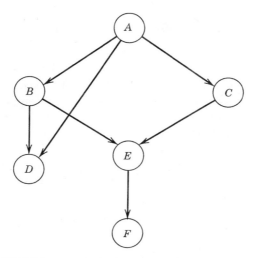

FIGURE 5.1   A directed acyclic graaph (DAG).

and let $G = (V, E)$ be a DAG. For each $v \in V$, let $c(v) \subseteq V$ be the set of all parents of v and $d(v) \subseteq V$ be the set of all descendents of v. Furthermore, for $v \in V$, let $a(v) \subseteq V$ be $V - (d(v) \cup \{v\})$, that is, the set of propositional variables in V excluding v and v's descendents. Suppose for every subset $W \subseteq a(v)$, W and v are conditionally independent given $c(v)$; that is, if $P(c(v)) > 0$, then

$$P(v \mid c(v)) = 0 \quad \text{or} \quad P(W \mid c(v)) = 0 \quad \text{or} \quad P(v \mid W \cup c(v)) = P(v \mid c(v)).$$

Then $C = (V, E, P)$ is called a causal network. The set $c(v)$ is called the set of causes (parents) of v.

If a variable v has no parents, then $c(v)$ is empty. Clearly, $P(v \mid \emptyset)$ (where $\emptyset$ represents an empty set of propositional variables) is simply $P(v)$. For the directed acyclic graph (DAG) in Figure 5.1, $c(E) = \{B, C\}$, $d(B) = \{D, E, F\}$, and $a(C) = \{A, B, D\}$.

We see now why, in chapter 2, we adopted the convention of using the logic symbol $\vee$ to represent the union of two events. In an expression such as $P(v \mid c(v) \cup W)$, the union symbol $\cup$ has an entirely different meaning than the union of two events. Since $c(v) \cup W$ stands for the event that all the variables in $c(v) \cup W$ are instantiated for values, the union symbol $\cup$ actually denotes the "and" of events, not the "or." Although the intersection symbol $\cap$ does not represent the "or" of events in this context, it still does not have its usual meaning as the "and" of events. The symbol $\emptyset$ does not represent the impossible event, but rather an empty set of propositional variables. Since, in a probability expression, a set of propositional variables stands for the event that all its members take specific values, the $P(\emptyset)$ is equal to 1 rather than 0.

FIGURE 5.2   The DAG for the causal network in Example 5.6.

Notice that an ordering of the variables has nothing to do with the defini-tion of a causal network. The definition has to do only with the structure of a DAG, a joint probability distribution, and conditional independences involving the DAG's structure and the probability distribution.

In section 5.5 we will explain why such networks are often called "causal." In the remainder of this section we develop some important properties of causal networks.

Example 5.6. Continuing Example 5.2, suppose P is defined as follows:

$$P(a_1,b_1) = .64 \qquad P(a_1,b_2) = .08 \qquad P(a_1,b_3) = .08$$
$$P(a_2,b_1) = .01 \qquad P(a_2,b_2) = .08 \qquad P(a_2,b_3) = .01$$
$$P(a_3,b_1) = .01 \qquad P(a_3,b_2) = .01 \qquad P(a_3,b_3) = .08.$$

Together with the DAG in Figure 5.2, this joint distribution defines a causal network, as can be ascertained by direct verification. Both $a(A)$ and $c(A)$ are the empty set. Therefore, easily for $W \subseteq a(A)$,

$$P(A \mid W \cup c(A)) = P(A \mid c(A)) = P(A \mid \emptyset) = P(A).$$

Furthermore, both $a(B)$ and $c(B)$ are the set $\{A\}$. Therefore, again easily for $W \subseteq a(B)$,

$$P(B \mid W \cup c(B)) = P(B \mid c(B)) = P(B \mid A).$$

Example 5.7. Together with the DAG in Figure 5.3, the joint distribution de-fined in Example 5.4 constitutes a causal network, which can be ascertained by direct verification. In this DAG, $c(B) = \emptyset$ and $a(B) = \{C\}$. Therefore, for $W \subseteq \{C\}$, we need to show that

$$P(B \mid W \cup c(B)) = P(B \mid c(B)).$$

That is, we need only show that

$$P(B \mid C) = P(B).$$

To this end

$$P(b_1) = \sum_{i,j} P(a_i, b_1, c_j)$$

$$= .0000099 + .008991 + .0000001 + .000999 = .01.$$

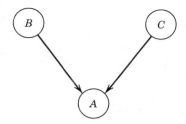

FIGURE 5.3 The DAG for the causal network in Example 5.7.

where we have computed the marginal value $P(b_1)$ from the joint probability distribution. Furthermore,

$$
\begin{aligned}
P(b_1 \mid c_1) &= \frac{P(b_1,c_1)}{p(c_1)} \\
&= \frac{\sum_i P(a_i,b_1,c_1)}{\sum_{i,j} P(a_i,b_j,c_1)} \\
&= \frac{.0000099 + .0000001}{.0000099 + .000495 + .0000001 + .000495} = .01,
\end{aligned}
$$

where again we have computed the marginal values from the joint distribution. The remaining verifications are left as an exercise. Since $c(C) = \emptyset$ and $a(C) = \{B\}$, we must also show $P(C \mid B) = P(C)$. However, this is implied by the fact that $P(B \mid C) = P(B)$, which we have already verified. Finally, since $c(A)$ and $a(A)$ both equal $\{B,C\}$, the verification for A is trivial, as in the previous example.

Example 5.8. Let

$a_1 = $ spouse is cheating

$a_2 = $ spouse is not cheating

$b_1 = $ spouse dines with another

$b_2 = $ spouse does not dine with another

$c_1 = $ spouse is reported seen dining with another

$c_2 = $ spouse is not reported seen dining with another

and

$$
\begin{array}{ll}
P(a_1,b_1,c_1) = .028 & P(a_1,b_1,c_2) = .042 \\
P(a_1,b_2,c_1) = .00003 & P(a_1,b_2,c_2) = .02997 \\
P(a_2,b_1,c_1) = .072 & P(a_2,b_1,c_2) = .108 \\
P(a_2,b_2,c_1) = .00072 & P(a_2,b_2,c_2) = .71928.
\end{array}
$$

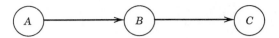

FIGURE 5.4 The DAG for the causal network in Example 5.8.

Together with the DAG in Figure 5.4, these assignments define a causal network, which again can be ascertained by direct verification. In this DAG, $a(C) = \{A, B\}$ and $c(C) = \{B\}$. Therefore we need to show for $W \subseteq \{A, B\}$ that

$$P(C \mid W \cup \{B\}) = P(C \mid B).$$

That is, we need to show that

$$P(C \mid A, B) = P(C \mid B).$$

To this end,

$$
\begin{aligned}
P(c_1 \mid a_1, b_1) &= \frac{P(a_1, b_1, c_1)}{P(a_1, b_1)} \\
&= \frac{P(a_1, b_1, c_1)}{\sum_i P(a_1, b_1, c_i)} \\
&= \frac{.028}{.028 + .042} = .4,
\end{aligned}
$$

and

$$
\begin{aligned}
P(c_1 \mid b_1) &= \frac{P(b_1, c_1)}{P(b_1)} \\
&= \frac{\sum_i P(a_i, b_1, c_1)}{\sum_{i,j} P(a_i, b_1, c_j)} \\
&= \frac{.028 + .072}{.028 + .042 + .072 + .108} = .4.
\end{aligned}
$$

The remaining verifications are left as an exercise. The verifications for A and B are again trivial.

The power of causal networks is obtained from the following theorems and lemmas:

Theorem 5.1. If $C = (V, E, P)$ is a causal network, then $P(V)$ is given by

$$P(V) = \prod_{\substack{v \in V \\ P(c(v)) > 0}} P(v \mid c(v)),$$

where $\prod_{v \in V, \ P(c(v)) > 0}$ means we are taking the product of all propositional variables in V for which $P(c(v)) > 0$.

Proof. By Theorem 3.7 it is possible to create an ancestral ordering of the vertices of V; that is, for every $v \in V$ the ancestors of v are ordered before v. Let $[v_1, v_2, \ldots, v_n]$ be such an ordering. For a given combination of values of the variables, there are two cases:

Case 1. $P(v_1, v_2, \ldots, v_n) > 0$. In this case we have for $1 \leq i \leq n$

$$P(v_1, v_2, \ldots, v_i) = \sum_{\substack{v_j \\ j > i}} P(v_1, v_2, \ldots, v_n) > 0.$$

($\sum_{v_j, \, j > i}$ means all the variables with labels greater than i, are running through all their possible values.) We can therefore repeatedly apply the definition of conditional probability to obtain the chain rule:

$$P(v_1, v_2, \ldots, v_n) = P(v_n \mid v_1, v_2, \ldots, v_{n-1}) P(v_{n-1} \mid v_1, v_2, \ldots, v_{n-2}) \ldots P(v_2 \mid v_1) P(v_1).$$

Since, for $1 \leq i \leq n$, all v_i's parents are numbered before v_i,

$$c(v_i) \subseteq \{v_1, v_2, \ldots, v_{i-1}\},$$

and hence

$$\{v_1, v_2, \ldots, v_{i-1}\} = \{v_1, v_2, \ldots, v_{i-1}\} \cup c(v_i).$$

Furthermore, since all v_i's descendents are ordered after v_i,

$$\{v_1, v_2, \ldots, v_{i-1}\} \subseteq a(v_i).$$

Therefore, by the definition of a causal network, we have

$$P(v_i \mid v_1, v_2, \ldots, v_{i-1}) = P(v_i \mid c(v_i))$$

(we know that $P(c(v_i)) > 0$ by the same argument that was used to show that $P(v_1, v_2, \ldots, v_i) > 0$ for all i). Thus

$$P(v_1, v_2, \ldots, v_n) = P(v_n \mid c(v_n)) P(v_{n-1} \mid c(v_{n-1})) \ldots P(v_2 \mid c(v_2)) P(v_1),$$

which, since v_1 is a root and therefore $c(v_1) = \emptyset$, proves case 1.

Case 2. $P(v_1, v_2, \ldots, v_n) = 0$. In this case either $P(v_1) = 0$ and the theorem is proved, or there exist some i where $2 \leq i \leq n$ such that

$$P(v_1, v_2, \ldots, v_i) = 0$$

and

$$P(v_1, v_2, \ldots, v_{i-1}) \neq 0.$$

For this i,

$$P(v_i \mid v_1, v_2, \ldots, v_{i-1}) = \frac{P(v_1, v_2, \ldots, v_i)}{P(v_1, v_2, \ldots, v_{i-1})} = 0.$$

Finally, by the same argument as in case 1,

$$P(v_i \mid c(v_i)) = P(v_i \mid v_1, v_2, \ldots, v_{i-1}) = 0,$$

which proves this case. $\square$

Thus, in a causal network, if one knows the conditional probability distribution of each propositional variable given its parents, one can compute the joint probability distribution of all the propositional variables in the network. This can reduce the complexity of determining the distribution enormously. For example, if there are ten propositional variables and each variable has two alternatives, there are 2^{10} values in the joint distribution. However, if there is one root, one variable has exactly one parent, and all other variables have exactly two parents, there are only 70 values in the conditional distributions. We will discuss this result in depth in the following sections. We will now investigate another important issue. Theorem 5.1 showed that if we know that a DAG and a joint probability distribution constitute a causal network, then the joint probability distribution can be retrieved from the conditional probability distributions of every variable given its parents. This does not imply, however, that if we arbitrarily specify a DAG and conditional probability distributions of every variable given its parents, we will necessarily have a causal network. We must show first that the specified conditional probability distributions determine a joint probability distribution of the variables in the DAG, second that the specified conditional distributions are indeed the conditional probabilities, relative to that joint distribution, of every variable given its parents, and third that this joint distribution together with the DAG satisfies the conditional independence assumptions in a causal network. The following lemmas and theorem obtain these results, thereby showing that a DAG, together with specified conditional distributions of every variable given its parents, determines a causal network. Before proving these lemmas and this theorem, we give some examples in which we determine the values in a joint probability distribution from the values in the conditional probability distributions.

Example 5.9. Consider the DAG in Figure 5.2 and the causal network in Example 5.6. The conditional probability of a propositional variable given its parents can be obtained from the joint distribution as follows:

$$P(b_1 \mid a_1) = \frac{P(a_1, b_1)}{P(a_1)}$$

$$= \frac{P(a_1, b_1)}{\sum_{i=1}^{3} P(a_1, b_i)}$$

$$= \frac{.64}{.64 + .08 + .08} = .8.$$

The other conditionals are computed in a similar fashion to obtain

$$P(a_1) = .8$$
$$P(a_2) = .1$$
$$P(a_3) = .1$$
$$P(b_j \mid a_i) = \begin{pmatrix} .8 & \text{if} \quad i = j \\ .1 & \text{if} \quad i \neq j. \end{pmatrix}$$

Theorem 5.1 shows how the joint distribution in Example 5.6 can be retrieved from these conditional distributions. That is,

$$P(A, B) = P(B \mid c(B))P(A \mid c(A))$$
$$= P(B \mid A)P(A).$$

For example,

$$P(a_1, b_1) = P(b_1 \mid a_1)P(a_1)$$
$$= .8 \times .8 = .64.$$

Verifying the remaining values is left as an exercise.

Notice in the above example that

$$\sum_{i=1}^{3} P(b_i \mid a_j) = 1 \qquad \text{for} \quad 1 \leq i \leq 3.$$

This is necessary, since b_1, b_2, and b_3 are mutually exclusive and exhaustive. When we get to specifying conditional distributions, instead of joint distributions, we must keep this in mind.

Example 5.10. Consider the DAG in Figure 5.3 and the causal network in Example 5.7. The conditional probability of a propositional variable given its parents can be obtained from the joint distribution to obtain

$$P(b_1) = .01 \qquad\qquad P(b_2) = .99$$
$$P(c_1) = .001 \qquad\qquad P(c_2) = .999$$
$$P(a_1 \mid b_1, c_1) = .99 \qquad P(a_2 \mid b_1, c_1) = .01$$
$$P(a_1 \mid b_1, c_2) = .9 \qquad P(a_2 \mid b_1, c_2) = .1$$
$$P(a_1 \mid b_2, c_1) = .5 \qquad P(a_2 \mid b_2, c_1) = .5$$
$$P(a_1 \mid b_2, c_2) = .01 \qquad P(a_2 \mid b_2, c_2) = .99.$$

Again Theorem 5.1 shows how the joint distribution in Example 5.7 can be retrieved from these conditional distributions. That is,

$$P(A, B, C) = P(A \mid c(A))P(B \mid c(B))P(C \mid c(C))$$
$$= P(A \mid B, C)P(B)P(C).$$

The verification of the values is left as an exercise.

Example 5.11. Consider the DAG in Figure 5.4 and the causal network in Example 5.8. The conditional probability of a propositional variable given its parents can be obtained from the joint distribution to obtain

$$P(a_1) = .1 \qquad\qquad P(a_2) = .9$$
$$P(b_1 \mid a_1) = .7 \qquad P(b_2 \mid a_1) = .3$$
$$P(b_1 \mid a_2) = .2 \qquad P(b_2 \mid a_2) = .8$$
$$P(c_1 \mid b_1) = .4 \qquad P(c_2 \mid b_1) = .6$$
$$P(c_1 \mid b_2) = .001 \qquad P(c_2 \mid b_2) = .999.$$

Again, due to Theorem 5.1, the joint distribution in Example 5.8 can be retrieved from these conditional distributions. That is,

$$P(A, B, C) = P(C \mid c(C))P(B \mid c(B))P(A \mid c(A))$$
$$= P(C \mid B)P(B \mid A)P(A).$$

The verification of the values is left is an exercise.

Next we prove that a DAG, together with a joint distribution obtained from conditional probability distributions of vertices given their parents, determines a causal network. The following lemmas and theorem accomplish this:

Lemma 5.1. Let V be a finite set of finite sets of alternatives (we are not yet calling the members of V propositional variables since we do not yet have a probability space), and let $G = (V, E)$ be a DAG. In addition, for $v \in V$, let $c(v) \subseteq V$ be the set of all parents of v, and let a conditional probability distribution of v given $c(v)$ be specified for every event in $c(v)$. That is, for every combination of alternatives in $c(v)$,

$$\hat{P}(v \mid c(v))$$

is specified for each alternative in v, and

$$\sum_v \hat{P}(v \mid c(v)) = 1.$$

Again, $\sum_v$ means the sum over all the alternatives for v. (Note that the values of variables in $c(v)$ are held fixed while we take this sum.) Then a joint probability distribution P of the vertices in V is uniquely determined by defining

$$P(V) = \prod_{v \in V} \hat{P}(v \mid c(v)).$$

(Since for each alternative, $v \in V$, and each combination of alternatives in $c(v)$, we are simply specifying a distinct real number, to be technically accurate we should denote the function $\hat{P}$ merely as $\hat{P}(v, c(v))$. However, Lemma

5.3 will prove that these values end up being conditional probabilities in P. Furthermore, in practice, when these distributions are created, they are conceived as being conditional probabilities. Therefore it is notationally more attractive to specify them as conditionals up front. This same convention will be adopted in the theorems and lemmas which follow.)

Proof. By Theorem 3.7, it is possible to obtain an ancestral ordering of the vertices in V. Let $[v_1, v_2, \ldots, v_n]$ be such an ordering. Then $c(v_1) = \emptyset$ and for $i > 1$,

$$c(v_i) \subseteq \{v_1, v_2, \ldots, v_{i-1}\}. \tag{5.1}$$

We need to show that

$$\sum_{\substack{v_i \\ 1 \leq i \leq n}} \hat{P}(v_n \mid c(v_n)) \hat{P}(v_{n-1} \mid c(v_{n-1})) \ldots \hat{P}(v_2 \mid c(v_2)) \hat{P}(v_1) = 1, \tag{5.2}$$

where the summation in (5.2) should be read to mean that we are summing over all the possible combinations of values of vertices in V. Now, from (5.1) we can see that the expression on the left in (5.2) is equal to

$$\sum_{v_1} \left(\hat{P}(v_1) \sum_{v_2} \left(\hat{P}(v_2 \mid c(v_2)) \sum_{v_3} \left(\hat{P}(v_3 \mid c(v_3)) \ldots \sum_{v_n} \hat{P}(v_n \mid c(v_n)) \right) \right) \right) \tag{5.3}$$

However, since for $1 \leq i \leq n$, $\sum_{v_i} \hat{P}(v_i \mid c(v_i)) = 1$ for every combination of alternatives in $c(v_i)$, the expression in (5.3)

$$= \sum_{v_1} \left(\hat{P}(v_1) \sum_{v_2} \left(\hat{P}(v_2 \mid c(v_2)) \sum_{v_3} \left(\hat{P}(v_3 \mid c(v_3)) \ldots 1 \right) \right) \right)$$

$$= \sum_{v_1} \left(\hat{P}(v_1) \sum_{v_2} \left(\hat{P}(v_2 \mid c(v_2)) 1 \right) \right)$$

$$= \sum_{v_1} \left(\hat{P}(v_1) 1 \right) = 1,$$

which proves the lemma. $\square$

Thus we see that a DAG and conditional distributions given parents yield a joint probability distribution of the propositional variables in the DAG. Two questions remain. The first is whether the specified conditional values are indeed the conditional probabilities in the resultant joint distribution. That is, is $\hat{P}(v \mid c(v))$, as specified, equal to $P(v \mid c(v))$? The second is whether the DAG and P constitute a causal network. To answer these questions, we first need the following lemma:

Lemma 5.2. Let the conditions of Lemma 5.1 hold and suppose P is the joint probability distribution obtained from the conditional distributions. Then, if

$[v_1, v_2, ..., v_n]$ is an ancestral ordering of the vertices in V, we have for $1 \leq i \leq n$ that the marginal distribution $P(v_1, v_2, ..., v_i)$ is given by

$$P(v_1, v_2, ..., v_i) = \hat{P}(v_i \mid c(v_i))\hat{P}(v_{i-1} \mid c(v_{i-1}))...\hat{P}(v_2 \mid c(v_2))\hat{P}(v_1).$$

Proof. Determining the marginal distribution, we have

$$P(v_1, v_2, ..., v_i) = \sum_{\substack{v_j \\ j>i}} \hat{P}(v_n \mid c(v_n))\hat{P}(v_{n-1} \mid c(v_{n-1}))...\hat{P}(v_2 \mid c(v_2))\hat{P}(v_1),$$

which, since $c(v_k) \subseteq \{v_1, v_2, ..., v_{k-1}\}$ for $1 \leq k \leq n$, is equal to

$$\left(\hat{P}(v_i \mid c(v_i))\hat{P}(v_{i-1} \mid c(v_{i-1}))...\hat{P}(v_1) \right) \left(\sum_{\substack{v_j \\ j>i}} \hat{P}(v_n \mid c(v_n))...\hat{P}(v_{i+1} \mid c(v_{i+1})) \right).$$

However,

$$\sum_{\substack{v_j \\ j>i}} \hat{P}(v_n \mid c(v_n))...\hat{P}(v_{i+1} \mid c(v_{i+1})) = 1$$

by the same argument used in Lemma 5.1, which proves the lemma.   □

We can now answer the two questions posed above. First we prove that the specified conditional distributions are indeed the conditional probabilities in the resultant joint distribution.

Lemma 5.3. Let the conditions of Lemma 5.1 hold and let P be the joint probability distribution obtained from the specified conditional distributions. Then for every $v \in V$ and every combination of alternatives in $c(v)$ such that $P(c(v)) > 0$, we have that

$$P(v \mid c(v)) = \hat{P}(v \mid c(v)),$$

the original specified conditional value.

Proof. Let $[v_1, v_2, ..., v_n]$ be an ancestral ordering of the vertices in V, let i be such that $1 \leq i \leq n$, and let $c(v_i)$ be such that $P(c(v_i)) > 0$. Since $c(v_i) \subseteq \{v_1, v_2, ..., v_{i-1}\}$, we have, due to Lemma 5.2, that

$$P(c(v_i)) = \sum_{v_j \notin c(v_i)} \hat{P}(v_{i-1} \mid c(v_{i-1}))...\hat{P}(v_2 \mid c(v_2))\hat{P}(v_1). \tag{5.4}$$

(note that this sum is simply 1 if $c(v_i)$ is empty. This is correct, since $P(\emptyset) = 1$)).

Next, since $\{v_i\} \cup c(v_i) \subseteq \{v_1, v_2, \ldots, v_i\}$, again due to Lemma 5.2, we have that

$$P(v_i \mid c(v_i)) = \frac{P(\{v_i\} \cup c(v_i))}{P(c(v_i))}$$

$$= \frac{\sum_{v_j \notin \{v_i\} \cup c(v_i)} \hat{P}(v_i \mid c(v_i)) \hat{P}(v_{i-1} \mid c(v_{i-1})) \ldots \hat{P}(v_2 \mid c(v_2)) \hat{P}(v_1)}{P(c(v_i))}$$

$$= \frac{\hat{P}(v_i \mid c(v_i)) \sum_{v_j \notin \{v_i\} \cup c(v_i)} \hat{P}(v_{i-1} \mid c(v_{i-1})) \ldots \hat{P}(v_2 \mid c(v_2)) \hat{P}(v_1)}{P(c(v_i))}$$

$$= \frac{\hat{P}(v_i \mid c(v_i)) \sum_{v_j \notin c(v_i)} \hat{P}(v_{i-1} \mid c(v_{i-1})) \ldots \hat{P}(v_2 \mid c(v_2)) \hat{P}(v_1)}{P(c(v_i))}.$$

The last equality is due to $v_i \notin c(v_j)$ for $j < i$. Together with (5.4), this completes the proof. $\square$

Notice that when we specify the conditional distributions, it is possible to specify values for $\hat{P}(v \mid c(v))$ for cases where $P(v \mid c(v))$ ends up being zero in the joint distribution and therefore the $P(v \mid c(v))$ does not exist. The corollary to Theorem 5.2 will show that such specified values are meaningless to the joint distribution; that is, the joint distribution is the same regardless of the values we give to them, and therefore in practice we need not specify any values for them. Indeed, care should be taken to recognize cases where $P(c(v))$ is equal to 0 so that we do not specify meaningless values and erroneously believe that they are conditional probabilities.

Since the conditional probability $P(v \mid c(v))$ is equal to $\hat{P}(v \mid c(v))$, the original specified value, for all variables v and all combinations of alternatives in $c(v)$, henceforth we will simply refer to the original specified value as $P(v \mid c(v))$. Next we prove that a DAG and the joint probability distribution of its vertices, obtained from conditional distributions given parents, constitute a causal network.

Theorem 5.2. Let the conditions of Lemma 5.1 hold and let P be the joint probability distribution obtained from the specified conditional distributions. Then the DAG and P, the joint probability distribution obtained from the specified conditional distributions of variables given their parents, constitute a causal network. Furthermore, the specifed conditional distributions are indeed conditional probabilities in this causal network for all cases where $P(c(v)) > 0$.

Proof. Lemma 5.1 proved that the joint probability distribution obtained is in fact a probability distribution. Lemma 5.3 proved that the specified conditional distributions are conditional probabilities in this distribution. Therefore, we need only show that the DAG and this distribution constitute a causal network. That is, we need to show, for each variable $v \in V$ and each subset

$W \subseteq a(v)$ such that $P(c(v) \cup W) > 0$, that

$$P(v \mid c(v) \cup W) = P(v \mid c(v)).$$

To that end, fix $v \in V$. Owing to Theorem 3.8 it is possible to create an ancestral ordering of the vertices in V so that only descendents of v are labeled after v. Let

$$[v_1, v_2, \dots, v_n]$$

be such an ordering. Let v be v_i in this ordering. Henceforth v will thus be denoted by v_i. We then have that $c(v_i)$, W, and $\{v_i\}$ are all subsets of $\{v_1, v_2, \dots, v_i\}$, and therefore

$$P(v_i \mid c(v_i) \cup W) \qquad \text{and} \qquad P(v_i \mid c(v_i))$$

can both be computed from the marginal distribution on $\{v_1, v_2, \dots, v_i\}$. However, due to Lemma 5.2, this marginal distribution is given by

$$P(v_i \mid c(v_i))P(v_{i-1} \mid c(v_{i-1}))\dots P(v_2 \mid c(v_2))P(v_1).$$

If we denote $c(v_i) \cup W$ by U, we then have

$$
\begin{aligned}
P(v_i \mid c(v_i) \cup W) &= P(v_i \mid U) \\[4pt]
&= \frac{P(\{v_i\} \cup U)}{P(U)} \\[6pt]
&= \frac{\sum_{v_j \notin U \cup \{v_i\}} P(v_i \mid c(v_i))P(v_{i-1} \mid c(v_{i-1}))\dots P(v_1)}{\sum_{v_j \notin U} P(v_i \mid c(v_i))P(v_{i-1} \mid c(v_{i-1}))\dots P(v_1)} \\[6pt]
&= \frac{P(v_i \mid c(v_i))(\sum_{v_j \notin U} P(v_{i-1} \mid c(v_{i-1}))\dots P(v_1))}{\sum_{v_j \notin U} P(v_i \mid c(v_i))P(v_{i-1} \mid c(v_{i-1}))\dots P(v_1)},
\end{aligned}
$$

since in the numerator we are not summing over v_i or over v_j for $v_j \in c(v_i)$.

Further, since $v_i \notin c(v_j)$ for $j < i$ and in the denominator we are also not summing over v_j for $v_j \in c(v_i)$, this last expression is equal to

$$\frac{P(v_i \mid c(v_i))(\sum_{v_j \notin U} P(v_{i-1} \mid c(v_{i-1}))\dots P(v_1))}{(\sum_{v_i} P(v_i \mid c(v_i)))(\sum_{v_j \notin U} P(v_{i-1} \mid c(v_{i-1}))\dots P(v_1))},$$

which clearly equals $P(v_i \mid c(v_i))$ and proves the theorem. $\qquad \square$

Corollary to Theorem 5.2. Let the conditions of Lemma 5.1 hold and let P be the joint probability distribution obtained from the specified conditional distributions. Then, if for some v and some combination of values of $c(v)$ we have that

$$P(c(v)) = 0,$$

the specified values of

$$\hat{P}(v \mid c(v))$$

for that combination do not enter into the determination of P.

Proof. Theorem 5.2 showed that (V, E, P) is a causal network, while Theorem 5.1 showed that the probability distribution in a causal network can be retrieved from the conditional probabilities of each variable given its parents. Since, whenever $P(c(v)) = 0$, the specified values of $\hat{P}(v \mid c(v))$ are not conditional probabilities, such specified values are not needed to determine P. $\quad\square$

We see by Theorem 5.2 that a DAG composed of propositional variables, together with a set of conditional distributions of variables given their parents, always determines a causal network. Another important question is whether every joint distribution of a set of propositional variables is contained in some causal network. Before proving that this is indeed the case, we mention that some authors define a causal network slightly differently than we did in Definition 5.4. They include the requirement that the set of parents $c(v)$ be minimal for each $v \in V$. That is, there is no proper subset $U \subset c(v)$ such that

$$P(v \mid U) = P(v \mid c(v)).$$

Our definition, however, is more general, and, as we shall now see, a causal network as defined here uniquely determines another causal network, containing the same joint distribution as the original causal network, in which the sets $c(v)$ are minimal. We need the following lemma to prove that result.

Lemma 5.4. Let V be a set of propositional variables. Suppose for some $v \in V$, $U \subseteq V$, and $W \subseteq V$ where $v \notin U$ and $v \notin W$,

$$P(v \mid U) = P(v \mid W).$$

Then

$$P(v \mid U \cap W) = P(v \mid U) = P(V \mid W).$$

Proof. Let $Y = U \cap W$ and $X = W - Y$. We have

$$\frac{P(\{v\} \cup U)}{P(U)} = \frac{P(\{v\} \cup W)}{P(W)}$$

$$P(\{v\} \cup U)P(W) = P(\{v\} \cup W)P(U)$$

$$\sum_X (P(\{v\} \cup U)P(W)) = \sum_X (P(\{v\} \cup W)P(U))$$

$$P(\{v\} \cup U)\sum_X P(W) = P(U)\sum_X P(\{v\} \cup W).$$

The last equality follows from the fact that $X \cap U = \varnothing$ and $v \notin X$. Computing the marginals by summing over X, we then have

$$P(\{v\} \cup U)P(Y) = P(U)P(\{v\} \cup Y)$$

$$\frac{P(\{v\} \cup U)}{P(U)} = \frac{P(\{v\} \cup Y)}{P(Y)}$$

$$P(v \mid U) = P(v \mid Y),$$

which proves the lemma. $\square$

We can now prove that every causal network determines another causal network in which the parent sets are minimal.

Theorem 5.3. Let $C = (V, E, P)$ be a causal network. Then there is a unique causal network $C' = (V, E', P)$, where $E' \subseteq E$, containing the same joint probability distribution, in which the parent sets are minimal.

Proof. Let $v \in V$, and let $\mathcal{W}$ be the set of all subsets $W \subseteq c(v)$ such that if $W \in \mathcal{W}$, then

$$P(v \mid W) = P(v \mid c(v))$$

whenever $P(c(v)) > 0$. Clearly, $\mathcal{W} \neq \varnothing$ since $c(v) \in \mathcal{W}$. Let $c'(v)$ be the intersection of all the sets in $\mathcal{W}$. By repeatedly applying Lemma 5.3, we have for each $v \in V$,

$$P(v \mid c'(v)) = P(v \mid c(v)). \tag{5.5}$$

If we then create E' by eliminating from E, for each v, all the arcs to v from vertices which are in $c(v)$ but not in $c'(v)$, the DAG (V, E') and the joint distribution P', determined by the conditional distributions, $P(v \mid c'(v))$ for $v \in V$, constitute a causal network by Theorem 5.2. Plainly, the parent sets in this new causal network are minimal. By Theorem 5.1 the conditional distributions, $P(v \mid c(v))$ for $v \in V$, determine P, and therefore by equality (5.5), $P' = P$, which completes the proof. $\square$

When constructing a causal network, our goal is to construct one in which the parent set is minimal. Notice that if we have a causal network (V, E, P), we can arbitrarily add arcs to E and the resultant structure is still a causal network containing the same probability distribution. However, by so doing, the structure of the DAG does not contain as much information. The conditional independencies in the probability distribution are implied by the absence of arcs. Therefore, to convey the most information in the structure of the DAG, we want the one in which the parent set is minimal. Note also that if we remove arcs from E, the resultant structure may not be a causal network containing the same distribution. Therefore arcs cannot be arbitrarily removed in the search for a structure in which the parent set is minimal.

Next we prove that any joint distribution has a causal network which contains it.

Theorem 5.4. Let P be a joint probability distribution of a finite set V of propositional variables. Then there is some causal network $C = (V, E, P)$ containing P.

Proof. Such a network can be created in the following way: First order the vertices arbitrarily. If $[v_1, v_2, \ldots, v_n]$ is that ordering, then create the arcs E in the DAG by putting

$$(v_k, v_i) \quad \text{in} \quad E \quad \text{if} \quad k < i.$$

That is, for each i, we set

$$c(v_i) = \{v_1, v_2, \ldots, v_{i-1}\}.$$

Now if $P(v_1, v_2, \ldots, v_n) \neq 0$, then by the chain rule

$$P(v_1, v_2, \ldots, v_n) = P(v_n \mid v_1, v_2, \ldots, v_{n-1}) P(v_{n-1} \mid v_1, v_2, \ldots, v_{n-2})$$
$$\times \ldots P(v_2 \mid v_1) P(v_1).$$
$$= P(v_n \mid c(v_n)) P(v_{n-1} \mid c(v_{n-1})) \ldots P(v_2 \mid c(v_2)) P(v_1).$$

If $P(v_1, v_2, \ldots, v_n) = 0$, then either $P(v_1) = 0$ or there exists some $i \geq 2$ such that

$$P(v_1, v_2, \ldots, v_i) = 0 \quad \text{and} \quad P(v_1, v_2, \ldots, v_{i-1}) > 0.$$

Then

$$P(v_i \mid c(v_i)) = P(v_i \mid v_1, v_2, \ldots, v_{i-1}) = 0$$

and the product of the conditional distributions still determines the value of $P(v_1, v_2, \ldots, v_n)$. Moreover, this product determines a causal network by Theorem 5.2, which proves the theorem. A causal network in which the parent sets are minimal could then be obtained by applying Theorem 5.3. $\square$

Before ending this section, we summarize the relationships among joint distributions, conditional distributions, DAGs, and causal networks:

1. A causal network by definition contains a unique DAG and a unique joint distribution on the variables in the DAG. Hence a causal network also determines unique conditional distributions of variables given their parents. Theorem 5.1 states that the joint distribution can be retrieved from these conditional distributions alone.

2. As Theorem 5.2 states, every DAG, whose vertices are propositional variables, together with a set of conditional distributions of variables given their parents, uniquely determines a causal network.

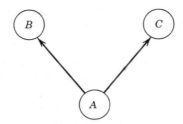

FIGURE 5.5   The DAG for Example 5.12.

3. A joint distribution does not uniquely determine a causal network. Example 5.15 (discussed a little later) gives an example of two causal networks which both contain the same joint distribution.

4. As Theorem 5.4 states, every joint distribution of a set of propositional variables has at least one causal network which contains it.

5. An arbitrary DAG, whose vertices are propositional variables, along with a joint distribution of those variables, is not necessarily a causal network. Notice that if we determine, from the joint distribution, the conditional distributions of variables given their parents, they will determine a joint distribution which, along with the DAG, constitutes a causal network by Theorem 5.2. However, the joint distribution they determine is not necessarily the original joint distribution from which the conditionals were computed. The following example shows that this is true.

Example 5.12. Let A, B, and C be the propositional variables whose joint distribution is given in Example 5.4, and let the DAG be the one in Figure 5.5. Then, if this were a causal network, we would have to have $P(B \mid A, C) = P(B \mid A)$. It is left as an exercise to verify that this is not the case for this distribution.

PROBLEMS 5.3

1. Perform the remaining verifications in Examples 5.7 and 5.8, thereby proving that the structures are indeed causal networks.

2. Retrieve the remaining values in the joint probability distribution from the conditional probability distributions in Examples 5.9, 5.10, and 5.11.

3. Complete Example 5.12 by showing that $P(B \mid A, C) \neq P(B \mid A)$ in the joint distribution in Example 5.4. Furthermore, show directly that the joint distribution $P(B \mid A)P(C \mid A)P(A)$ is not equal to the joint distribution in Example 5.4.

5.4. CAUSAL NETWORKS AND EXPERT SYSTEMS

In general, in expert systems, we are interested in how the realization of values by some propositional variables affects the probabilities of others. For example, in Example 5.2, we would be interested in the probability of each of the individuals being the murderer given that it became known for certain that person 2 was the last holder of the weapon. In Example 5.3 Mr. Holmes would be interested in the probability of his residence having been burglarized given that his burglar alarm had sounded. Second, he would want to know how that probability would change if he learned there were an earthquake. In Example 5.8, a wife would be interested in the probability of her spouse cheating given that it is reported that he was seen dining with another.

In expert systems applications, we assume that the propositional variables of interest have some underlying joint distribution. As mentioned and illustrated above, our goal is to determine how the realization of values by some of the variables affects the probabilities of the others. For example, if $\{A,B,C,D,E\}$ are the variables of interest, and it is known that A equals a_i and B equals b_j, in some cases we would want to determine the conditional distributions

$$P(C \mid a_i,b_j), \qquad P(D \mid a_i,b_j), \qquad \text{and} \qquad P(E \mid a_i,b_j).$$

In other cases we may wish to determine joint conditional distributions. For example, we might want to determine the conditional distribution

$$P(C,D \mid a_i,b_j).$$

Such cases will be discussed in detail in chapter 8.

There are actually two tasks involved: The first is to determine the joint distribution of the variables of interest, and the second is to compute the conditional distributions given certain variables realize specific values. To accomplish the first task by determining the probability of every event in the joint distribution is in general not possible. For, if there are n variables, each with only two possible values, there would be 2^n values in the joint distribution. This is an unmanageable number even for a relatively small value of n. Even if we could determine and store all 2^n values, it would not in general be possible to accomplish the second task by direct computation on values in the joint distribution. The reason is that it would take an exponentially large number of calculations to perform this computation. For example, to compute $P(c_k \mid a_i,b_j)$ from the joint distribution directly, we would need to evaluate

$$\frac{\sum_{m,p} P(a_i,b_j,c_k,d_m,e_p)}{\sum_{k,m,p} P(a_i,b_j,c_k,d_m,e_p)}.$$

Since it is not possible to determine the joint distribution by determining the probability of every event and it is also not possible to directly compute the desired conditional distributions from the joint distribution, a different method must be found to achieve both these tasks. Recall that Theorem 5.4 proves that every joint distribution has at least one causal network which contains it. If we could determine such a network, containing the joint distribution of interest, in which the number of parents of each variable in the network were small, then the number of values in the conditional distributions of the variables given their parents would be small. If we could then determine the values in these conditional distributions, we would have totally specified the joint distribution due to Theorem 5.1, thus accomplishing the first task, namely the determination of the joint distribution. Furthermore, the structure of the DAG in a causal network implies many of the conditional independencies in the joint distribution (it implies by definition the fact that a variable v is independent of any subset $W \subseteq a(v)$ given $c(v)$; there are, as we shall see in the next chapter, other independencies which can be deduced). This property of causal networks will be exploited in the following chapters to compute the desired conditional distributions given that certain variables have realized specific values, thus accomplishing the second task, namely the determination of the desired conditional distributions. The techniques in these chapters will not require every value in the joint distribution, but rather only the values in the conditional distributions of variables given their parents.

In the remainder of this chapter we show how to accomplish the first task, namely the determination of the joint distribution. As mentioned previously, all the values in the joint distribution cannot be determined directly. Additionally, the techniques in the following chapters require that the joint distribution be represented in a causal network. Therefore we will develop a method for determining the joint distribution by 1) for a given joint distribution, deriving a causal network containing that distribution; and 2) for that causal network, determining the values in the conditional distributions of variables given their parents.

5.5. CAUSAL NETWORKS AND THE NOTION OF CAUSALITY

Let V be a set of propositional variables with some unknown joint probability distribution. A DAG containing V is created according to the human notion of causality by directing an arc from one propositional variable to another if it is perceived that the value assumed by the first variable is a direct cause of the value assumed by the second variable. Many feel that a DAG created in this manner, along with the joint probability distribution, constitutes a causal network. Recall from section 5.3 that not every DAG and joint distribution constitute a causal network. The reason that many feel that constructing a DAG in this manner yields a causal network is that they believe that the conditional independencies in Definition 5.4 are conditional independencies which humans feel exist among causes and effects. Stated in terms of causes and effects, these conditional independencies are the following: given the direct causes (parents) of a variable, the value which that variable assumes is conditionally independent of all other variables except its effects (descendents).

Therefore, when a human creates a DAG by specifying all the perceived direct causes, he is stating that the joint probability distribution on the variables contains, relative to that DAG, all the conditional independencies in Definition 5.4. Thus, if the human is correct, the DAG, together with the joint probability distribution, constitutes a causal network. We finally see why some call such networks "causal." We will discuss this further later in this section and in the next section. First we examine the networks in the examples in the previous section to determine whether the conditional independencies in Definition 5.4 are indeed ones which we feel exist among cause sand effects. The DAGs in those networks were created according to the notion of causality. We will assume that the DAGs imply all the conditional independencies in Definition 5.4, and investigate whether we feel that these conditional independencies exist among the variables.

In Example 5.6 a possible cause of someone being the last holder of the weapon would be that he is the actual murderer. Hence we have created an arc from A to B. This is the only perceived cause/effect relation among the variables in this example. Thus we have the DAG in Figure 5.2. The question is whether the conditional independencies implied by this DAG agree with our notions. A network which adequately represents all the variables of interest in this situation would have more vertices in the graph and more causes of B. However, in this simple example we have only two; consequently, there are no conditional independencies to consider. Let us investigate something else here. A simple calculation, using the values in the joint distribution, shows that A is conditionally dependent on B even though B would not be considered a cause of A and there is no arc from B to A. This agrees with our notions. The fact that someone last held the weapon certainly would not cause him to be the murderer. However, that fact is evidence that he did indeed commit the murder. Therefore it agrees with our notions that A is conditionally dependent on B.

In Example 5.7 there are two causes of the burglar alarm sounding. Of course, a burglary could cause it to sound. However, in the past Mr. Holmes has noticed that earthquakes have occasionally triggered his alarm. Accordingly, with lower probability, an earthquake could also cause the alarm to sound. The arcs are in the direction of perceived causation. For example, a burglary could cause the burglar alarm to sound, but the sounding of the burglar alarm would not cause a burglar to materialize. Our notions would indicate that neither B nor C would cause the other. That is, that an earthquake should not cause a burglar to appear or vice versa. Therefore we have placed no arc between B and C. Again the question is whether the conditional independencies implied by the resultant DAG agree with our notions. First of all, since $c(B) = \emptyset$ and $C \in a(B)$, the DAG implies that $P(B \mid C) = P(B)$; that is, that B and C are independent. This agrees with our notions. For example, if Mr. Holmes heard there was an earthquake, he would not start expressing concern over being burglarized. This is the only conditional independency implied by the DAG. Let us again investigate something else. If Mr. Holmes'

burglar alarm sounded, he would feel that it is likely that he had been burglarized. However, when he found out that there had been an earthquake, he
would think perhaps the earthquake had caused the alarm to sound, and he
would conclude that it is less likely that he had been burglarized. Thus our
notions would not indicate that B and C are independent given A. In fact,
this conditional independence is not implied by the DAG. Indeed, a calculation using the values in the joint distribution shows that B and C are not
conditionally independent given A. That is, $P(B \mid A,C) \neq P(B \mid A)$.

Notice, in this last example, that the structure of the DAG does not imply that B and C are dependent given A; rather it only implies that B and
C *may* be dependent given A. It was necessary to perform a calculation using the joint distribution to determine that this dependency does exist. This
does not mean that the structure of the DAG is inadequate. The structure
of the DAG implies important conditional independencies. It does not imply
every dependency and independency in the joint distribution. Indeed, since
if the propositional variables are mutually independent the DAG can have
an arbitrary structure, without some additional assumptions (beyond those in
Definition 5.4) the structure of the DAG could never imply any dependencies. Pearl, Verma, and Geiger (Pearl [1986b]; Verma & Pearl [1988]; Geiger
& Pearl [1988]) have investigated the problem of determining exactly what independencies are implied by the structure of the DAG. In Geiger, Verma, &
Pearl [1989] they obtain an efficient algorithm which identifies all the independencies implied by the structures of the DAG. We will discuss some of their
results in chapter 6.

In Example 5.8, the act of cheating may cause the spouse to dine with another, and dining with another may cause him to be seen dining with another.
However, dining with another would not be a cause of his cheating. Notice in
this example that the act of cheating could indeed be a cause of the spouse
being seen dining with another. However, there is no arc from A to C. This
method of DAG construction is at the heart of creating a DAG according to
the notion of causality. B, the act of dining with another, is the only direct
cause of C, the act of being seen dining with another. A causes C indirectly
by causing B. Thus, in a DAG created according to the notion of causality,
there are only arcs from direct causes to effects. Again we investigate whether
the conditional independencies implied by the DAG agree with our notions.
Since $A \in a(C)$ and $c(C) = B$, the DAG implies that $P(C \mid A,B) = P(C \mid B)$;
that is, that C is independent of A given B. This agrees with our notions for
the following reason. Once we know that the spouse is dining with another, we
ascertain the likelihood of his being seen dining. Learning that he is cheating
would not change that likelihood. The idea is that the direct cause B shields C
from the influence of A. That is, once we are certain of B, knowing the value
of A would have no effect on the probability of values of C. This is why we
only draw arcs from direct causes.

Again, rephrasing the independence assumptions in Definition 5.4 in terms
of human notions concerning causality, if we eliminate a variable v's descen-

dents (i.e., the events that v causes either directly or indirectly), then the probabilities of the values at v are fixed once we know the values of all the parents (direct causes) of v. Knowledge of any of the other values in the DAG, namely indirect causes of v and other events caused by these indirect causes, cannot change the probabilities of values of v. We eliminate v's descendents in the conditional independence assumptions because this also agrees with the human notion of causation. Given that a spouse is cheating, knowledge that he is seen dining with another still increases the probability that he actually did dine with another.

The following more complex example from Cooper [1984] better illustrates these principles:

Example 5.13. Suppose that metastatic cancer is a cause of a brain tumor and can also cause an increase in total serum calcium. Suppose further that either a brain tumor or an increase in total serum calcium could cause a patient to fall into a coma, and that a brain tumor could cause papilledema. Let

a_1 = metastatic cancer present	a_2 = metastatic cancer not present
b_1 = total serum calcium increased	b_2 = total serum calcium not increased
c_1 = brain tumor present	c_2 = brain tumor not present
d_1 = coma present	d_2 = coma not present
e_1 = papilledema present	e_2 = papilledema not present.

Then the DAG in Figure 5.6 embodies the causal relations described above.

Again, we investigate whether the conditional independencies implied by this DAG agree with our notions. First, the DAG implies that B, increased total serum calcium, and C, a brain tumor, are independent given A, metastatic cancer. Now A is the only direct cause of both B and C. Therefore knowing that a patient has increased total serum calcium makes it more probable that he has metastatic cancer (since cancer could be the cause) and therefore makes it more probable that he has a brain tumor. Accordingly, we would not feel that B and C are independent. That is, we would not feel that $P(C \mid B) = P(C)$. However, once it becomes known for certain that the patient has or does not have metastatic cancer, knowledge of increased total serum calcium cannot change the probability of metastatic cancer and hence should have no effect on the probability of a brain tumor. Consequently, we would feel that B and C are independent given A. That is, we would feel that $P(C \mid A, B) = P(C \mid A)$, the independence implied by the DAG. There are other conditional independencies implied by the DAG. For example, D is independent of both A and E given B and C. It is left as an exercise to investigate whether the other independencies agree with our notions. Finally, note that although D is a descendent of B, it is also a descendent of C and is therefore not included in the set of variables which we have required to be independent given C's parent A. This agrees with our notions. Since a brain

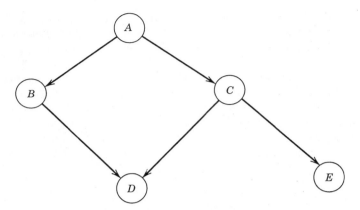

FIGURE 5.6   The DAG for Example 5.13.

tumor can cause a coma directly, even if we know metastatic cancer is present, knowledge of a coma should increase the probability of a brain tumor.

If we accept the arguments in this section that DAGs created according to the notion of causality embody human notions of conditional independence among causes and effects, then we can construct causal networks as follows. Suppose a set of propositional variables is given and that a joint probability distribution on the variables exists. We have an expert, who is familiar with the cause/effect relationships among the variables, identify what he perceives to be all the direct causes of each variable. Assuming he does not identify a circular path of causes, we then create a DAG according to his identifications. As mentioned at the beginning of this section, if we assume that the expert is implying, by his identification of the direct causes, that he knows that the joint distribution contains, relative to the DAG, all the conditional independencies in Definition 5.4, then this DAG, together with the joint distribution, constitutes a causal network. As mentioned at the end of the last section, by Theorem 5.1 we need next only determine the conditional probabilities of every variable given its parents; the joint distribution will then be totally specified. Thus, if we assume for the moment that we are able to determine those conditional probabilities, we will have completely constructed the desired causal network.

Before proceeding, we pose a question: What would happen if the expert incorrectly identified the causes? That is, what if the conditional independencies which he feels exist do not actually exist among the variables? Theorem 5.2 resolves this problem. Due to that theorem, the DAG, together with the joint distribution determined from the conditional probabilities of variables given their parents, still constitutes a causal network. Hence any manipulatory technique based on the assumptions in a causal network still apply. But what of the fact that the joint distribution in this causal network may not be the "correct" joint distribution? If we take the frequentist point of view that objective probabilities exist in the particular application, then even if we have accurately determined all the conditional probability distributions, the joint distribution which they determine, along with the DAG, may not be the correct one because the independence assumptions implied by the DAG may not necessarily be correct. Theorem 5.1 holds only if we assume that the joint distribution has all the conditional independencies implied by the DAG. A subjectivist would not view this as a problem, since the joint distribution represents the expert's beliefs and therefore it is a correct distribution by definition. Actually, in order to represent his beliefs, if we obtained the structure of the network from him, we would be bound to obtain the probabilities from him also. However, if he felt that the sequences were infinitely exchangeable and that his beliefs could be affected by the frequencies, then he would feel that the probabilities, obtained by augmenting his original probabilistic assessments with information from a data base, represent his beliefs in light of the information in the data base. Therefore we could augment his initial assessments with information in a data base (we will see how to do this in chapter 10). A frequentist would say that this joint distribution is the expert's estimate of the objective distribution, and that it may be possible to obtain better estimates from empirical data or from other experts. Regardless of the point of view, this distribution does represent the expert's knowledge, and hence it can be used to make deductions consistent with those of the expert.

When the expert identifies the direct causes, it is understood that he only identifies causes and does not include extraneous variables. That is to say, if the variables in $c(v)$ are identified as the direct causes of v, then there is no proper subset $U \subset c(v)$ such that

$$P(v \mid U) = P(v \mid c(v)).$$

Thus we ordinarily assume that the parent set is minimal as defined in the previous section.

Srinivas et al. [1989] have noted that in some cases there are problems in obtaining the structure of the DAG solely from the expert. The expert may have only partial knowledge about the domain. Furthermore, there may be a "knowledge acquisition bottleneck" when constructing the DAG in this manner. They develop a method for constructing the DAG from information supplied by the expert and from empirical data from the domain.

Identifying the direct causes, and thereby determining the DAG, is a natural task. Humans are good at making judgments concerning local causes and effects. Looking again at Example 5.13, a physician could readily identify that metastatic cancer causes increased total serum calcium and brain tumors and that brain tumors cause severe headaches. A more difficult problem is the determination of the values in the conditional distributions of variables given their parents. However, this task is much simpler than the task of determining the joint probability distribution directly.

As mentioned previously, to fully specify the joint probability distribution for every possible event would take a forbiddingly large number of values even for a fairly small number of variables. Aside from this problem, in general these values are very difficult to ascertain, especially if they are obtained from an expert rather than a data base. In Example 5.7 it is a simple matter for Mr. Holmes to determine the causal relations among the three variables in his system. However, it would be asking a bit much to have Mr. Holmes state that the joint probability of his residence being burglarized, of an earthquake, and of his burglar alarm sounding is .0000099. Humans do not ordinarily make judgments concerning a large number of variables simultaneously. If it seems too much to ask Mr. Holmes to specify the value of .0000099 for these three variables, imagine the absurdity if there were 10, 20, or 100 variables.

On the other hand, it is much more reasonable to ask a human to state the probability of a variable given its direct causes. Mr. Holmes might judge that the probability of his being burglarized is fairly low, around .01, the probability of an earthquake is substantially lower, around .001, and, since both these events are likely to sound his alarm, that the probability of his alarm sounding given both these events is very high, around .99. Thus, once the direct causes of a variable are identified, the conditional probabilities given the direct causes can often be judged by an expert. These judgments can be augmented with information in a data base. It is true that the data base would have no more difficulty determining the joint distribution directly than determining the conditional ones. However, in practice, data bases are often unavailable or contain insufficient data. The expert's judgments play a central role in obtaining the probability values. In any case, even if the data base is adequate, it would usually take an astronomical number of calculations to determine the joint distribution directly. In contrast, if the number of direct causes of each variable is small, there are a manageable number of values in the conditional probability distributions of variables given the direct causes. We see then that by asking the expert to identify the cause/effect relations among variables, we have changed the problem from one involving the determination of an astronomical number of nebulous values to a manageable number of meaningful values.

If a variable has a large number of direct causes, the number of needed values would again become large. But, as we have already seen, sparse, irregular causal networks are often appropriate.

We look again at Example 5.13 to see another example of the determination of the conditional probability values.

Example 5.14. After the DAG for Example 5.13 (Figure 5.6) is created, the conditional probabilities can be obtained from an expert along with possible augmentation from a data base. They may be the following (for the sake of brevity, we are now only listing the probability of one alternative; since in this example there are only two alternatives for each variable, the other alternative is uniquely determined):

$$P(a_1) = .2$$

$$P(b_1 \mid a_1) = .8 \qquad P(b_1 \mid a_2) = .2$$

$$P(c_1 \mid a_1) = .2 \qquad P(c_1 \mid a_2) = .05$$

$$P(d_1 \mid b_1, c_1) = .8 \qquad P(d_1 \mid b_1, c_2) = .9$$

$$P(d_1 \mid b_2, c_1) = .7 \qquad P(d_1 \mid b_2, c_2) = .05$$

$$P(e_1 \mid c_1) = .8 \qquad P(e_1 \mid c_2) = .6.$$

Again, notice that an expert would have an easier time judging these probabilities than the joint probabilities of all five variables. Further note that $P(d_1 \mid b_1, c_2) > P(d_1 \mid b_1, c_1)$. Although this relationship may not be plausible for this particular example, we made these assignments to illustrate that such a relationship is possible.

We have seen in this section that causal networks can be used to determine joint distributions based on the human notion of causality. We shall see in the next section that their use extends beyond this.

PROBLEMS 5.5

1. Investigate the other conditional independencies implied by the DAG for Example 5.13 (Figure 5.6) to determine whether they agree with your notions.

2. Consider the following small piece of fictitious medical knowledge [Lauritzen & Spiegelhalter, 1988]: Tuberculosis and lung cancer can cause shortness of breath (dyspnea) with equal likelihood. The same is true for a positive chest Xray (i.e., a positive chest Xray is also equally likely given either tuberculosis or lung cancer). Bronchitis is another cause of dyspnea. A recent visit to Asia increases the likelihood of tuberculosis, while smoking is a possible cause of both lung cancer and bronchitis. Create a DAG according to the cause/effect relations described in this knowledge base, and investigate whether you feel that the conditional independencies implied by the DAG agree with your notions.

3. Create a causal network for a mini-expert system. The system could incorporate uncertain reasoning into one of the expert systems created in chapter 1. For example, it could be an expert system for diagnosing the problem in an automobile. The money which this author has wasted in garages is testimony to the fact that mechanics reason under uncertainty.

5.6. BEYOND THE NOTION OF CAUSALITY

Looking again at the definition of a causal network (Definition 5.4), we see that it has only to do with conditional independencies and has nothing to do with the human notion of causality. Since it is often straightforward for an expert to identify direct causes and probabilities of effects given direct causes, causal networks are often created based on the notion of causality. On the other hand, perhaps the reason he identifies the causes in a certain manner is that his experience with the independencies among the variables has led him to believe that these are causes. In other words, the observed independencies could well determine the identification of the causes.

The definition and determination of "real" causes are beyond our scope and can only be left to philosophers. In practice, however, if a DAG is created in such a manner that each variable is conditionally independent of all others, except its descendents, given its parents, then that DAG, together with the joint distribution, constitutes a causal network. Therefore any manipulatory technique, based only on the assumptions in the definition of a causal network, can be applied. That is to say, there is no need for someone to actually feel that the arcs are in the direction of causation. The arcs can be determined according to which conditional independencies are easiest to identify or according to which conditional probabilities are easiest to ascertain.

For example, sometimes it is easier for a person to ascertain the probability of a cause given an effect than that of an effect given a cause. Consider the following situation: If a physician had worked at the same clinic for a number of years, he would have seen a large population of similar people with certain symptoms. Since his job is to reason from the symptoms to determine the likelihood of diseases, through the years he may have become adept at judging the probabilities of diseases given symptoms for the population of people who attend this clinic. On the other hand, a physician does not have the task of looking at a person with a known disease and judging whether a symptom is present. Hence it is not likely that the physician would have acquired the ability from his experience to judge the probabilities of symptoms given diseases. He does ordinarily learn something about this probability in medical school. However, if a particular disease were rare in the population, whereas a particular symptom of the disease were common, and the physician had not studied the disease in medical school, he would certainly not be able to determine the probability of the symptom given the disease. On the other hand, he could readily determine the probability of the disease given the symptom for this particular population. We see then that in this case it is easier to ascertain the probabilities of causes given effects. Notice that these conditional probabilities are only valid for the population of people who attend that particular clinic, and thus we would not want to incorporate them into an expert system. We could, however, use the definition of conditional probability to determine the probabilities of symptoms given diseases from these conditional probabilities obtained from the physician. These latter probabilities could then be used in an expert system which would be applicable at any clinic. Many of the ideas in this paragraph are from Shachter and Heckerman [1987].

Causal networks often incorporate the notion of time. That is, for a given vertex v, the parents of v often precede v in time. This is consistent with the scientific notion of a state. If we give a sufficiently detailed present-state description of a system, information concerning previous states cannot provide any additional assistance in predicting future states. This observation was taken from Pearl [1986a]. But, as we have seen, the independence also often applies to events occurring at the same time. Given its parents, a variable is independent of all other variables in the network, except its descendents. For example, in Example 5.13, increased serum calcium becomes independent of a brain tumor once we know the patient has metastatic cancer. Furthermore, as we shall see in the examples at the end of this section, the parents of a vertex need not precede a vertex in time. Regardless of whether a particular parent–child relationship incorporates the notion of time, it is the probabilistic dependencies and conditional independencies which are the fundamental considerations in the construction of the network, not the perceived causal relationships.

As mentioned in chapter 2, many use psychological modeling in an effort to create systems which reason in the same way as humans. At the end of chapter 4, we offered the argument that a causal network, created according to the notion of causality, is a structure which represents human uncertain knowledge at some level. However, it is again the notion of independency rather than causality which is fundamental to that argument. Pearl [1986a] a proponent of that argument, states that

> the notions of dependence and conditional independence are more basic to human reasoning than are the numerical values attached to probability judgments... Moreover, the nature of probabilistic dependency between propositions is similar in many respects to that of connectivity in a graph. For instance, we find it plausible to say that a proposition B affects proposition C directly, while A influences C indirectly, via B. Similarly, we find it natural to identify a set of direct justifications for C to sufficiently shield C from all other influences and describe them as the direct neighbors of C [Doyle, 1979]. These graphical metaphors suggest that the fundamental structure of human knowledge can be represented by dependency graphs and that mental tracing of links in these graphs are the basic steps in querying and updating that knowledge.

Pearl's argument does not assume that the parents of a vertex be conceived as "causes." Instead it is only necessary that they shield the vertex from all other influences (in the sense of probabilistic independency).

When creating a causal network by identifying the conditional independencies directly, there is again the question of whether these identifications are "correct;" that is, whether these conditional independencies really exist in some objective sense. Again, they are sufficient to make deductions consistent with those of the expert, and Theorem 5.2 guarantees that the resultant structure is a causal network.

Next these concepts are illustrated with some examples:

FIGURE 5.7   The DAG for the causal network in Example 5.15.

Example 5.15. Looking again at Examples 5.6 and 5.9, if we compute $P(B)$ and $P(A \mid B)$ from the joint distribution, we obtain

$$P(b_1) = .66 \qquad P(b_2) = .17 \qquad P(b_3) = .17$$
$$P(a_1 \mid b_1) = .97 \qquad P(a_2 \mid b_1) = .015 \qquad P(a_3 \mid b_1) = .015$$
$$P(a_1 \mid b_2) = .47 \qquad P(a_2 \mid b_2) = .47 \qquad P(a_3 \mid b_2) = .06$$
$$P(a_1 \mid b_3) = .47 \qquad P(a_2 \mid b_3) = .06 \qquad P(a_3 \mid b_3) = .47.$$

Together with the DAG in Figure 5.7, these conditional distributions constitute a causal network which contains the same joint distribution as the causal network described in Example 5.6. Thus, as far as determining a causal network containing that joint distribution, it makes no difference as to whether we perceive A as the cause of B, or B as the cause of A. Indeed, the notion of causality is not even needed. We can place the arc in the direction for which it is easiest to obtain the conditional probabilities.

Example 5.16. Let

b_1 = patient has a cold b_2 = patient does not have a cold
c_1 = patient has hay fever c_2 = patient does not have hay fever
a_1 = patient is sneezing a_2 = patient is not sneezing.

Our notions of causality would imply that B and C are independent (i.e., a cold does not cause hay fever and hay fever does not cause a cold) and both A and B can cause C. If we construct a DAG according to these notions, we'd obtain the DAG in Figure 5.8(a). Suppose we then determine the following conditional probabilities (again, the probability of only one alternative is now given for the sake of brevity):

$$P(b_1) = .01 \qquad\qquad P(c_1) = .02$$
$$P(a_1 \mid b_1,c_1) = .99 \qquad P(a_1 \mid b_1,c_2) = .9$$
$$P(a_1 \mid b_2,c_1) = .8 \qquad P(a_1 \mid b_2,c_2) = .05.$$

As mentioned before, although in such a case B and C are independent, they are often not conditionally independent given A. In this case, $P(B \mid C,A) \neq P(B \mid A)$. This agrees with our notions. If a patient is sneezing, we would believe there is a good chance he has a cold. However, when we learn that he has hay fever, since hay fever could explain the sneezing, we feel that it is less likely that he has a cold. If we then compute $P(B \mid A)$ and $P(C \mid B,A)$ from

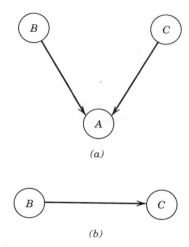

(a)

(b)

FIGURE 5.8   The DAGs for the causal networks in Example 5.16.

the joint distribution obtained from the conditional distributions above, these new conditional distributions, along with the DAG in Figure 5.8(b), constitute a causal network which represents the joint conditional probability distribution of B and C given A. The actual determination of these values is left as an exercise. In this new network the presence of a cold would "cause" the likelihood of hay fever to be lessened. Thus, in a subpopulation in which everyone was sneezing, one might perceive a causal relationship between colds and hay fever.

Example 5.17. [Pearl, 1986a]. Suppose we are to toss two fair coins simultaneously and then ring a bell if either of them turns up heads. Let

$$a_1 = \text{first coin turns up heads}$$

$$a_2 = \text{first coin turns up tails}$$

$$b_1 = \text{second coin turns up heads}$$

$$b_2 = \text{second coin turns up tails}$$

$$c_1 = \text{bell rings}$$

$$c_2 = \text{bell does not ring.}$$

Our notions in this situation are that either coin could cause the bell to ring by turning up heads, that the bell ringing follows the coin tosses in time and therefore could not cause either coin to land a certain way, and that the coins are independent; that is, neither has any effect on the way the other lands. Therefore, creating a DAG according to the notion of causality, we obtain the DAG in Figure 5.9(a). The conditional probabilities of variables given their parents are (again, only the probability of one event is given, since there are only two alternatives)

$$P(a_1) = 1/2 \qquad\qquad P(b_1) = 1/2$$
$$P(c_1 \mid a_1, b_1) = 1 \qquad P(c_1 \mid a_1, b_2) = 1$$
$$P(c_1 \mid a_2, b_1) = 1 \qquad P(c_1 \mid a_2, b_2) = 0.$$

The causal network, determined by these conditional distributions and the DAG in Figure 5.9(a), clearly contains the following joint distribution:

$$P(a_1, b_1, c_1) = 1/4 \qquad P(a_1, b_1, c_2) = 0$$
$$P(a_1, b_2, c_1) = 1/4 \qquad P(a_1, b_2, c_2) = 0$$
$$P(a_2, b_1, c_1) = 1/4 \qquad P(a_2, b_1, c_2) = 0$$
$$P(a_2, b_2, c_1) = 0 \qquad P(a_2, b_2, c_2) = 1/4.$$

Next we create the DAG shown in Figure 5.9(b) and compute the conditional probabilities of variables given their parents using the above joint distribution:

$$P(c_1) = 3/4$$
$$P(a_1 \mid c_1) = 2/3 \qquad P(a_1 \mid c_2) = 0$$
$$P(b_1 \mid a_1, c_1) = 1/2 \qquad P(b_1 \mid a_1, c_2) = 0$$
$$P(b_1 \mid a_2, c_1) = 1 \qquad P(b_1 \mid a_2, c_2) = 0.$$

It is left as an exercise to verify that the joint distribution can be retrieved from these conditional distributions, and that therefore we have another causal network containing that joint distribution. However, this network has nothing to do with our usual notions of causality. The bell ringing actually follows the coin tosses in time; yet it is a "cause" of their results. It is a "cause" in the following sense. Given that we know that the bell rung and the first coin turned up tails, the second coin must have turned up heads. Thus the ringing of the bell and the first coin turning up tails "caused" the second coin to turn up heads. Again, the definition of a causal network has only to do with conditional independencies and has nothing to do with "real" causes. Consequently, in this example, if we are using a manipulatory technique which relies only on the properties of a causal network, we can use either causal network equally as well.

In this chapter we have seen how to obtain causal networks which contain a desired joint probability distribution. In the following two chapters we will see how to use these causal networks to compute the conditional probabilities of remaining variables given that certain variables have realized specific values.

PROBLEMS 5.6

1. In Example 5.16 determine the values in the conditional joint distribution of B and C given A.

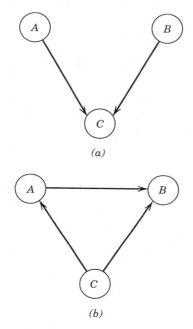

FIGURE 5.9   The DAGs for the causal networks in Example 5.17.

2. Verify that the joint distribution can be retrieved from the second set of conditional distributions in Example 5.17.

COMPUTER PROBLEM 5.6

Using your favorite programming language, create a causal network. Initially the program should prompt for the set of propositional variables and their possible values and the arcs in the DAG. After creating the graph and performing an error check to verify that it is acyclic, the program should prompt for the conditional probabilities of each variable given its parents. There are no outputs. This program will be used in the programs written in the following chapters.

5.7. A ROSE BY ANY OTHER NAME...

Since the notion of causality is not fundamental to a causal network, the term "causal" is somewhat misleading. As mentioned at the start of this chapter, these networks are also called Bayesian or belief networks. However, these terms seem to imply that the probabilities must be obtained from the beliefs of an expert. That is, they seem to exclude the possibility of identifying the independencies and/or obtaining the probability values using a data base (as noted in section 2.4, if we are to use the techniques in this book, we must acknowledge that probabilities obtained from a data base become our beliefs). Of the current names, "influence diagram" may be most appropriate, since the set of parents shields a vertex from the influence of all other vertices in the network (except its descendents). However, this name is ordinarily used when the network includes decision vertices (as discussed in chapter 9). Furthermore, the word "influence" has ambiguous connotations of its own. Since the fundamental assumption in these networks is that their structures represent conditional independencies, we suggest that perhaps the most appropriate name for them is *independence networks*.

CHAPTER 6

PROBABILITY PROPAGATION IN SINGLY CONNECTED CAUSAL NETWORKS

We saw how to construct causal networks in the previous chapter. Next we must develop a method for determining how the realization of specific values of some variables affects the probabilities of the remaining variables. Since Pearl's efforts have been to model human reasoning, he calls these probabilities "beliefs," and his method the propagation of belief. As mentioned in chapter 2, we will simply call the methods, in both this chapter and the next, the propagation of probabilities.

Recall from the last chapter that Pearl [1986a] argues that

> the fundamental structure of human knowledge can be represented by dependency graphs and that the mental tracing of links in these graphs are the basic steps in querying and updating the knowledge.

Thus Pearl feels that to propagate probabilities in a manner which is consistent with the way humans reason, we must perform the updates locally, via communication between vertices in the DAG, rather than globally. Pearl [1988] states further that

> ...there is a growing interest in reasoning models that permit unsupervised parallelism. The interest is motivated both by technological advances in parallel computation and by the need to develop viable models of human reasoning. The speed and ease with which people perform some low-level interpretive functions, such as recognizing scenes, reading text, and even understanding stories, strongly suggest that such processes involve a significant amount of parallelism and that most of the processing is done at the knowledge level itself [Shastri and Feldman, 1984]...

We can model such phenomena by viewing a belief network not merely as a passive code for storing factual knowledge but also as a computational architecture for reasoning about that knowledge. This means that the links in the network should be treated as the only mechanisms that direct and propel the data through the process of querying and updating belief. Accordingly, we assume that each node in the network is given a separate processor, which maintains the parameters of belief for the host variable and manages the communication links to and from neighboring, conceptually related variables.

Thus Pearl argues that local communication between vertices models the way humans reason.

When we construct a causal network according to conditional independencies specified by a human (which may or may not be according to the notion of causality), the structure of the network represents the human's beliefs concerning the conditional independencies among the variables. Therefore, the argument that the structure of the network represents the structure of human knowledge appears to be valid at some level. Moreover, it seems reasonable that if we determined the conditional probabilities of remaining variables given that certain variables are instantiated, these conditional probabilities would be consistent with the judgments of a human. For example, if Mr. Holmes learned that his alarm had sounded, he would conclude that it is highly probable that his residence had been burglarized. If he then learned that there had been an earthquake, he would conclude that it is less probable that he had been burglarized. These are the same judgments which we would reach if we computed the corresponding conditional probabilities in a causal network. Furthermore, this conjecture, that we can reach the same judgments as a human by computing the conditional probabilities in the causal network, can be tested. That is, we can give a human the information in a causal network, tell him that certain events have happened, and evaluate his judgments.

On the other hand, the conjecture that humans reach their judgments by traversing the arcs in the graph is more provocative and less testable. Alternatively, we could claim that when a variable is instantiated, the human mind creates a new network based on that instantiation. For example, when Mr. Holmes learns that his alarm went off, perhaps his mind creates a new network with an arc from the earthquake to the burglar. Then, when Mr. Holmes learns of the earthquake, he traverses that arc directly to conclude that it is less probable that he has been burglarized. If we accept this conjecture, and if we developed a system which created a new network each time a variable is instantiated, then we could argue that this system models human reasoning. However, mathematically both systems would yield identical results. As noted in chapter 2, the fundamental goal in expert systems is to obtain the best possible results, not necessarily to model the human reasoning process. Indeed, if we want to create systems which can go beyond human capabilities, computers may have to do things in ways different from the ways of humans. If we take this approach and are unconcerned with the modeling of human reasoning, it is immaterial as to whether Pearl's conjecture is correct. The important thing is that by conceiving the propagation problem in terms of human reasoning, he was able to devise a mathematically sound propagation method.

In addition, even if we take the approach that we are unconcerned with modeling the way humans reason, there are still advantages to updating probabilities locally. Pearl [1986a] notes further that such a scheme satisfies a primary goal of rule-based systems, namely the separation of the control mechanism or inference engine from the knowledge base. Furthermore, such a scheme can be implemented easily in object-oriented languages, the updating of vertices can be done in any order and in parallel, and the communication between vertices can be implemented using a blackboard architecture [Lesser & Erman, 1977]. Therefore, even if the propagation method is not meant to model the way a human reasons, a method which updates probabilities locally has implementation advantages. Finally, regardless of the method used to construct the network and regardless of whether the aim is to model the way humans reason, the act of updating locally stays within the framework of the network builder. This is particularly advantageous if the network is built and analyzed dynamically.

In the case of singly connected networks, Pearl has accomplished his goal, namely a local probability propagation scheme. Pearl [1986a] has devised a method which makes no ad hoc adjustments or unfounded assumptions of conditional independence. In this chapter we explain Pearl's method. A simplified version, which works only in trees, will be covered first. Before that, some necessary background material is discussed in the following section. In the next chapter we will develop a method which works in an arbitrary network, thus removing the single connectivity requirement.

6.1. D-SEPARATION

As mentioned in the previous chapter, Pearl, Verma, and Geiger [Pearl, 1986b; Verma & Pearl, 1988; Geiger & Pearl, 1988; Geiger, Verma, & Pearl, 1989] have investigated the problem of determining exactly what independencies are implied by the structure of the DAG in a causal network. Their fundamental result is discussed in this section. This result will be needed to develop the propagation schemes in the following sections. Before giving that result, we offer an alternative notation for representing conditional independencies and prove a preliminary theorem. This notation, taken from Pearl [1986b], simplifies the notation in the proofs of the theorems in this section.

Definition 6.1. Let $(\Omega, \mathcal{F}, P)$ be a probability space and let E_1, E_2, and E_3 be three events in $\mathcal{F}$ such that E_1 is conditionally independent of E_2 given E_3. That is,

$$P(E_1 \mid E_2 \wedge E_3) = P(E_1 \mid E_3).$$

This relation will also be denoted as follows:

$$I_P(E_1, E_3, E_2).$$

The following theorem enables us to investigate the conditional independencies implied by the structure of the DAG in a causal network:

Theorem 6.1. Let X, Y, W, and Z be disjoint sets of propositional variables, each of which is defined on the same probability space $(\Omega, \mathcal{F}, P)$. Then the following four implications hold:

Symmetry $\qquad I_P(X, Z, Y)$ if and only if $I_P(Y, Z, X)$ $\qquad\qquad$ (a)

Decomposition $\quad I_P(X, Z, Y \cup W)$ implies $I_P(X, Z, Y)$ $\qquad\qquad$ (b)

Weak union $\qquad I_P(X, Z, Y \cup W)$ implies $I_P(X, Z \cup Y, W)$ $\qquad\qquad$ (c)

Contraction $\qquad I_P(X, Z \cup Y, W)$ and $I_P(X, Z, Y)$ implies $I_P(X, Z, Y \cup W)$ $\quad$ (d)

(Recall from the previous chapter that if X is a set of propositional variables, $P(X)$ means the probability that each variable in X assumes one of its values.)

Proof. In order to keep the background necessary to the understanding of this proof to a minimum, the proof will be for the case where the variables are finite. The continuous case is a simple extension.

(a) By the definition of conditional probability, this result follows immediately for arbitrary events, not just the case in which each event is the instantiation of a set of propositional variables.

(b) We need show that

$$P(Y \cup W \mid Z \cup X) = P(Y \cup W \mid Z) \qquad (6.1)$$

implies that

$$P(Y \mid Z \cup X) = P(Y \mid Z).$$

To that end, (6.1) implies that

$$\frac{P(Y \cup W \cup Z \cup X)}{P(Z \cup X)} = \frac{P(Y \cup W \cup Z)}{P(Z)}$$

$$P(Y \cup W \cup Z \cup X)P(Z) = P(Y \cup W \cup Z)P(Z \cup X)$$

$$\sum_W (P(Y \cup W \cup Z \cup X)P(Z)) = \sum_W (P(Y \cup W \cup Z)P(Z \cup X)),$$

where $\sum_W$ means we are summing over all possible combinations of values of variables in W. We then have further that

$$P(Z)\sum_W (P(Y \cup W \cup Z \cup X)) = P(Z \cup X)\sum_W (P(Y \cup W \cup Z))$$

$$P(Z)P(Y \cup Z \cup X) = P(Z \cup X)P(Y \cup Z).$$

Finally, since by assumption $P(Z \cup X)$ and $P(Z)$ are both nonzero, we have

$$\frac{P(Y \cup Z \cup X)}{P(Z \cup X)} = \frac{P(Y \cup Z)}{P(Z)}$$

$$P(Y \mid Z \cup X) = P(Y \mid Z).$$

(c) We need show that

$$P(X \mid Z \cup Y \cup W) = P(X \mid Z) \qquad\qquad (6.2)$$

implies that

$$P(X \mid Z \cup Y \cup W) = P(X \mid Z \cup Y).$$

Plainly, it suffices to show that

$$P(X \mid Z \cup Y) = P(X \mid Z).$$

To that end, (6.2) implies that

$$\frac{P(X \cup Z \cup Y \cup W)}{P(Z \cup Y \cup W)} = \frac{P(X \cup Z)}{P(Z)}.$$

The proof can now easily be completed by cross multiplying and summing over W as in the proof of (b).

(d) We need show that

$$P(X \mid Z \cup Y \cup W) = P(X \mid Z \cup Y) \qquad \text{and} \qquad P(X \mid Z \cup Y) = P(X \mid Z)$$

$$(6.3)$$

implies that

$$P(X \mid Z \cup Y \cup W) = P(X \mid Z).$$

The result follows directly from (6.3). $\square$

Before proving the main result about the conditional independencies implied by the structure of the DAG in a causal network, we investigate these independencies via some examples.

Example 6.1. Consider the DAG in Figure 6.1(a). If this DAG were part of causal network $C = (V, E, P)$, its structure would imply that

$$I_P(\{B\}, \{A\}, \{C\}),$$

since $c(B) = \{A\}$ and $a(B) = \{A, C\}$. If the structure is meant to represent the notion of causality, this makes sense. A is the only cause of B and C. If we did not know A's value, knowledge of B's value would tell us something about A's value and therefore something about C's value. Thus we would not expect B and C to be independent. However, once we know A's value, knowledge of B's value can tell us nothing about A and hence should tell us nothing about

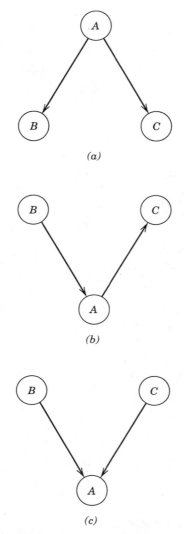

FIGURE 6.1 The DAGs for Examples 6.1, 6.2, and 6.3, respectively.

C. Accordingly, we would expect B and C to be independent given A. Recall that we had this situation in Example 5.13 in the previous chapter.

Example 6.2. Consider the DAG in Figure 6.1(b). If this DAG were part of causal network $C = (V, E, P)$, its structure would not imply that B and C are independent. Indeed, if this network is meant to represent the notion of causality, we would not expect that B and C are independent. Since B is a cause of A and A is a cause of C, we would expect B's value to tell us something about C. However, if we ascertained A's value, B's value could no longer tell us anything about A's value and therefore we would not expect it to tell us anything about C's value. Thus we would expect that B and C are independent given A. This independence also follows immediately from the definition of a causal network. From that definition, we have

$$I_P(\{C\},\{A\},\{B\}),$$

since $c(C) = \{A\}$ and $a(C) = \{A,B\}$. Note that this is the DAG in Example 5.8 in the previous chapter.

Example 6.3. Consider the DAG in Figure 6.1(c). If this DAG were part of causal network $C = (V,E,P)$, its structure would imply that

$$I_P(\{B\},\emptyset,\{C\}),$$

since $c(B) = \emptyset$ and $a(B) = C$. Again, if this network is meant to represent the notion of causality, we'd expect this, since neither variable has any causes. However, we would not expect that B and C are independent given A. This DAG is found Examples 5.7 and 5.16 in the previous chapter. Recall that given that Mr. Holmes' burglar alarm sounded, the occurrence of an earthquake would make it less likely that his residence had been burglarized. One can verify from the joint distributions in both Examples 5.7 and 5.16 that, indeed, $I_P(\{B\},\{A\},\{C\})$ cannot be implied by this DAG.

In order to discuss these examples further, we need the following definition, which is purely graph theoretic. It is presented here, rather that in the chapter on graph theory (Chapter 3), because its intuitive meaning can only be explained here.

Definition 6.2. Let $G = (V,E)$ be a DAG with $u \in V$ and $v \in V$ (refer to Figure 6.2).

1. If $(v,u) \in E$ and $(v,w) \in E$, it is said that the arcs (v,u) and (v,w) meet tail-to-tail at v.
2. If $(u,v) \in E$ and $(v,w) \in E$, it is said that the arcs (u,v) and (v,w) meet head-to-tail at v.
3. If $(u,v) \in E$ and $(w,v) \in E$, it is said that the arcs (u,v) and (w,v) meet head-to-head at v.

Looking back now at these three examples we see in the first of these that B and C are connected by a chain containing arcs which meet tail-to-tail at A, and B and C are independent given A. In the second example, B and C are connected by a chain containing arcs which meet head-to-tail at A, and B and C are independent given A. In the third example, B and C are connected by a chain containing arcs which meet head-to-head at A, and B and C are not independent given A. One might speculate from this that whenever two

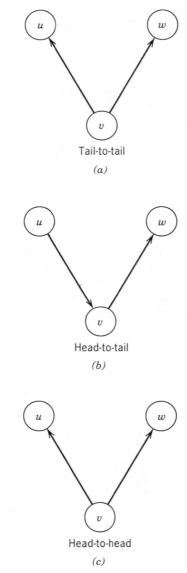

FIGURE 6.2 An illustration of Definition 6.2.

variables, B and C, are connected by a chain containing arcs which meet tail-to-tail or head-to-tail at A the structure of the DAG implies that B and C are independent given A, but when they are connected by a chain containing arcs which meet head-to-head at A, the structure of the DAG cannot imply that they are independent given A. This certainly does not follow, since two variables can be connected by more than one chain. For example, in Figure

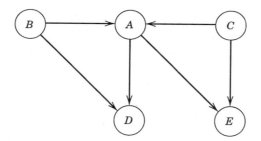

FIGURE 6.3   A DAG in which one chain meets tail-to-tail at A while another meets head-to-head at A.

6.3 the chain $[B, A, C]$ contains arcs which meet head-to-head at A while that chain $[B, D, A, E, C]$ contains arcs which meet tail-to-tail at A. Therefore, if the previous statement were correct, we would have to conclude contradictory results. But what of the case where the only chain between B and C has arcs which meet tail-to-tail at A? In the DAG in Figure 6.4 the only chain between B and C is $[B, A, D, C]$ and that chain contains arcs which meet tail-to-tail at A. Does the structure of the DAG imply that B and C are independent given A? This does not follow immediately from Example 6.2 because the argument in that example holds only when the DAG contains exactly three vertices. However, if the DAG is meant to represent the notion of causality, it does agree with our notions that B and C would be independent given A. Once the value of A is ascertained, B's value can tell us nothing more about A, therefore it can tell us nothing more about D, and in turn it can tell us nothing more about C. Our main result of this section is a theorem which shows that the structure of the DAG does indeed imply that B and C are independent given A.

Example 6.4. Consider the DAG in Figure 6.5. If this DAG were part of a causal network $C = (V, E, P)$, constructed according to the notion of causality, we would expect that B and C are independent given A, their common cause. However, this does not follow immediately from the definition of a causal network, since B has two parents, D and A, and C has two parents, A and E. We will use Theorem 6.1 to prove that indeed $I_P(\{B\}, \{A\}, \{C\})$. We have, owing to the definition of a causal network, that

$$I_P(\{D\}, \varnothing, \{A, C\}).$$

Theorem 6.1(c) then implies that

$$I_P(\{D\}, \{A\}, \{C\}).$$

Furthermore, owing to the definition of a causal network,

$$I_P(\{B\}, \{D, A\}, \{C\}).$$

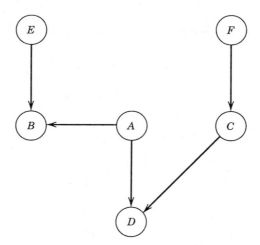

FIGURE 6.4 A DAG in which the only chain between B and C meets tail-to-tail at A.

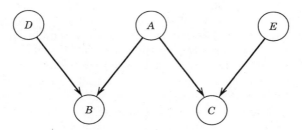

FIGURE 6.5 The DAG for Example 6.4.

We therefore have, owing to Theorem 6.1(d), that

$$I_P(\{B,D\},\{A\},\{C\}),$$

and finally, owing to Theorem 6.1(b), that

$$I_P(\{B\},\{A\},\{C\}).$$

The result in Example 6.4 follows immediately from the general theorem proved at the end of this section. We proved it directly in the example to illustrate how Theorem 6.1 can be used to determine conditional independencies from the DAG in a causal network. That theorem will be used repeatedly in the proof of the general theorem at the end of this section. Before that, we first consider two more examples.

Example 6.5. Consider the DAG in Figure 6.6. If that DAG were in a causal network $C = (V,E,P)$, constructed according to the notion of causality, we

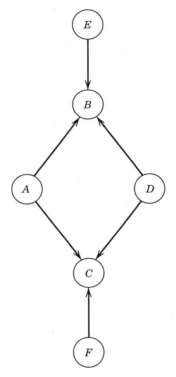

FIGURE 6.6   The DAG for Example 6.5.

would expect B and C to be independent given both their common causes, A and D, but not independent given only one of them, say A. For, if only A is given, knowledge of B's value still tells us something about D and therefore something about C. We will see that this is indeed the case. In other words, the general theorem along with its converse will show that the structure of the DAG implies that B and C are independent given both A and D, but not given only one of them. Of course, in this case, we can use the definition of a causal network to conclude that B and C are independent given both A and D.

Example 6.6. Consider the DAG in Figure 6.7. If that DAG were part of a causal network $C = (V, E, P)$, constructed according to the notion of causality, we would expect that B and C are not independent given F, since B's value still tells us something about A, which in turn tells us something about C. However, we would expect B and C to be independent given A, their common cause. On the other hand, we would not expect B and C to be independent given both A and E, because E's value tells us something about D. Thus, once we know something about D, information about one of D's causes tells us

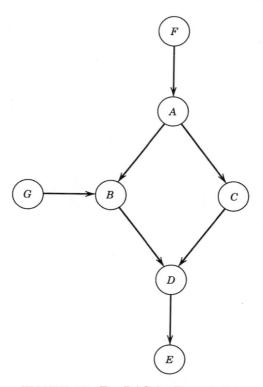

FIGURE 6.7 The DAG for Example 6.6.

something about the other cause. This is similar to the situation in Examples 5.7 and 5.16. We will see that the general theorem along with its converse shows that the structure of the DAG only implies the independencies which we would expect.

Note that in all these examples the general theorem shows only that the structure of the DAG implies independences. It does not imply any dependencies. For example, for some particular distribution, B and C may be independent given both A and E in Example 6.6.

We have belabored the independencies through so many examples to impart an intuitive feel for the results in the general theorem at the end of this section. In order to prove that theorem, we need the following definitions, which should now have some intuitive appeal:

Definition 6.3. Let $G = (V, E)$ be a DAG, $W \subseteq V$, and u and v be vertices in $V-W$. Then a chain ρ, between u and v, is blocked by W if one of the following is true:

1. There is a vertex $w \in W$ on the chain ρ, such that the arcs, which determine that w is on ρ, meet tail-to-tail at w.
2. There is a vertex $w \in W$ on the chain ρ such that the arcs, which determine that w is on ρ, meet head-to-tail at w.
3. There is a vertex $x \in V$, for which x and none of x's descendents are in W, on the chain ρ such that the arcs, which determine that x is on ρ, meet head-to-head at x.

Although this definition is made for DAGs in general, it only has application when the DAG is in a causal network. The idea in the definition is that if a chain between two vertices, u and v, is blocked by W, it is not possible for u and v to be dependent given W via that chain. For example, in Example 6.5, if W contains A, then B and C cannot be conditionally dependent given W via the chain $[B, A, C]$. But this is not enough. If W does not contain D, they could still be conditionally dependent given W via the chain $[C, D, B]$. That is to say, given W, C still tells us something about B through this chain. Thus all chains between two vertices must be blocked by W in order for the structure to imply that the vertices are conditionally independent given W. This is in fact the requirement in the theorem which will be proved shortly. Before that, consider again Example 6.6. If W contains D or E but not B, then the chain $[G, B, D, C]$ is not blocked by W. This is what we would expect, since G and C should not, in general, be conditionally independent given D or E (G tells us something about B, and we've already seen that B and C should not, in general, be conditionally independent given D or E). However, if W also contained B, then the chain is blocked by W. We'd also expect this. Once B's value is known for certain, G can tell us nothing more about B, and therefore nothing more about C.

We can now define d-separation and prove our main result.

Definition 6.4. Let $G = (V, E)$ be a DAG, $W \subseteq V$, and u and v vertices in $V-W$. Then u and v are d-separated by W if every chain between u and v is blocked by W.

In the DAG in Figure 6.8, B and C are d-separated by $\{A\}$; B and C are d-separated by $\{A, D\}$; B and C are not d-separated by $\{A, G\}$.

Definition 6.5. Let $G = (V, E)$ be a DAG, and W_1, W_2, and W_3 disjoint subsets of V. Then W_1 and W_2 are d-separated by W_3 if for every $u \in W_1$ and $v \in W_2$, u and v are d-separated by W_3.

In the graph in Figure 6.8, $\{B, D\}$ and $\{C\}$ are d-separated by $\{A\}$; $\{B, D\}$ and $\{C\}$ are not d-separated by $\{A, E\}$.

We finally prove the main result of this section, that is, the theorem by Verma and Pearl [1988] concerning the conditional independencies implied by the structure of the DAG in a causal network.

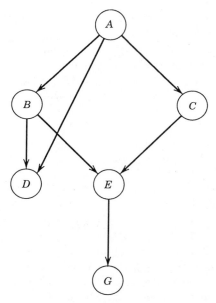

FIGURE 6.8   A DAG.

Theorem 6.2. [Verma & Pearl, 1988]. Let $C = (V, E, P)$ be a causal network with $G = (V, E)$ the DAG in C, and let X, Y, and Z be subsets of V such that X and Y are d-separated by Z in G. Then

$$I_P(X, Z, Y).$$

That is, X and Y are conditionally independent given Z.

Proof. Unless otherwise stated, in all that follows, d-separation means d-separation in G. The theorem will be proved by induction on the number of vertices in V. It is trivially true if V has one vertex. By way of induction, assume it is true for all causal networks for which V has less than n vertices. Suppose V has n vertices. By Theorem 3.7, it is possible to obtain an ancestral ordering of the vertices in V. Let $[v_1, v_2, \ldots, v_n]$ be such an ordering, and let X, Y, and Z be subsets of V such that X and Y are d-separated by Z. The theorem will now be proved by considering three cases:

Case 1. v_n does not appear in X, Y, or Z. Let $W = \{v_1, v_2, \ldots, v_{n-1}\}$, G_W the subgraph induced on W by G, and E'' the arcs in G_W. In this case we have that X, Y, and Z are all subsets of W. Suppose Z does not d-separate X and Y in G_W. Then there exist some $x \in X$ and $y \in Y$ and a chain in G_W, between x and y, which is not blocked in G_W by Y. However, adding a vertex (namely v_n) and arcs (namely those in $E - E''$) to a graph cannot cause an existing chain to become blocked by Z. Thus the chain is not blocked in G, and Z does not d-separate X and Y in G, a contradiction. Therefore X and Y are d-separated in G_W by Z.

Now by Theorem 5.1 and Lemma 5.2, the marginal distribution P'' on W is uniquely determined by the conditional distributions, $P(v_i \mid c(v_i))$ for $1 \leq i \leq n - 1$. Therefore, by Theorem 5.2, (W, E'', P'') is a causal network. Thus, since X and Y are d-separated in G_W by Z, we have by the induction assumption that

$$I_{P''}(X, Z, Y).$$

However, since P'' is just a marginal distribution of P, P'' and P are equal on any event on which they both are defined. Therefore trivially any conditional independencies which exist in P'' also exist in P. Therefore

$$I_P(X, Z, Y).$$

Case 2. $v_n \in X$ or $v_n \in Y$. Owing to Theorem 6.1(a), without loss of generality we can assume $v_n \in X$. Let $X' = X - \{v_n\}$, $B = c(v_n)$, and $R = W - B$. Since v_n is the last vertex in an ancestral ordering, v_n has no descendents. Therefore $R \subseteq a(v_n)$, and we have by the definition of a causal network that

$$I_P(\{v_n\}, B, R). \tag{6.4}$$

Now let R_x, R_y, R_z, R_o be a partitioning of the vertices in R and B_x, B_y, B_z, B_o a partitioning of the vertices in B such that

$$X' = B_x \cup R_x$$
$$Y = B_y \cup R_y$$
$$Z = B_z \cup R_z.$$

Since $B = c(v_n)$, there is an arc from every vertex in B to v_n. Therefore if B_y is not empty, there is an arc from a vertex $y \in Y$ to v_n. This single arc would constitute a chain between y and v_n which is not blocked by Z, which contradicts the fact that $\{v_n\}$ and Y are d-separated by Z. Therefore B_y is empty. We thus have by (6.4) that

$$I_P(\{v_n\}, B_x \cup B_o \cup B_z, R_x \cup R_z \cup Y \cup R_o).$$

Applying Theorem 6.1(b) and (c) and noting that $X' = B_x \cup R_x$ and $Z = B_z \cup R_z$, we then have

$$I_P(\{v_n\}, X' \cup B_o \cup Z, Y). \tag{6.5}$$

Now since there is an arc from every vertex in B_o to v_n, and v_n and Y are d-separated by Z, B_o and Y must also be d-separated by Z. For, if this were not the case, there would exist a $b \in B_o$ and a $y \in Y$ and a chain between b and y which is not blocked by Z. However, since there is an arc from b to v_n, there would then be a chain between v_n and y which is not blocked by Z, a contradiction. We therefore have that Y and B_o are d-separated by Z. Thus, since Y and X' are also d-separated by Z, we have trivially that Y and

$X' \cup B_o$ are d-separated by Z. Since v_n is not in Y, $X' \cup B_o$, or Z, case 1 therefore implies that

$$I_P(X' \cup B_o, Z, Y).$$

Together with (6.5) and Theorem 6.1(d), this implies that

$$I_P(\{v_n\} \cup X' \cup B_o, Z, Y).$$

Finally, noting that $X = X' \cup \{v_n\}$ and applying Theorem 6.1(b), we have

$$I_P(X, Z, Y).$$

Case 3. $v_n \in Z$. Let $Z' = Z - \{v_n\}$, and let B and R be defined as in case 2. Furthermore, let R_x, R_y, R_z, R_o be a partitioning of the vertices in R and B_x, B_y, B_z, B_o be a partitioning of the vertices in B such that

$$X = B_x \cup R_x$$

$$Y = B_y \cup R_y$$

$$Z' = B_z \cup R_z.$$

Again, since v_n has no descendents, we have by the definition of a causal network that

$$I_P(\{v_n\}, B, R)$$

and therefore

$$I_P(\{v_n\}, B_x \cup B_y \cup B_z \cup B_o, R_x \cup R_y \cup R_z \cup R_o). \tag{6.6}$$

Now either B_y is empty and B_o and Y are d-separated by Z', or B_x is empty and B_o and X are d-separated by Z'. For, if B_y were not empty, there would be an arc from some $y \in Y$ to v_n, and therefore trivially there would be a chain between y and v_n, in which the arc touching v_n points toward v_n, which is not blocked by Z'. Furthermore, if B_o and Y were not d-separated by Z', there would be some $y \in Y$ and $b \in B_o$ and a chain between y and b which is not blocked by Z'. However, by adding the arc from b to v_n to this chain, we would have a chain between y and v_n, in which the arc touching v_n points toward v_n, which is not blocked by Z'. Thus if both conditions were not true, we would need to have some $y \in Y$ such that there would be a chain between y and v_n, in which the arc touching v_n points toward v_n, which is not blocked by Z'. Similarly, if both conditions were not true for X, we would need to have some $x \in X$ such that there would be a chain between x and v_n, in which the arc touching v_n points toward v_n, which is not blocked by Z'. Thus, if neither X nor Y had both conditions true, we would have a chain between x and y, containing two successive arcs meeting head-to-head at v_n, which is not blocked by Z'. However, since $Z = Z' \cup \{v_n\}$, this implies that there would be a chain between x and y which is not blocked by Z, which contradicts the fact that X and Y are d-separated by Z. Thus the two conditions are true either for X or for Y. Without loss of generality, assume B_y is empty and B_o and Y are d-separated by Z'.

Now suppose X and Y are not d-separated by Z'. Then there would be some $x \in X$ and $y \in Y$ and a chain between x and y which is not blocked by Z'. Since X and Y are d-separated by $Z = Z' \cup \{v_n\}$, this would imply that there must be two successive arcs on the chain meeting either head-to-tail or tail-to-tail at v_n. However, since v_n has no descendents and therefore has only incoming arcs, this is not possible. Therefore X and Y are d-separated by Z' in G.

Since B_o and Y are d-separated by Z', and X and Y are d-separated by Z', trivially $X \cup B_o$ and Y are d-separated by Z'. Therefore, owing to case 1, we have

$$I_P(X \cup B_o, Z', Y).$$

Furthermore, since B_y is empty, (6.6) and an argument analogous to that in case 2 implies

$$I_P(\{v_n\}, X \cup B_o \cup Z', Y).$$

We therefore have, owing to Theorem 6.1(d), that

$$I_P(\{v_n\} \cup X \cup B_o, Z', Y).$$

Applying Theorem 6.1(b) and (c) and noting that $Z = Z' \cup \{v_n\}$ finally yields

$$I_P(X, Z, Y). \quad \square$$

Geiger and Pearl [1988] have obtained an important converse to this theorem. They show that, given a DAG $G = (V, E)$, if for all P such that (V, E, P) constitute a causal network it is the case that $I_P(X, Z, Y)$, then X and Y must be d-separated by Z. Thus d-separation identifies all independencies implied by the structure of the DAG.

Theorem 6.2 has only to do with the mathematical definition of a causal network; it has nothing to do with the notion of causality. Consequently, this theorem can be applied to any causal network, regardless of the method used to create the network. In the next two sections we will use this theorem to develop Pearl's [1986a] propagation technique.

PROBLEMS 6.1

1. Complete the proof of (c) in Theorem 6.1.

2. For the graph in Figure 6.8, determine all the vertex pairs which are d-separated 1) by $\{A\}$; 2) by $\{A, G\}$.

COMPUTER PROBLEM 6.1

Write a program which determines all the vertex pairs which are d-separated by a given subset of vertices. The inputs should be the vertices and arcs of the graph and the subset of vertices; the output should be a set of vertex pairs (see Geiger, Verma, & Pearl [1989]).

6.2. PROBABILITY PROPAGATION IN TREES

Section 6.2.1 contains the theoretical development of Pearl's method for propagating probabilities in trees, section 6.2.2 illustrates the method, section 6.2.3 contains a high-level algorithm for the method, and section 6.2.4 discusses virtual evidence. Since section 6.2.1 contains a number of laborious theorems, a reader interested primarily in applying the method should read it only for an understanding of the statements of the theorems. We will develop the theory under the assumption that $P(V) > 0$ for all combinations of values of variables in V. The method is still applicable when we remove this restriction; however, it would obscure the clarity of the theorems to address the special cases, which result from allowing probability values of 0. Therefore we do not allow such values in this initial development of the theory.

6.2.1. Theoretical Development of Probability Propagation in Trees

Thus far we have used lowercase letters near the end of the alphabet (e.g., u, v, and w) to denote variables whose values can be propositional variables, uppercase letters near the front of the alphabet (e.g., A, B, and C) to represent actual propositional variables, and the lowercase letter, corresponding to the uppercase letter used to represent a propositional variable, to represent a value that a propositional variable can take (e.g., b_i for $1 \leq i \leq k$ represents the values B can take if there are k values in the range of B). In the theory which follows, it will often be necessary to refer to the values a propositional variable can take. Accordingly, instead of using letters such as v to represent a variable whose value can be an actual propositional variable, we will now use uppercase letters such as A and B both for that purpose and to represent actual propositional variables.

As mentioned previously, in the definitions, lemmas, and theorems which follow it is assumed that all probabilities are positive.

Definition 6.6. Let $C = (V, E, P)$ be a causal network in which the DAG is a tree, and let W be a subset of variables which are instantiated (i.e., the actual values assumed by these variables are known). For $B \in V$, we will denote

W_B^- to mean the subset of W in the tree rooted at B
W_B^+ to mean $W - W_B^-$.

Example 6.7. Suppose in the graph in Figure 6.9, $W = \{F, E, J, D\}$. Then $W_B^- = \{J, D\}$ and $W_B^+ = \{F, E\}$.

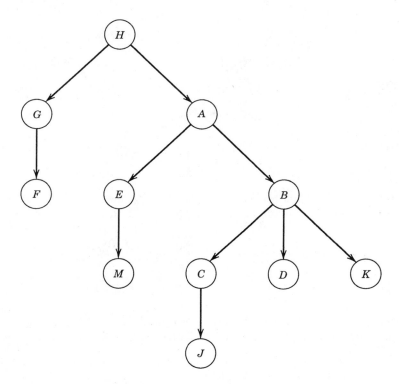

FIGURE 6.9   A DAG which is a tree.

Lemma 6.1. Let $C = (V, E, P)$ be a causal network in which the DAG is a tree and let W be a subset of instantiated variables. In addition, let $B \in V - W$ be a propositional variable with k possible values, and b_i, such that $1 \leq i \leq k$, be the possible values of B. Then for $1 \leq i \leq k$,

$$P(b_i \mid W) = \beta_B P(W_B^- \mid b_i) P(b_i \mid W_B^+),$$

where β_B is a constant which depends on B and the instantiated values in W but not on i.

Proof. First recall that the event W is the event that all the variables in W take specific values. Therefore, if W is the empty set, $P(W)$ and $P(W \mid b_i)$ are both equal to 1. The same is true for W_B^+ and W_B^- if either of these sets is empty. The proof below is therefore valid even if any of these sets are empty.

The proof is as follows: We have that

$$P(b_i \mid W) = P(b_i \mid W_B^- \cup W_B^+)$$

$$= \frac{P(W_B^- \cup W_B^+ \cup \{b_i\})}{P(W_B^- \cup W_B^+)}$$

$$= \frac{P(W_B^- \cup W_B^+ \mid b_i)P(b_i)}{P(W_B^- \cup W_B^+)}$$

$$= \frac{P(W_B^- \mid b_i)P(W_B^+ \mid b_i)P(b_i)}{P(W_B^- \cup W_B^+)}.$$

The last inequality is due to Theorem 6.2, since W_B^- and W_B^+ are d-separated by $\{B\}$ if $B \notin W$. We therefore have, by the definition of conditional probability, that

$$P(b_i \mid W) = \frac{P(W_B^- \mid b_i)P(b_i \mid W_B^+)P(W_B^+)}{P(W_B^- \cup W_B^+)},$$

which proves the theorem if we let $\beta_B = P(W_B^+)/P(W_B^- \cup W_B^+)$. $\square$

Definition 6.7. Let $C = (V, E, P)$ be a causal network in which the graph is a tree, let W be a subset of instantiated variables, and let $B \in V$ have k possible values. Then we define for $1 \le i \le k$,

$$\lambda(b_i) = \begin{cases} P(W_B^- \mid b_i) & \text{if} \quad B \notin W \\ 1 & \text{if} \quad B \in W \text{ and } b_i \text{ is the instantiated value} \\ 0 & \text{if} \quad B \in W \text{ and } b_i \text{ is not the instantiated value} \end{cases}$$

$$\pi(b_i) = P(b_i \mid W_B^+).$$

The entire vector of values, $\lambda(b_i)$ for $1 \le i \le k$, is called B's λ value and is denoted $\lambda(B)$. Likewise, the vector of values, $\pi(b_i)$ for $1 \le i \le k$, is called B's π value and is denoted $\pi(B)$. $\lambda(B)$ and $\pi(B)$ clearly depend on W. However, we will omit that reference for the sake of brevity.

Theorem 6.3. Let $C = (V, E, P)$ be a causal network in which the graph is a tree, W a subset of instantiated variables, and $B \in V$ a variable with k possible values. Then

1. If $B \notin W$ and B is a leaf (i.e., a vertex with no children), then

$$\lambda(b_i) = 1 \quad \text{for} \quad 1 \le i \le k.$$

2. If B is the root, then

$$\pi(b_i) = P(b_i) \quad \text{for} \quad 1 \le i \le k.$$

Proof

1. If B is a leaf and $B \notin W$, then W_B^- is the empty set and therefore
$$P(W_B^- \mid b_i) = 1 \qquad \text{for} \quad 1 \le i \le k.$$

2. If B is the root, then W_B^+ is the empty set and therefore
$$P(b_i \mid W_B^+) = P(b_i) \qquad \text{for} \quad 1 \le i \le k. \qquad \square$$

Theorem 6.4. Let $C = (V, E, P)$ be a causal network in which the graph is a tree, let W be a subset of instantiated variables, and let $B \in V$ have k possible values. Then for $1 \le i \le k$,

$$P(b_i \mid W) = \alpha_B \lambda(b_i) \pi(b_i),$$

where α_B is a constant depending on B and the instantiated values in W, but not on i.

Proof

Case 1. $B \notin W$. This case follows trivially from Lemma 6.1 and Definition 6.7 by letting $\alpha_B = \beta_B$.

Case 2. $B \in W$. Let b_j be the instantiated value of B. Then

$$P(b_i \mid W) = \begin{pmatrix} 1 & \text{if} \quad i = j \\ 0 & \text{if} \quad i \ne j. \end{pmatrix}$$

Due to Definition 6.7, $\lambda(b_j) = 1$ and $\lambda(b_i) = 0$ for $i \ne j$. The proof therefore follows if we let $\alpha_B = 1/\pi(b_j)$. $\quad \square$

Definition 6.8. Let $G = (V, E)$ be DAG. For $B \in V$, we will denote the subset of all children of B by $s(B)$.

In the graph in Figure 6.9, $s(B) = \{C, D, K\}$.

Lemma 6.2. Let $C = (V, E, P)$ be a causal network in which the graph is a tree, W a subset of instantiated variables, and $B \in V - W$ a variable with k possible values. Then for $1 \le i \le k$, we have

$$\lambda(b_i) = \prod_{C \in s(B)} P(W_C^- \mid b_i),$$

where, if B is a leaf and therefore there are no terms in the product, it is understood that the product represents the value 1.

Proof

Case 1. W_B^- is empty. In this case W_C^- is also empty for all children C

of B. Therefore for $1 \leq i \leq k$ and for all children C of B,

$$P(W_B^- \mid b_i) = 1 \qquad \text{and} \qquad P(W_C^- \mid b_i) = 1.$$

Case 2. W_B^- is not empty. Since $B \notin W$, we have

$$W_B^- = \bigcup_{C \in s(B)} W_C^-,$$

where for some $C \in s(B)$, W_C^- may be empty. Then, since $\{B\}$ d-separates all the subsets W_C^-, for $C \in s(B)$, from each other, the lemma follows by repeatedly applying Theorem 6.2, and noting that $P(W_C^- \mid b_i) = 1$ if W_C^- is empty. $\square$

Notice that the first result in Theorem 6.3 follows as a corollary of Lemma 6.2. However, because of its importance, that result is placed in a separate theorem.

Definition 6.9. Let $C = (V, E, P)$ be a causal network in which the graph is a tree, W a subset of instantiated variables, $B \in V$ a variable with k possible values, and $C \in s(B)$ a child of B with m possible values. Then for $1 \leq i \leq k$, we define

$$\lambda_C(b_i) = \sum_{j=1}^{m} P(c_j \mid b_i) \lambda(c_j).$$

The entire vector of values, $\lambda_C(b_i)$ for $1 \leq i \leq k$, is called the λ message from C to B and is denoted $\lambda_C(B)$.

The reason $\lambda_C(B)$ is called a message is that its value is obtained from information about C (namely C's λ value) and is meaningful to B. We will see in Theorem 6.5 why this message is meaningful to B.

Lemma 6.3. Let $C = (V, E, P)$ be a causal network in which the graph is a tree, W a subset of instantiated variables, $B \in V$ a variable with k possible values, and $C \in s(B)$ a child of B. Then for $1 \leq i \leq k$, we have

$$P(W_C^- \mid b_i) = \sigma_C \lambda_C(b_i),$$

where σ_C is a constant which depends on C and on the instantiated values in W, but not on i.

Proof

Case 1. $C \notin W$. Suppose C has m possible values. Since the events c_j, for $1 \leq j \leq m$, are mutually exclusive and exhaustive we have, owing to Lemma 2.1, that

$$P(W_C^- \mid b_i) = \sum_{j=1}^{m} P(W_C^- \mid b_i, c_j) P(c_j \mid b_i). \qquad (6.7)$$

Since in this case $C \notin W$, W_C^- and $\{B\}$ are d-separated by $\{C\}$. Therefore by

Theorem 6.2,

$$P(W_C^- \mid b_i, c_j) = P(W_C^- \mid c_j)$$
$$= \lambda(c_j),$$

which, together with (6.7), completes the proof if we let $\sigma_C = 1$.

Case 2. $C \in W$. Let c_n be the instantiated value of C, and $X = W_C^- - \{C\}$. We then have, by the definition of conditional probability,

$$P(W_C^- \mid b_i) = P(X \cup \{c_n\} \mid b_i)$$
$$= \frac{P(b_i \mid X \cup \{c_n\}) P(X \cup \{c_n\})}{P(b_i)}$$
$$= \frac{P(b_i \mid c_n) P(X \cup \{c_n\})}{P(b_i)}.$$

The last equality is due to X and $\{B\}$ being d-separated by $\{C\}$. Therefore, again by the definition of conditional probability,

$$P(W_C^- \mid b_i) = \frac{P(c_n \mid b_i) P(b_i) P(X \cup \{c_n\})}{P(c_n) P(b_i)}$$
$$= P(c_n \mid b_i) P(X \mid c_n).$$

If we now let $\sigma_C = P(X \mid c_n)$, this case is proved by noting that $\lambda(c_n) = 1$ and $\lambda(c_j) = 0$ for $j \neq n$. $\square$

Lemma 6.4. Let $C = (V, E, P)$ be a causal network in which the graph is a tree, W a subset of instantiated variables, $B \in V$ a variable with k possible values, and $C \in s(B)$ a child of B such that W_C^- is empty. Then for $1 \leq i \leq k$, we have

$$\lambda_C(b_i) = 1.$$

Proof. Since in this case $C \notin W$, looking at the proof in Lemma 6.3 we see that $\sigma_C = 1$. Therefore for $1 \leq i \leq k$,

$$P(W_C^- \mid b_i) = \lambda_C(b_i).$$

Since W_C^- is empty, $P(W_C^- \mid b_i) = 1$ for $1 \leq i \leq k$, which completes the proof. $\square$

Theorem 6.5. Let $C = (V, E, P)$ be a causal network in which the graph is a tree, W a subset of instantiated variables, and $B \in V$ a variable with k possible values. Then for $1 \leq i \leq k$, we have

$$\lambda(b_i) = \begin{cases} \rho_B \displaystyle\prod_{C \in s(B)} \lambda_C(b_i) & \text{if } B \notin W \\ 1 & \text{if } B \in W \text{ and } B \text{ is instantiated for } b_i \\ 0 & \text{if } B \in W \text{ and } B \text{ is not instantiated for } b_i, \end{cases}$$

where ρ_B is a constant depending on B and the instantiated values in W but not on i, and $\lambda_C(b_i)$ is defined in Definition 6.9. It is understood that if $s(B)$ is empty and therefore there are no terms in the product, the product represents the value 1.

Proof. The proof follows immediately from Definition 6.7 if $B \in W$. Hence suppose $B \notin W$. Lemmas 6.2 and 6.3 imply that

$$\lambda(b_i) = \prod_{C \in s(B)} \sigma_C \lambda_C(b_i).$$

The proof is now completed by letting $\rho_B = \prod_{C \in s(B)} \sigma_C$. $\square$

Definition 6.10. Let $C = (V, E, P)$ be a causal network in which the graph is a tree, W a subset of instantiated variables, $B \in V$ a variable which is not the root, and $A \in V$ the father of B. Suppose A has m possible values. Then we define for $1 \le j \le m$,

$$
\pi_B(a_j) = \begin{cases}
1 & \text{if } A \text{ is instantiated for } a_j \\
0 & \text{if } A \text{ is instantiated, but not for } a_j \\
\pi(a_j) \prod_{\substack{C \in s(A) \\ C \ne B}} \lambda_C(a_j) & \text{if } A \text{ is not instantiated,}
\end{cases}
$$

where $\lambda_C(a_j)$ is defined in Definition 6.9. Again, if there are no terms in the product, it is meant to represent the value 1. The entire vector of values, $\pi_B(a_j)$ for $1 \le j \le m$, is called the π message from A to B and is denoted $\pi_B(A)$.

The reason $\pi_B(A)$ is called a message is that its value is obtained from information about A (namely A's π value and its λ messages from its other children) and is meaningful to B. We will see in Theorem 6.6 why this message is meaningful to B.

Theorem 6.6. Let $C = (V, E, P)$ be a causal network in which the graph is a tree, W a subset of instantiated variables, $B \in V$ a variable, which is not the root, with k possible values, and $A \in V$ B's parent. Suppose A has m possible values. Then for $1 \le i \le k$,

$$\pi(b_i) = \mu_B \sum_{j=1}^{m} P(b_i \mid a_j) \pi_B(a_j),$$

where μ_B is a constant which depends on B and on the instantiated values in W but not on i, and $\pi_B(a_j)$ is defined in Definition 6.10.

Proof

Case 1. $A \notin W$. Since the events a_j, for $1 \leq j \leq m$, are mutually exclusive and exhaustive, by Lemma 2.1 we have for $1 \leq i \leq k$ that

$$\pi(b_i) = P(b_i \mid W_B^+)$$

$$= \sum_{j=1}^{m} P(b_i \mid W_B^+ \cup \{a_j\}) P(a_j \mid W_B^+)$$

$$= \sum_{j=1}^{m} P(b_i \mid a_j) P(a_j \mid W_B^+). \tag{6.8}$$

The last equality is due to W_B^+ and $\{B\}$ being d-separated by $\{A\}$ in this case. Let W' equal W_B^+, and let C be a descendent of A which is not in the tree rooted at B. Since clearly $W_C'^- = W_C^-$, we have by Definition 6.7, if c_n is a possible value of C, that

$$\lambda'(c_n) = P(W_C'^- \mid c_n) = P(W_C^- \mid c_n) = \lambda(c_n). \tag{6.9}$$

Next, Theorem 6.5 implies that for $1 \leq j \leq m$,

$$\lambda'(a_j) = \rho_A \prod_{C \in s(A)} \lambda_C'(a_j).$$

Since there are no variables in W' from the tree rooted at B, we have due to Lemma 6.4 that $\lambda_B'(a_j) = 1$ for $1 \leq j \leq m$. Therefore

$$\lambda'(a_j) = \rho_A \prod_{\substack{C \in s(A) \\ C \neq B}} \lambda_C'(a_j).$$

However, (6.9) and Definition 6.9 imply that for $C \in s(A)$ and $C \neq B$,

$$\lambda_C'(a_j) = \lambda_C(a_j)$$

for $1 \leq j \leq m$. We thus have for $1 \leq j \leq m$,

$$\lambda'(a_j) = \rho_A \prod_{\substack{C \in s(A) \\ C \neq B}} \lambda_C(a_j).$$

Next it is easily seen that $W_A'^+ = W_A^+$. Therefore for $1 \leq j \leq m$,

$$\pi'(a_j) = \pi(a_j).$$

Owing to Theorem 6.4 we therefore have for $1 \leq j \leq m$,

$$P(a_j \mid W') = \alpha_A \pi(a_j) \rho_A \prod_{\substack{C \in s(A) \\ C \neq B}} \lambda_C(a_j) = \alpha_A \rho_A \pi_B(a_j).$$

Recalling that $W' = W_B^+$, we thus have for $1 \leq j \leq m$,

$$P(a_j \mid W_B^+) = \alpha_A \rho_A \pi_B(a_j),$$

which together with (6.8) proves this case if we let $\mu_B = \alpha_A \rho_A$.

Case 2. $A \in W$. Since $\{A\}$ d-separates $W_B^+ - \{A\}$ from B, we have

$$\pi(b_i) = P(b_i \mid W_B^+)$$
$$= P(b_i \mid a_n),$$

where a_n is the instantiated value of A. Since, in this case $\pi_B(a_n) = 1$, and for $j \neq n$, $\pi_B(a_j) = 0$, the proof follows by letting $\mu_B = 1$. $\square$

We need one final theorem:

Theorem 6.7. Let $C = (V, E, P)$ be a causal network in which the graph is a tree, W a subset of instantiated variables, $B \in V$ a variable which is not the root and which is not instantiated, and $A \in V$ the parent of B. Suppose A has m possible values. Then for $1 \leq j \leq m$,

$$\pi_B(a_j) = \tau_B \frac{P(a_j \mid W)}{\lambda_B(a_j)},$$

where τ_B is a constant which depends on B and on the instantiated values in W, but not on j.

Proof. Suppose B has k possible values. Since B is not instantiated, then by Definition 6.7, for $1 \leq i \leq k$,

$$\lambda(b_i) = P(W_B^- \mid b_i),$$

which is positive, since we are assuming that all probabilities are positive. Therefore, owing to Definition 6.9 and, again, since all probabilities in the joint distribution are assumed to be positive, we have for $1 \leq j \leq m$ that

$$\lambda_B(a_j) > 0.$$

The theorem will now be proved by considering two cases (case 2 will not be needed in Pearl's propagation method, but is included for completeness):

Case 1. $A \notin W$. By Theorems 6.4 and 6.5, we then have for $1 \leq j \leq m$,

$$P(a_j \mid W) = \alpha_A \rho_A \left(\prod_{C \in s(A)} \lambda_C(a_j) \right) \pi(a_j).$$

The proof now follows due to Definition 6.10 by letting $\tau_B = 1/\alpha_A \rho_A$.

Case 2. A is instantiated for a_n. Then, clearly for $j \neq n$,

$$P(a_j \mid W) = 0$$

and

$$P(a_n \mid W) = 1.$$

If we let $\tau_B = \lambda_B(a_n)$, the theorem follows owing to Definition 6.10. $\square$

6.2.2. Illustrating the Method of Probability Propagation in Trees

We are rewarded for our efforts developing the theory in section 6.2.1 with an elegant method for determining the probabilities of remaining variables given that certain variables have realized specific values. This method has the desirable property, mentioned in section 6.1, of determining the probabilities only through communication between neighboring vertices. Additionally, this method has the property that it does not matter in what order the intervertex communication is performed. The result is that we can compute the a priori probabilities of each variable (i.e., the probability based on the instantiation of no variables) using the method. This would leave the network in a certain state, say S_1. When a variable becomes instantiated, the method can be applied to the network, in state S_1, to determine the conditional probability of each variable given the instantiation of the one variable. This would leave the network in a new state, say S_2. When a second variable becomes instantiated, the method can be applied to the network, in state S_2, to determine the conditional probability of each variable given the instantiation of the two variables. This process can continue for as many instantiations as desired. The resultant conditional probabilities do not depend on the order in which the variables are instantiated. In this subsection, the process will be illustrated by centering on a specific example. In section 6.2.3, a high-level algorithm for the method is given.

The only theory we need for the propagation method are Definitions 6.9 and 6.10, Lemma 6.4, and Theorems 6.3 through 6.7. The purpose of the other theory in this chapter was to obtain these results. Accordingly, we will only refer to the definitions, lemma, and theorems just identified. For illustration, we will consider the following elaboration of Example 5.8. Let the propositional variables A, B, and C be defined as in that example, and let D be a propositional variable with the following possible values:

$$d_1 = \text{strange man/lady calls on the phone}$$

$$d_2 = \text{no strange man/lady calls on the phone.}$$

Furthermore, let the DAG in Figure 6.10 represent the causal relationships among the propositional variables. The conditional probabilities are stored in Figure 6.10 so that the figure represents the entire causal network.

For illustration, we show first how to compute the a priori probability of each variable using the propagation method. Of course, since Theorem 5.2 implies that the joint distribution is just the product of the specified conditional distributions, we could compute the probabilities in the joint distribution directly from the conditional distributions. The probability of each

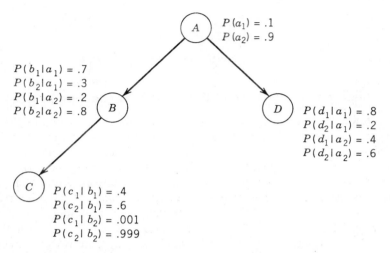

$P(a_1) = .1$
$P(a_2) = .9$

$P(b_1|a_1) = .7$
$P(b_2|a_1) = .3$
$P(b_1|a_2) = .2$
$P(b_2|a_2) = .8$

$P(d_1|a_1) = .8$
$P(d_2|a_1) = .2$
$P(d_1|a_2) = .4$
$P(d_2|a_2) = .6$

$P(c_1|b_1) = .4$
$P(c_2|b_1) = .6$
$P(c_1|b_2) = .001$
$P(c_2|b_2) = .999$

FIGURE 6.10   The causal network for the cheating spouse example.

variable could then be computed as a marginal value. However, this is exactly what we wish to avoid. As mentioned in chapter 5, in general this procedure is not computationally feasible. That is why it was necessary to devise the propagation method.

Recall that each variable passes a λ message to its parent and a π message to each of its children. The graph is therefore actually a weighted graph with these values resident on the arcs. Furthermore, recall that each variable has a λ value and a π value associated with it. The entire network thus consists of the conditional probabilities stored at each variable (see Figure 6.10), the λ and π values stored at each variable, and the λ and π messages stored at each arc. The π and λ values and the π and λ messages are depicted in Figure 6.11.

Theorem 6.5 gave a formula for computing a variable's λ value from the λ messages it receives from its children, Theorem 6.6 gave a formula for computing a variable's π value from the π message it receives from its parent, Theorem 6.4 gave a formula for computing the conditional probability of a variable from its λ value and π value, and Theorem 6.3 gave a formula for the π value of the root and a formula for the λ values of the leaves. We will now see how to compute the a priori probabilities of each variable by using these formulas and propagating the λ and π messages. Using the notation of section 6.2.1, we will denote the set of instantiated vertices by W. When computing the a priori probabilities, W is the empty set. We compute $P(B \mid W)$ (i.e., $P(B)$) first.

By Theorem 6.3,

$$\lambda(c_1) = 1 \quad \text{and} \quad \lambda(c_2) = 1$$
$$\lambda(d_1) = 1 \quad \text{and} \quad \lambda(d_2) = 1.$$

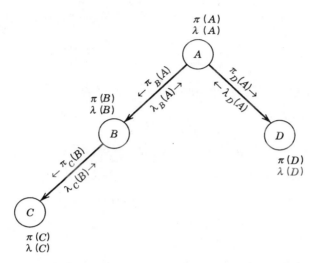

FIGURE 6.11 The DAG from Figure 6.10 with λ and π values and λ and π messages indicated.

Therefore, by Definition 6.9,

$$\lambda_C(b_1) = P(c_1 \mid b_1)\lambda(c_1) + P(c_2 \mid b_1)\lambda(c_2)$$
$$= .4(1) + .6(1) = 1$$
$$\lambda_C(b_2) = P(c_1 \mid b_2)\lambda(c_1) + P(c_2 \mid b_2)\lambda(c_2)$$
$$= .001(1) + .999(1) = 1$$
$$\lambda_D(a_1) = P(d_1 \mid a_1)\lambda(d_1) + P(d_2 \mid a_1)\lambda(d_2)$$
$$= .8(1) + .2(1) = 1$$
$$\lambda_D(a_2) = P(d_1 \mid a_2)\lambda(d_1) + P(d_2 \mid a_2)\lambda(d_2)$$
$$= .4(1) + .6(1) = 1.$$

We could have used Lemma 6.4 to determine immediately that all those λ messages have value 1. However, we are now demonstrating a general technique for determining the values of λ messages.

When B receives its λ message from its only child C, we can compute B's λ values using Theorem 6.5 as follows:

$$\lambda(b_1) = \rho_B\lambda_C(b_1) = \rho_B$$
$$\lambda(b_2) = \rho_B\lambda_C(b_2) = \rho_B.$$

We will see that it is not necessary to determine the value of the constant ρ_B.

Using Theorem 6.3 we can compute the π value of the root A as follows:

$$\pi(a_1) = P(a_1) = .1$$
$$\pi(a_2) = P(a_2) = .9.$$

Using Definition 6.10 we can compute the π message which B receives from A:

$$\pi_B(a_1) = \pi(a_1)\lambda_D(a_1) = .1(1) = .1$$
$$\pi_B(a_2) = \pi(a_2)\lambda_D(a_2) = .9(1) = .9.$$

Next, using Theorem 6.6, we can compute B's π value:

$$\pi(b_1) = \mu_B\left(P(b_1 \mid a_1)\pi_B(a_1) + P(b_1 \mid a_2)\pi_B(a_2)\right)$$
$$= \mu_B\left((.7)(.1) + (.2)(.9)\right)$$
$$= \mu_B(.25)$$
$$\pi(b_2) = \mu_B\left(P(b_2 \mid a_1)\pi_B(a_1) + P(b_2 \mid a_2)\pi_B(a_2)\right)$$
$$= \mu_B\left((.3)(.1) + (.8)(.9)\right)$$
$$= \mu_B(.75).$$

Owing to Definition 6.7, we could eliminate the constant μ_B by normalization. However, we shall see that this is unnecessary.

Finally, using Theorem 6.4, we have, since $W = \emptyset$,

$$P(b_1) = P(b_1 \mid W) = \alpha_B\lambda(b_1)\pi(b_1)$$
$$= \alpha_B\rho_B\mu_B(.25)$$
$$= \alpha(.25)$$
$$P(b_2) = P(b_2 \mid W) = \alpha_B\lambda(b_2)\pi(b_2)$$
$$= \alpha_B\rho_B\mu_B(.75)$$
$$= \alpha(.75),$$

where α is defined to be $\alpha_B\rho_B\mu_B$. Since $P(b_1) + P(b_2) = 1$, we then can get rid of the normalizing constant α to obtain

$$P(b_1) = \frac{\alpha(.25)}{\alpha(.25) + \alpha(.75)} = .25$$

$$P(b_2) = \frac{\alpha(.75)}{\alpha(.25) + \alpha(.75)} = .75.$$

Notice that the constant α, which is the product of all our other constants, cancels out of the expression for $P(B)$. Since our only interest, relative to B, is its probability, and since the constants do not enter into the determination of that probability, there is no reason to concern ourselves with the constants when computing that probability. In the same manner, it is easily seen that these constants are meaningless to the probabilities of other variables in the network. Therefore we will henceforth drop the constants α_B, ρ_B, and μ_B when applying Theorems 6.4, 6.5, and 6.6 respectively. We need only include one constant, α, in the final expression for the probability of a variable and eliminate that constant by normalization at the end.

We compute one more a priori probability, namely $P(A)$, to illustrate the principles further. We already know that

$$\lambda_D(A) = (1, 1),$$

where we are using a shorthand vector notation to state that $\lambda_D(a_1)$ is 1 and $\lambda_D(a_2)$ is 1. By Definition 6.9,

$$\lambda_B(a_1) = P(b_1 \mid a_1)\lambda(b_1) + P(b_2 \mid a_1)\lambda(b_2)$$
$$= (.7)1 + (.3)1 = 1$$
$$\lambda_B(a_2) = P(b_1 \mid a_2)\lambda(b_1) + P(b_2 \mid a_2)\lambda(b_2)$$
$$= (.2)1 + (.8)1 = 1.$$

We therefore have

$$\lambda(a_1) = \lambda_B(a_1)\lambda_D(a_1)$$
$$= (1)(1) = 1$$
$$\lambda(a_2) = \lambda_B(a_2)\lambda_D(a_2)$$
$$= (1)(1) = 1.$$

Finally,

$$P(a_1) = P(a_1 \mid W) = \alpha\lambda(a_1)\pi(a_1)$$
$$= \alpha(1)(.1) = \alpha(.1)$$
$$P(a_2) = P(a_2 \mid W) = \alpha\lambda(a_1)\pi(a_1)$$
$$= \alpha(1)(.9) = \alpha(.9).$$

After normalization, we have $P(A) = (.1, .9)$.

There was really no need to compute the a priori probability of A, since A is the root and therefore its a priori probability is stored in the network. Again, it was done to illustrate the method. It is left as an exercise to compute the remaining a priori probabilities. When this is done, the network is left in the state depicted in Figure 6.12. This is the state of the network before any variables are instantiated. Notice that all λ values and λ messages are 1. Before any variables are instantiated, all λ messages are always 1 due to Lemma 6.4 and, thus, by Theorem 6.5, all λ values are also equal to 1 (up to an unnecessary multiplicative constant). We will incorporate this fact into

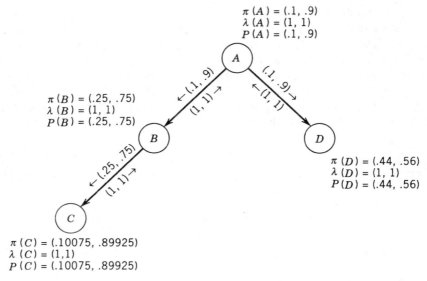

$\pi(A) = (.1, .9)$
$\lambda(A) = (1, 1)$
$P(A) = (.1, .9)$

$\pi(B) = (.25, .75)$
$\lambda(B) = (1, 1)$
$P(B) = (.25, .75)$

$\pi(D) = (.44, .56)$
$\lambda(D) = (1, 1)$
$P(D) = (.44, .56)$

$\pi(C) = (.10075, .89925)$
$\lambda(C) = (1,1)$
$P(C) = (.10075, .89925)$

FIGURE 6.12   The state of the causal network in the cheating spouse example before any variables are instantiated.

our actual propagation algorithm in section 6.2.3. The calculations were done above for the sake of illustration.

Suppose B is now instantiated for b_1. Owing to Theorem 6.5, this would change B's λ value to $(1,0)$, and, owing to Definition 6.10, change B's π message to C to $(1,0)$. The question is how this affects the probabilities of the remaining variables in the network. That is, if W is now $\{B\}$, what is $P(E \mid W)$ for every variable E in the network? Remember, owing to Theorem 6.4, the probability of every variable in the network can be computed from the variable's λ value and its π value. So all we need determine is how the instantiation of B affects the λ and π values of the other variables. We shall see that it is through the propagation of new λ and π messages.

B's parent A obtains a new λ value owing to its new λ message from B and Theorem 6.5. However, A's π value is unchanged. In the same way, if A had a parent, A's new λ values would cause A to send a new λ message to that parent, thus changing the parent's λ value. However, the parent's π value would remain unchanged. If A's parent had a parent, it too would receive a new λ message, and so on up the tree until the root was reached. Thus, when a variable is instantiated, the λ values of all its ancestors are changed owing to receiving new λ messages, but their π values are unchanged. B's new λ message to A also has an effect on A's other children. They each receive a new π message owing to Definition 6.10 and consequently their π values are changed owing to Theorem 6.6. If A's children had children, the change in A's children's π values would in turn cause a change in the π messages they sent to their children. Thus A's children's children would receive new π values. This propagation continues down the tree until leaves are reached. Finally, B's new π message to C causes C's π value to change owing to Theorem 6.6. If C had children, the propagation of π messages would again continue down the tree until leaves were reached.

Hence we see that the new conditional probabilities, based on the instantiation of B, can be determined from the propagation of λ and π messages. After all propagation is finished (i.e., λ messages are absorbed at the root and π messages at the leaves), the network reaches a new state based on the instantiation of B. We have thus accomplished the goal of computing the conditional probability based on the instantiation of one variable by intervertex communication. Before discussing the instantiation of more than one variable, we illustrate the propagation scheme further by actually instantiating B:

Suppose B is instantiated for b_1. Then $W = \{B\}$ and we have

$$\lambda(B) = (1,0)$$

$$P(B \mid W) = (1,0).$$

B sends a new π message to C:

$$\pi_C(B) = (1,0).$$

C therefore obtains a new π value:

$$\pi(c_1) = P(c_1 \mid b_1)\pi_C(b_1) + P(c_1 \mid b_2)\pi_C(b_2)$$
$$= .4(1) + .001(0) = .4$$
$$\pi(c_2) = P(c_2 \mid b_1)\pi_C(b_1) + P(c_2 \mid b_2)\pi_C(b_2)$$
$$= .6(1) + .999(0) = .6.$$

$P(C \mid W)$ is then obtained as follows:

$$P(c_1 \mid W) = \alpha\lambda(c_1)\pi(c_1)$$
$$= \alpha(1)(.4) = \alpha(.4)$$
$$P(c_2 \mid W) = \alpha\lambda(c_2)\pi(c_2)$$
$$= \alpha(1)(.6) = \alpha(.6).$$

Normalizing, we obtain $P(C \mid W) = (.4, .6)$.

B sends a new λ message to A:

$$\lambda_B(a_1) = P(b_1 \mid a_1)\lambda(b_1) + P(b_2 \mid a_1)\lambda(b_2)$$
$$= .7(1) + .3(0) = .7$$
$$\lambda_B(a_2) = P(b_1 \mid a_2)\lambda(b_1) + P(b_2 \mid a_2)\lambda(b_2)$$
$$= .2(1) + .8(0) = .2.$$

A therefore obtains a new λ value:

$$\lambda(a_1) = \lambda_B(a_1)\lambda_C(a_1)$$
$$= .7(1) = .7$$
$$\lambda(a_2) = \lambda_B(a_2)\lambda_C(a_2)$$
$$= .2(1) = .2.$$

$P(A \mid W)$ is then obtained as follows:

$$P(a_1 \mid W) = \alpha\lambda(a_1)\pi(a_1)$$
$$= \alpha(.7)(.1) = \alpha(.07)$$
$$P(a_2 \mid W) = \alpha\lambda(a_2)\pi(a_2)$$
$$= \alpha(.2)(.9) = \alpha(.18).$$

Normalizing, we have

$$P(a_1 \mid W) = \frac{.07}{.07 + .18} = .28$$
$$P(a_2 \mid W) = \frac{.18}{.07 + .18} = .72.$$

A sends a new π message to D. We could use Definition 6.10 to determine this π message. However, if A had many children, this would call for multiplying many λ messages. The purpose of Theorem 6.7 is to avoid this in the case where all probabilities are known to be positive. On account of that theorem, we can use $P(A \mid W)$, which is now known, to determine the new π message to D. We have

$$\pi_D(a_1) = \tau_D \frac{P(a_1 \mid W)}{\lambda_D(a_1)} = \tau_D \frac{.28}{1} = \tau_D(.28)$$
$$\pi_D(a_2) = \tau_D \frac{P(a_2 \mid W)}{\lambda_D(a_2)} = \tau_D \frac{.72}{1} = \tau_D(.72).$$

As is the case with the other constants, the constant τ_D has no bearing on our final results. We will therefore drop this constant just as we did with the others. Thus

$$\pi_D(A) = (.28, .72).$$

D's new π value is now given by

$$\pi(d_1) = P(d_1 \mid a_1)\pi_D(a_1) + P(d_1 \mid a_2)\pi_D(a_2)$$
$$= .8(.28) + .4(.72) = .512$$
$$\pi(d_2) = P(d_2 \mid a_1)\pi_D(a_1) + P(d_2 \mid a_2)\pi_D(a_2)$$
$$= .2(.28) + .6(.72) = .488.$$

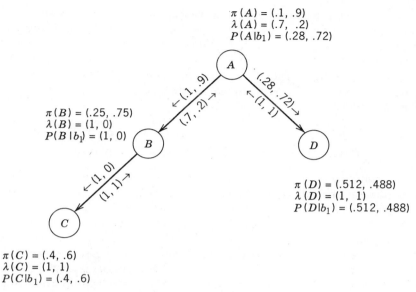

$\pi(A) = (.1, .9)$
$\lambda(A) = (.7, .2)$
$P(A|b_1) = (.28, .72)$

$\pi(B) = (.25, .75)$
$\lambda(B) = (1, 0)$
$P(B|b_1) = (1, 0)$

$\pi(D) = (.512, .488)$
$\lambda(D) = (1, 1)$
$P(D|b_1) = (.512, .488)$

$\pi(C) = (.4, .6)$
$\lambda(C) = (1, 1)$
$P(C|b_1) = (.4, .6)$

FIGURE 6.13   The state of the causal network in the cheating spouse example after B is instantiated.

Computing $P(D \mid W)$ and normalizing, we finally obtain

$$P(D \mid W) = (.512, .488).$$

The network will now be in a new state based on the instantiation of only B. That state is depicted in Figure 6.13.

Next we investigate the instantiation of more than one variable. Suppose the network is in its initial state (depicted in Figure 6.12) and C and D are both instantiated. C sends a new λ message to B, which causes B to send a new λ message to A. D also sends a new λ message to A. Each of these messages does not affect the other. Rather they each cause A to post a new π message to the other. The new π message B receives from A has nothing to do with the λ message B sends to A. The messages simply cross on the arc from A to B. The order of this updating is thus immaterial to B and, if B had other children, to those children. Each new λ message also changes A's λ value due to Theorem 6.5. However, since A's λ value is simply the product of its λ messages, again the order in which these new λ messages are posted is immaterial to A and, if A had any ancestors, to A's ancestors and those ancestors' other descendents. When C is instantiated, if C had children, C would post a new π message to those children. Owing to Definition 6.10, this message is now fixed. Therefore the instantiation of D will not affect this message. Hence the order in which C and D are instantiated also does not matter to C's descendents.

We see then that the order in which variables are instantiated and therefore the order in which new λ and π messages are posted are immaterial. Thus we have achieved the goal of being able to instantiate variables individually in whatever order the information is received. We will illustrate this by instantiating another variable in the cheating spouse example. However, first we summarize the propagation method in a high-level algorithm.

6.2.3. An Algorithm for Probability Propagation in Trees

We assume that a causal network is given and that we wish to compute the conditional probability of each variable in the network given that certain variables are instantiated. The following algorithm accomplishes this by instantiating each variable in sequence. As we have seen, the order in which the variables are instantiated does not matter. For a given variable B in the network, $P'(B)$ is used to denote the conditional probability of E based on the variables thus far instantiated. Initially, $P'(B)$ is the a priori probability. Lemma 6.4 and Theorems 6.3 through 6.7 can be used to establish the validity of this algorithm in the same way as they were used in the illustrations in the previous section.

PROBABILITY PROPAGATION IN TREES

Operative Formulas

1. If B is a child of A, B has k possible values, and A has m possible values, then for $1 \leq j \leq m$ the λ message from B to A is given by

$$\lambda_B(a_j) = \sum_{i=1}^{k} P(b_i \mid a_j)\lambda(b_i).$$

2. If B is a child of A and A has m possible values, then for $1 \leq j \leq m$ the π message from A to B is given by

$$\pi_B(a_j) = \begin{cases} 1 & \text{if } A \text{ is instantiated for } a_j \\ 0 & \text{if } A \text{ is instantiated, but not for } a_j \\ \dfrac{P'(a_j)}{\lambda_B(a_j)} & \text{if } A \text{ is not instantiated,} \end{cases}$$

where $P'(a_j)$ is defined to be the current conditional probability of a_j based on the variables thus far instantiated.

3. If B is a variable with k possible values, $s(B)$ is the set of B's children, then for $1 \leq i \leq k$ the λ value of B is given by

$$
\lambda(b_i) = \begin{cases} \displaystyle\prod_{C \in s(B)} \lambda_C(b_i) & \text{if } B \text{ is not instantiated} \\ 1 & \text{if } B \text{ is instantiated for } b_i \\ 0 & \text{if } B \text{ is instantiated, but not for } b_i. \end{cases}
$$

4. If B is a variable with k possible values, A is the parent of B, and A has m possible values, then for $1 \leq i \leq k$ the π value of B is given by

$$
\pi(b_i) = \sum_{j=1}^{m} P(b_i \mid a_j) \pi_B(a_j).
$$

5. If B is a variable with k possible values, then for $1 \leq i \leq k$, $P'(b_i)$, the conditional probability of b_i based on the variables thus far instantiated, is given by

$$
P'(b_i) = \alpha \lambda(b_i) \pi(b_i).
$$

The causal network is first initialized to compute the a priori probabilities (i.e., the probabilities based on the instantiation of no variables) of all variables as follows:

Initialization

A. Set all λ messages and λ values to 1.

B. If the root A has m possible values, then for $1 \leq j \leq m$, set

$$
\pi(a_j) = P(a_j).
$$

C. For all children B of the root A do
   Post a new π message to B using operative formula 2.

   {A propagation flow will then begin due to updating procedure C.}

When a variable is instantiated or a λ or π message is received by a variable, one of the following updating procedures is used:

Updating

A. If a variable B is instantiated for b_j, then
   begin
      1. Set $P'(b_j) = 1$ and for $i \neq j$ set $P'(b_i) = 0$;
      2. Compute $\lambda(B)$ using operative formula 3;

  3. Post a new λ message to B's parent using operative formula 1;

  4. Post new π messages to B's children using operative formula 2
end.

B. If a variable B receives a new λ message from one of its children and if B is not already instantiated, then
  begin
    1. Compute the new value of $\lambda(B)$ using operative formula 3;
    2. Compute the new value of $P'(B)$ using operative formula 5;
    3. Post a new λ message to B's parent using operative formula 1;
    4. Post new π messages to B's other children using operative formula 2
  end.

C. If a variable B receives a new π message from its parent and if B is not already instantiated, then
  begin
    1. Compute the new value of $\pi(B)$ using operative formula 4;
    2. Compute the new value of $P'(B)$ using operative formula 5;
    3. Post new π messages to B's children using operative formula 2
  end.

In the above algorithm, operative formula 2 is based on Theorem 6.7. and therefore assumes that all probabilities are positive. If this is not the case, Definition 6.10 would be used in operative formula 2 to compute $\pi_B(a_j)$ when A is not instantiated.

Notice in the above algorithm that variables which have already been instantiated are always dead ends as far as propagation is concerned. This is due to the fact that if a variable B is instantiated, $P'(B)$, $\lambda(B)$, and B's π messages to its children all become fixed. Since $\lambda(B)$ becomes fixed, B's λ message to its parent also becomes fixed. Thus we see that nothing can change at B and no new messages can be sent from B once B is instantiated. Therefore an instantiated variable is a dead end. We could deduce this fact directly by noticing that a variable d-separates its children (and their descendents) from each other and from its ancestors.

It is left as an exercise to use the above algorithm to redo the initialization of the network and the instantiation of B in the cheating spouse example. We illustrate the updating method by assuming that B is already instantiated (i.e., the network is in the state depicted in Figure 6.13) and then instantiating D for d_2:

When D is instantiated for d_2, we have

A.1. $P'(d_1) = 0$ and $P'(d_2) = 1$.

A.2. $\lambda(d_1) = 0$ and $\lambda(d_2) = 1$.

A.3. $\lambda_D(a_1) = P(d_1 \mid a_1)\lambda(d_1) + P(d_2 \mid a_1)\lambda(d_2)$

$\quad\quad\quad = .8(0) + .2(1) = .2$

$\quad\lambda_D(a_2) = P(d_1 \mid a_2)\lambda(d_1) + P(d_2 \mid a_2)\lambda(d_2)$

$\quad\quad\quad = .4(0) + .6(1) = .6.$

A.4. D has no children to receive new π messages.

When A receives a new λ message from D,

B.1. $\lambda(a_1) = \lambda_B(a_1)\lambda_D(a_1) = (.7)(.2) = .14$

$\quad\lambda(a_2) = \lambda_B(a_2)\lambda_D(a_2) = (.2)(.6) = .12.$

B.2. $P'(a_1) = \alpha\lambda(a_1)\pi(a_1) = \alpha(.14)(.1) = \alpha(.014)$

$\quad P'(a_2) = \alpha\lambda(a_2)\pi(a_2) = \alpha(.12)(.9) = \alpha(.108).$

Normalization yields $P'(A) = (.1148, .8852)$.

B.3. A has no parent to which to send a new λ message.

B.4. $\pi_B(a_1) = \dfrac{P'(a_1)}{\lambda_B(a_1)} = \dfrac{.1148}{.7} = .164$

$\quad\quad \pi_B(a_2) = \dfrac{P'(a_2)}{\lambda_B(a_2)} = \dfrac{.8852}{.2} = 4.426.$

When B receives a new π message from A,

C.1, C.2, and C.3. None of these steps execute, since B is already instantiated.

Propagation ends.

Notice that we could, before executing step B.4, observe that B is already instantiated and therefore not bother to compute a π message to B. However, this would violate the principle that we are treating each vertex as an autonomous processor. The state of the network after the instantiation of both B and D is shown in Figure 6.14.

Further familiarity with this algorithm can be obtained by working the exercises at the end of this section. Before that, we address another issue. Thus far, we have assumed that the values of particular variables become known for certain. That is, the variables are instantiated. But what about the case where we have evidence that a variable takes a certain value, but the evi-

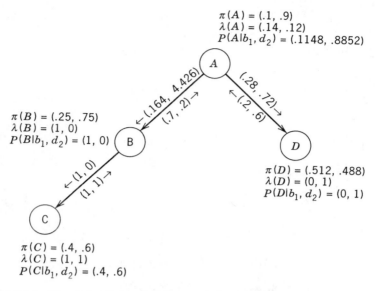

FIGURE 6.14 The state of the causal network in the cheating spouse example after both B and D are instantiated.

dence is not absolutely conclusive? This is discussed in the following subsection.

6.2.4. Virtual Evidence

To illustrate the concept of virtual evidence we will continue the murder trial example (Example 5.6). This continuation is also from Pearl [1986a]. A and B are defined as in Example 5.6 and the causal network is the one in Figure 6.15(a). First we perform the initialization procedure:

A. $\lambda(A) = (1, 1, 1)$

$\lambda(B) = (1, 1, 1)$

$\lambda_B(A) = (1, 1, 1).$

B. $\pi(A) = (.8, .1, .1)$

C. $\pi_B(a_j) = \pi(a_j)(1) = \pi(a_j)$ for $1 \le j \le 3$.

Therefore

$$\pi_B(A) = (.8, .1, .1).$$

{Updating procedure C now takes over.}

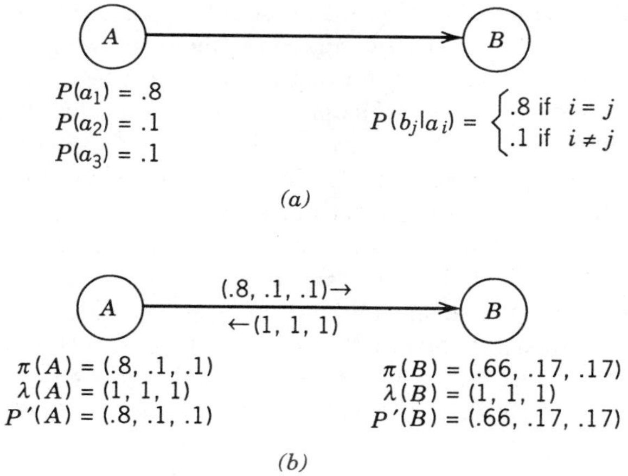

(a)

$\pi(A) = (.8, .1, .1)$
$\lambda(A) = (1, 1, 1)$
$P'(A) = (.8, .1, .1)$

$\pi(B) = (.66, .17, .17)$
$\lambda(B) = (1, 1, 1)$
$P'(B) = (.66, .17, .17)$

(b)

FIGURE 6.15 The state of the causal network in the murder trial example after initialization.

When B receives a new π message from A,

C.1. $(\pi b_1) = P(b_1 \mid a_1)\pi_B(a_1) + P(b_1 \mid a_2)\pi_B(a_2) + P(b_1 \mid a_3)\pi_B(a_3)$

$= .8(.8) + .1(.1) + .1(.1) = .66$

$\pi(b_2) = P(b_2 \mid a_1)\pi_B(a_1) + P(b_2 \mid a_2)\pi_B(a_2) + P(b_2 \mid a_3)\pi_B(a_3)$

$= .1(.8) + .8(.1) + .1(.1) = .17$

$\pi(b_3) = P(b_3 \mid a_1)\pi_B(a_1) + P(b_3 \mid a_2)\pi_B(a_2) + P(b_3 \mid a_3)\pi_B(a_3)$

$= .1(.8) + .1(.1) + .8(.1) = .17.$

C.2. $P'(b_1) = \alpha\lambda(b_1)\pi(b_1) = \alpha(1)(.66) = \alpha(.66)$

$P'(b_2) = \alpha\lambda(b_2)\pi(b_2) = \alpha(1)(.17) = \alpha(.17)$

$P'(b_3) = \alpha\lambda(b_3)\pi(b_3) = \alpha(1)(.17) = \alpha(.17).$

Normalization yields $P(B) = (.66, .17, .17)$.

C.3. B has no children to which to send a new π message.

Propagation ends.

The state of the network after initialization is depicted in Figure 6.15(b).
Suppose the murder weapon is then sent to a lab to be analyzed for fingerprints. The fingerprint analysis cannot determine conclusively who last held

the weapon. It is possible, however, to judge the probability of these particular test results given that a specific person last held the weapon. For example, the probability of these results given that person 1 last held the weapon might be .8, the probability of these results given that person 2 last held the weapon might be .6, while the probability given person 3 was the last holder might be .5. Thus we see that variable B is not instantiated for certain, but rather we have evidence for the particular values of B. We could have included a vertex C for the lab results in our original network with an arc from B to C. However, this would have called for judging the probability of every possible lab result given B, an incomprehensible and unnecessary task. This situation can better be handled by deferring any reference to the evidence until the time the evidence is received. At that time we add a virtual vertex C to the network, as depicted in Figure 6.16(a). C is a variable which can take exactly one value, c_1, where

$$c_1 = \text{these lab results occurred.}$$

Since variables which are leaves do not affect the probabilities of other variables unless they are instantiated, we can think of the variable C as having always been in the network but having not yet been instantiated for c_1. Then when the evidence arrives, we instantiate C for c_1, and propagate in the usual fashion, hence staying within the natural framework of probability propagation. We will now do this:

When C is instantiated for c_1,
A.1. $P'(c_1) = 1$.
A.2. $\lambda(c_1) = 1$.

A.3. $\lambda_C(b_1) = P(c_1 \mid b_1)\lambda(c_1) = .8(1) = .8$

$\lambda_C(b_2) = P(c_1 \mid b_2)\lambda(c_1) = .6(1) = .6$

$\lambda_C(b_3) = P(c_1 \mid b_3)\lambda(c_1) = .5(1) = .5.$

A.4. A virtual vertex never has any children to receive new π messages.

Notice that the λ message to B is simply the conditional probabilities of C given B. Therefore, when a virtual vertex is instantiated, we need only post the conditional probabilities as the λ message to the parent. There is no need to go through the calculations for instantiation.

When B receives a new λ message from C,

B.1. $\lambda(b_1) = \lambda_C(b_1) = .8$

$\lambda(b_2) = \lambda_C(b_2) = .6$

$\lambda(b_3) = \lambda_C(b_3) = .5.$

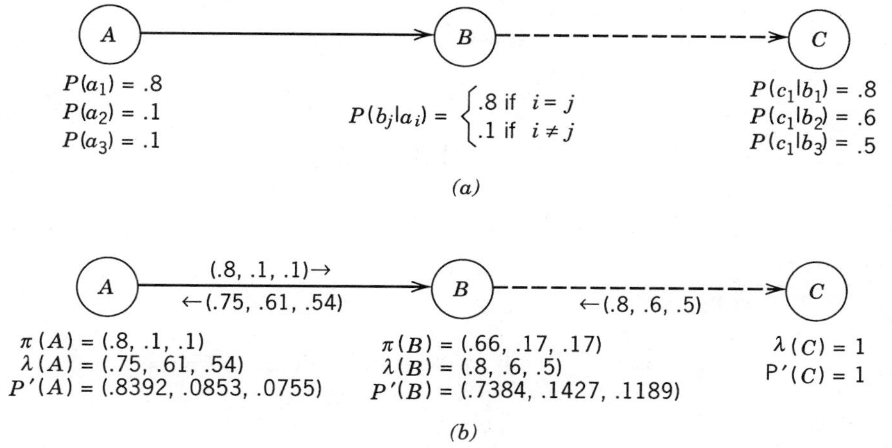

$P(a_1) = .8$
$P(a_2) = .1$
$P(a_3) = .1$

$P(b_j|a_i) = \begin{cases} .8 \text{ if } i = j \\ .1 \text{ if } i \neq j \end{cases}$

$P(c_1|b_1) = .8$
$P(c_1|b_2) = .6$
$P(c_1|b_3) = .5$

(a)

$(.8, .1, .1) \rightarrow$
$\leftarrow (.75, .61, .54)$

$\leftarrow (.8, .6, .5)$

$\pi(A) = (.8, .1, .1)$
$\lambda(A) = (.75, .61, .54)$
$P'(A) = (.8392, .0853, .0755)$

$\pi(B) = (.66, .17, .17)$
$\lambda(B) = (.8, .6, .5)$
$P'(B) = (.7384, .1427, .1189)$

$\lambda(C) = 1$
$P'(C) = 1$

(b)

FIGURE 6.16 The state of the causal network in the murder trial example after the lab report on the fingerprints is received.

B.2. $P'(b_1) = \alpha\lambda(b_1)\pi(b_1) = \alpha(.8)(.66) = \alpha(.528)$

$P'(b_2) = \alpha\lambda(b_2)\pi(b_2) = \alpha(.17)(.6) = \alpha(.102)$

$P'(b_3) = \alpha\lambda(b_3)\pi(b_3) = \alpha(.17)(.5) = \alpha(.085).$

Normalization yields $P'(B) = (.7384, .1427, .1189)$.

B.3. $\lambda_B(a_1) = P(b_1 \mid a_1)\lambda(b_1) + P(b_2 \mid a_1)\lambda(b_2) + P(b_3 \mid a_1)\lambda(b_3)$

$= .8(.8) + .1(.6) + .1(.5) = .75$

$\lambda_B(a_2) = P(b_1 \mid a_2)\lambda(b_1) + P(b_2 \mid a_2)\lambda(b_2) + P(b_2 \mid a_2)\lambda(b_3)$

$= .1(.8) + .8(.6) + .1(.5) = .61$

$\lambda_B(a_3) = P(b_1 \mid a_3)\lambda(b_1) + P(b_2 \mid a_3)\lambda(b_2) + P(b_3 \mid a_3)\lambda(b_3)$

$= .1(.8) + .1(.6) + .8(.5) = .54.$

B.4. B has no other children to receive new π messages.

When A receives a new λ message from B,

B.1. $\lambda(a_1) = \lambda_B(a_1) = .75$

$\lambda(a_2) = \lambda_B(a_2) = .61$

$\lambda(a_3) = \lambda_B(a_3) = .54.$

B.2. $P'(a_1) = \alpha\lambda(a_1)\pi(a_1) = \alpha(.75)(.8) = \alpha(.6)$

   $P'(a_2) = \alpha\lambda(a_2)\pi(a_2) = \alpha(.61)(.1) = \alpha(.061)$

   $P'(a_3) = \alpha\lambda(a_3)\pi(a_3) = \alpha(.54)(.1) = \alpha(.054).$

Normalization yields $P'(A) = (.8392, .0853, .0755)$.

Propagation ends.

The causal network is now left in the state depicted in Figure 6.16(b).
We will instantiate one more virtual vertex in order to illustrate the propagation method further. Suppose person 1 now gives a strong alibi in his favor, and we judge that the probability of his giving this alibi given that he is the murderer is .1, while the probability of his giving this alibi given that one of the others is the murderer is 1. We then have the network indicated in Figure 6.17(a), where

$$d_1 = \text{person 1 gives this alibi.}$$

Note that the virtual vertex C is no longer included in the network, since it can have no further bearing. We now instantiate D for d_1:

When D is instantiated,

$$\lambda_D(A) = (.1, 1, 1).$$

Recall that this is the only step that is necessary when a virtual vertex is instantiated.

When A receives a new λ message from D,

B.1. $\lambda(a_1) = \lambda_B(a_1)\lambda_D(a_1) = .75(.1) = .075$

   $\lambda(a_2) = \lambda_B(a_2)\lambda_D(a_2) = .61(1) = .61$

   $\lambda(a_2) = \lambda_B(a_2)\lambda_D(a_2) = .54(1) = .54.$

B.2. $P'(a_1) = \alpha\lambda(a_1)\pi(a_1) = \alpha(.075)(.8) = \alpha(.06)$

   $P'(a_2) = \alpha\lambda(a_2)\pi(a_2) = \alpha(.61)(.1) = \alpha(.061)$

   $P'(a_3) = \alpha\lambda(a_3)\pi(a_3) = \alpha(.54)(.1) = \alpha(.054).$

Normalization yields $P'(A) = (.3429, .3486, .3085)$.

B.3. A has no parent to receive a new λ message.

B.4. $\pi_B(a_1) = \dfrac{P'(a_1)}{\lambda_B(a_1)} = \dfrac{.3429}{.75} = .4572$

$\pi_B(a_2) = \dfrac{P'(a_2)}{\lambda_B(a_2)} = \dfrac{.3486}{.61} = .5715$

$\pi_B(a_3) = \dfrac{P'(a_3)}{\lambda_B(a_3)} = \dfrac{.3085}{.54} = .5713.$

When B receives a new π message from A,

C.1. $\pi(b_1) = P(b_1 \mid a_1)\pi_B(a_1) + P(b_1 \mid a_2)\pi_B(a_2) + P(b_1 \mid a_3)\pi_B(a_3)$

$\qquad = .8(.4572) + .1(.5715) + .1(.5713) = .48$

$\quad \pi(b_2) = P(b_2 \mid a_1)\pi_B(a_1) + P(b_2 \mid a_2)\pi_B(a_2) + P(b_2 \mid a_3)\pi_B(a_3)$

$\qquad = .1(.4572) + .8(.5715) + .1(.5713) = .56$

$\quad \pi(b_3) = P(b_3 \mid a_1)\pi_B(a_1) + P(b_3 \mid a_2)\pi_B(a_2) + P(b_3 \mid a_3)\pi_B(a_3)$

$\qquad = .1(.4572) + .1(.5715) + .8(.5713) = .56$

C.2. $P'(b_1) = \alpha\lambda(b_1)\pi(b_1) = .8(.48) = \alpha(.384)$

$\quad P'(b_2) = \alpha\lambda(b_2)\pi(b_2) = .6(.56) = \alpha(.336)$

$\quad P'(b_3) = \alpha\lambda(b_3)\pi(b_3) = .5(.56) = \alpha(.28).$

Normalization yields $P(B) = (.384, .336, .28)$.

C.3. B has no children to receive new π messages.

Propagation ends.

The state of the network after person 1's strong alibi is depicted in Figure 6.17(b).

PROBLEMS 6.2

1. Using the propagation algorithm, perform the initialization of the causal network in the cheating spouse example.

2. Starting with the initialized causal network (Figure 6.12) in the cheating spouse example, use the propagation algorithm to instantiate D for d_2 and B for b_1 in sequence. Notice the results are the same as those when B is instantiated first.

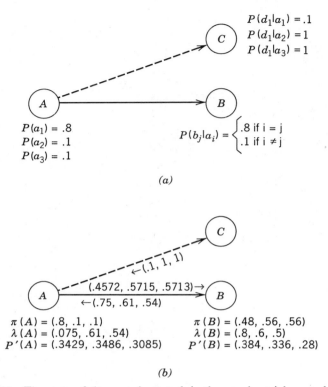

$P(d_1|a_1) = .1$
$P(d_1|a_2) = 1$
$P(d_1|a_3) = 1$

$P(a_1) = .8$
$P(a_2) = .1$
$P(a_3) = .1$

$P(b_j|a_i) = \begin{cases} .8 \text{ if } i = j \\ .1 \text{ if } i \neq j \end{cases}$

(a)

←(.1, 1, 1)
(.4572, .5715, .5713)→
←(.75, .61, .54)

$\pi(A) = (.8, .1, .1)$
$\lambda(A) = (.075, .61, .54)$
$P'(A) = (.3429, .3486, .3085)$

$\pi(B) = (.48, .56, .56)$
$\lambda(B) = (.8, .6, .5)$
$P'(B) = (.384, .336, .28)$

(b)

FIGURE 6.17 The state of the causal network in the murder trial example after person 1 gives a strong alibi.

3. In the cheating spouse example, suppose a husband goes out for dinner and his wife has a hunch that the dinner is with another woman (perhaps he took suspiciously meticulous care in preparing for the dinner). Say that the wife judges that the probability of this meticulous care given that he is dining with another is .9, while the probability of such care given that he is not dining with another is .1. Starting with the initialized causal network (Figure 6.12), use the propagation algorithm to determine how this virtual evidence affects the probabilities of all variables in the network. Next instantiate variable D for d_1 and again use the propagation algorithm to compute the effect on the variables in the network.

4. Create and initialize a causal network which represents the relationships described in Problem 2.1.2. Propagate the information that James tested positive to determine the probability of James having syphilis.

5. Continuation of Problem 4: Several weeks later James acquires a canker sore. Suppose syphilis often causes such sores, and that the probability of acquiring such a sore given that syphilis is present is .8, while the probability of acquiring one given that syphilis is not present is .1. Determine the new

probability of the individual having syphilis. Notice that a vertex for the canker sore should have been included in the network in the first place (this is not like the situation involving the lab test where there are many possible values of the variable which are not germane to the problem), but that the results are the same if the canker sore is treated as virtual evidence. (The values in this problem are entirely fictitious.)

COMPUTER PROBLEM 6.2

Write a program which performs probability propagation in trees. There should be two user interfaces: one for the knowledge engineer and one for the end user. The inputs from the knowledge engineer should include a set of propositional variables and their alternatives, a cause (parent) for each propositional variable which has a cause, and the probability of each variable given its cause. The output to the knowledge engineer should be the a priori probabilities of all the variables. The input from the end user should either be 1) evidence that a variable in the network has been instantiated and its value, or 2) a variable in the network and virtual evidence for that variable (i.e., the probability of the evidence given each alternative of the variable). The output to the end user should be the new probability of each variable based on the evidence. The user should be allowed to enter as many items of evidence as desired. The output should always be based on the accumulation of all the evidence. You should use the program from Computer Problem 5.6. Test the program on the causal network created as an exercise in Problem 5.5.3 if that network is a tree or on one of the tree-structured causal networks we have discussed.

6.3. PROBABILITY PROPAGATION IN SINGLY CONNECTED NETWORKS

The method developed in the previous section works only for causal networks in which the DAG is a tree. However, this is not the case in many causal networks. Looking back at the simple examples in Chapter 5, the DAG was not a tree in Examples 5.7, 5.13, and 5.16. The most prevalent deviation from a tree structure is that a variable can have more than one parent. For example, in medical applications, a symptom is often caused by more than one disease. If the symptom is known to be present and it is learned that one of the diseases is also present, that disease "explains" the symptom, thereby making the other diseases less probable.

This section extends the propagation method to allow for multiple parents. But, this extension does require that the DAG must be singly connected. That is, there can be at most one chain between any two variables. The development of the theory is very similar to that for trees, and is left as a major exercise. Included here is the following high-level algorithm for the method. An example follows the algorithm. For the sake of notational simplicity, the algorithm is given for the case where there are exactly two parents. The case where there are more than two parents is a straightforward generalization.

PROBABILITY PROPAGATION IN SINGLY CONNECTED NETWORKS

Operative Formulas

1. If B is a child of A, B has k possible values, A has m possible values, and B has one other parent D, with n possible values, then for $1 \leq j \leq m$ the λ message from B to A is given by

$$\lambda_B(a_j) = \sum_{p=1}^{n} \pi_B(d_p) \left(\sum_{i=1}^{k} P(b_i \mid a_j, d_p) \lambda(b_i) \right).$$

2. If B is a child of A and A has m possible values, then for $1 \leq j \leq m$ the π message from A to B is given by

$$\pi_B(a_j) = \begin{cases} 1 & \text{if} \quad A \text{ is instantiated for } a_j \\ 0 & \text{if} \quad A \text{ is instantiated, but not for } a_j \\ \dfrac{P'(a_j)}{\lambda_B(a_j)} & \text{if} \quad A \text{ is not instantiated,} \end{cases}$$

where $P'(a_j)$ is defined to be the current conditional probability of a_j based on the variables thus far instantiated.

3. If B is a variable with k possible values, $s(B)$ is the set of B's children, then for $1 \leq i \leq k$ the λ value of B is given by

$$\lambda(b_i) = \begin{cases} \displaystyle\prod_{C \in s(B)} \lambda_C(b_i) & \text{if} \quad B \text{ is not instantiated} \\ 1 & \text{if} \quad B \text{ is instantiated for } b_i \\ 0 & \text{if} \quad B \text{ is instantiated, but not for } b_i. \end{cases}$$

4. If B is a variable with k possible values and exactly two parents, A and D, A has m possible values, and D has n possible values, then for $1 \leq i \leq k$ the π value of B is given by

$$\pi(b_i) = \sum_{j=1}^{m} \sum_{p=1}^{n} P(b_i \mid a_j, d_p) \pi_B(a_j) \pi_B(d_p).$$

5. If B is a variable with k possible values, then for $1 \leq i \leq k$, $P'(b_i)$, the conditional probability of b_i based on the variables thus far instantiated, is given by

$$P'(b_i) = \alpha \lambda(b_i) \pi(b_i).$$

The causal network is first initialized to compute the a priori probabilities (i.e., the probabilities based on the instantiation of no variables) of all variables as follows:

Initialization

A. Set all λ values, λ messages, and π messages to 1.
B. For all roots A, if A has m possible values, then for $1 \leq j \leq m$, set

$$\pi(a_j) = P(a_j).$$

C. For all roots A for all children B of A, do
   Post a new π message to B using operative formula 2.

{A propagation flow will then begin due to updating procedure C.}

When a variable is instantiated, or a λ or π message is received by a variable, one of the following updating procedures is used:

Updating

A. If a variable B is instantiated for b_j, then
   begin
      1. Set $P'(b_j) = 1$ and for $i \neq j$, set $P'(b_i) = 0$;
      2. Compute $\lambda(B)$ using operative formula 3;
      3. Post new λ messages to all B's parents using operative formula 1;
      4. Post new π messages to all B's children using operative formula 2
   end.
B. If a variable B receives a new λ message from one of its children, then if B is not already instantiated,
   begin
      1. Compute the new value of $\lambda(B)$ using operative formula 3;
      2. Compute the new value of $P'(B)$ using operative formula 5;
      3. Post λ messages to all B's parents using operative formula 1;
      4. Post new π messages to B's other children using operative formula 2
   end.
C. If a variable B receives a new π message from a parent, then
   begin
      If B is not already instantiated, then
        begin
          1. Compute the new value of $\pi(B)$ using operative formula 4;

      2. Compute the new value of $P'(B)$ using operative formula 5;

      3. Post new π messages to all B's children using formula 2

   end;

  If $\lambda(B) \neq (1,1,\ldots,1)$, then

      4. Post new λ messages to B's other parents using operative formula 1

  end.

As in the case of trees, if all probabilities were not necessarily positive, then Definition 6.10 would be used to compute $\pi_B(a_j)$ in operative formula 2 when A is not instantiated.

The formulas in the above algorithm can easily be generalized to the case where a vertex has more than two parents. For example, if B had another parent E, with r possible values, then operative formula 1 would be as follows:

$$\lambda_B(a_j) = \sum_{s=1}^{r} \sum_{p=1}^{n} \pi_B(e_s)\pi_B(d_p) \left(\sum_{i=1}^{k} P(b_i \mid a_j, d_p, e_s)\lambda(b_i) \right).$$

Notice that instantiated variables are not dead ends as they are in trees. Even if a variable is already instantiated, if it receives a new π message from one parent, the λ messages to its other parents can change. This is what we would expect, since a variable does not d-separate its parents. However, notice in operative formula 1 that if $\lambda(B) = (1,1,\ldots,1)$, then the new λ messages to other parents will be unchanged (except for a multiplicative constant). Thus we do not bother to start unnecessary propagation by sending a new λ message in this case. This would be the case if B and none of B's descendents were instantiated, and is what we would expect, since a variable's parents are d-separated by the empty set (and by their descendents through other children) in a singly connected causal network, and therefore, by Theorem 6.2, the parents are independent in the space at large (and in the space including only descendents through other children). The examples in the last chapter bear these results out in the case where a causal network is created according to the notion of causation. For example, Mr. Holmes would not suspect that he had or had not been burglarized if he heard there were an earthquake. However, once he knew his burglar alarm had sounded, the news of the earthquake would relax him a bit. As in the case of trees, an instantiated variable will send no new π message to its children when it receives a new π message from a parent, and it will send no new λ messages to its parents when it receives a new λ message from a child. Notice further that, in the case of trees, this algorithm reduces to the algorithm given previously for propagation in trees. Last of all, note that π messages are also initialized to 1 in the initialization procedure. This is so that during initialization, when updating procedure

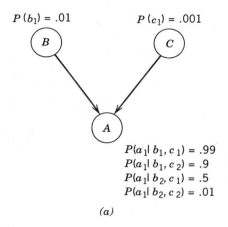

$P(b_1) = .01$ $P(c_1) = .001$

$P(a_1 | b_1, c_1) = .99$
$P(a_1 | b_1, c_2) = .9$
$P(a_1 | b_2, c_1) = .5$
$P(a_1 | b_2, c_2) = .01$

(a)

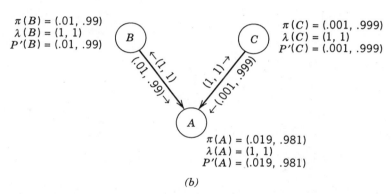

$\pi(B) = (.01, .99)$
$\lambda(B) = (1, 1)$
$P'(B) = (.01, .99)$

$\pi(C) = (.001, .999)$
$\lambda(C) = (1, 1)$
$P'(C) = (.001, .999)$

$\pi(A) = (.019, .981)$
$\lambda(A) = (1, 1)$
$P'(A) = (.019, .981)$

(b)

FIGURE 6.18 The state of causal network for the burglar alarm example before any variables are instantiated.

C computes a new π value based on the new π message received from one parent, there will always be a π message on the arcs from the other parents.

To illustrate this method, we apply it to Example 5.7. That is, the example involving Mr. Holmes, his burglar alarm, and earthquakes. The causal network for this example is depicted in Figure 6.18(a).

First we perform the initialization:

A. Set all λ values, all λ messages, and all π messages to 1.

B. $\pi(B) = P(B) = (.01, .99)$

$\pi(C) = P(C) = (.001, .999)$.

C. $\pi_A(b_1) = \dfrac{P'(b_1)}{\lambda_A(b_1)} = \dfrac{.01}{1} = .01.$

Similarly,

$$\pi_A(b_2) = .99, \qquad \pi_A(c_1) = .001, \qquad \pi_A(c_2) = .999.$$

{Updating procedure C now takes over.}

When A receives a new π message from B (note $\pi_A(C)$ still equals (1,1) at this point),

C.1. $\pi(a_1) = P(a_1 \mid b_1,c_1)\pi_A(b_1)\pi_A(c_1) + P(a_1 \mid b_1,c_2)\pi_A(b_1)\pi_A(c_2)$
$\qquad + P(a_1 \mid b_2,c_1)\pi_A(b_2)\pi_A(c_1) + P(a_1 \mid b_2,c_2)\pi_A(b_2)\pi_A(c_2)$
$\qquad = (.99)(.01)(1) + (.9)(.01)(1)$
$\qquad\quad + (.5)(.99)(1) + (.01)(.99)(1) = .5238.$

Similarly,
$$\pi(a_2) = 1.4762.$$

C.2. $P'(a_1) = \alpha\lambda(a_1)\pi(a_1) = \alpha(.5238)$
$\quad\; P'(a_2) = \alpha\lambda(a_2)\pi(a_2) = \alpha(1.4762).$

Normalization yields $P'(A) = (.2619, .7381)$.

C.3. A has no children to which to send new π messages.

When A receives a new π message from C,

C.1. $\pi(a_1) = P(a_1 \mid b_1,c_1)\pi_A(b_1)\pi_A(c_1) + P(a_1 \mid b_1,c_2)\pi_A(b_1)\pi_A(c_2)$
$\qquad + P(a_1 \mid b_2,c_1)\pi_A(b_2)\pi_A(c_1) + P(a_1 \mid b_2,c_2)\pi_A(b_2)\pi_A(c_2)$
$\qquad = (.99)(.01)(.001) + (.9)(.01)(.999)$
$\qquad\quad + (.5)(.99)(.001) + (.01)(.99)(.999) = .019.$

Similarly,
$$\pi(a_2) = .981.$$

C.2. $P'(a_1) = \alpha\lambda(a_1)\pi(a_1) = \alpha(.019)$
$\quad\; P'(a_2) = \alpha\lambda(a_2)\pi(a_2) = \alpha(.981).$

Normalization yields $P'(A) = (.019, .981)$.

C.3. A has no children to which to send new π messages.

Propagation ends.

The state of the causal network after initialization is shown in Figure 6.18(b).

Suppose now Mr. Holmes is sitting in his office and his wife calls, informing him that the burglar alarm has sounded. He would most likely rush home believing that he has probably been burglarized. We shall determine if Mr. Holmes is correct in his belief by instantiating A for a_1 and propagating.

When A is instantiated,

A.1. $P'(a_1) = 1$ and $P'(a_2) = 0$.
A.2. $\lambda(a_1) = 1$ and $\lambda(a_2) = 0$.

A.3. $\lambda_A(b_1) = \pi_A(c_1)(P(a_1 \mid b_1,c_1)\lambda(a_1) + P(a_2 \mid b_1,c_1)\lambda(a_2))$

$\qquad + \pi_A(c_2)(P(a_1 \mid b_1,c_2)\lambda(a_1) + P(a_2 \mid b_1,c_2)\lambda(a_2))$

$\qquad = .001(.99(1) + .01(0)) + .999(.9(1) + .1(0))$

$\qquad = .9.$

Similarly,

$$\lambda_A(b_2) = .01$$
$$\lambda_A(c_1) = .505$$
$$\lambda_A(c_2) = .019.$$

A.4. A has no children to receive new π messages.

When B receives a new λ message from A,

B.1. $\lambda(b_1) = \lambda_A(b_1) = .9$

$\qquad \lambda(b_2) = \lambda_A(b_2) = .01.$

B.2. $P'(b_1) = \alpha\lambda(b_1)\pi(b_1) = \alpha(.9)(.01) = \alpha(.009)$

$\qquad P'(b_2) = \alpha\lambda(b_2)\pi(b_2) = \alpha(.01)(.99) = \alpha(.0099).$

Normalization yields $P'(B) = (.476, .524)$.

B.3. B has no parents to receive new λ messages.
B.4. B has no other children to receive new π messages.

When C receives a new λ message from A,

B.1. $\lambda(c_1) = \lambda_A(c_1) = .505$
   $\lambda(c_2) = \lambda_A(c_2) = .019.$

B.2. $P'(c_1) = \alpha\lambda(c_1)\pi(c_1) = \alpha(.505)(.001) = \alpha(.000505)$
   $P'(c_2) = \alpha\lambda(c_2)\pi(c_2) = \alpha(.019)(.999) = \alpha(.018981).$

Normalization yields $P'(C) = (.026, .974)$.

B.3. C has no parents to receive new λ messages.
B.4. C has no other children to receive new π messages.

Propagation ends.

The network is now in the state depicted in Figure 6.19(a). Notice that there is about a 50/50 chance that Mr. Holmes has been burglarized. So he certainly has reason to be concerned. Suppose next that on his way home, Mr. Holmes hears a radio broadcast that there has been an earthquake. He then relaxes, figuring that there is a good chance that the earthquake caused the alarm to sound. Next we instantiate C for c_1 and propagate to see if Mr. Holmes should really relax.

When C is instantiated,

A.1. $P'(c_1) = 1$ and $P'(c_2) = 0$.
A.2. $\lambda(c_1) = 1$ and $\lambda(c_2) = 0$.
A.3. C has no parents to receive new λ messages.
A.4. $\pi_A(C) = (1, 0)$.

When A receives a new π message,

C.1, C.2, and C.3. These steps are skipped because A has already been instantiated.

C.4. $\lambda_A(b_1) = \pi_A(c_1)(P(a_1 \mid b_1, c_1)\lambda(a_1) + P(a_2 \mid b_1, c_1)\lambda(a_2))$
   $+ \pi_A(c_2)(P(a_1 \mid b_1, c_2)\lambda(a_1) + P(a_2 \mid b_1, c_2)\lambda(a_2))$
   $= 1(.99(1) + .01(0)) + 0(.9(1) + .1(0)) = .99.$

Similarly,

$$\lambda_A(b_2) = .5.$$

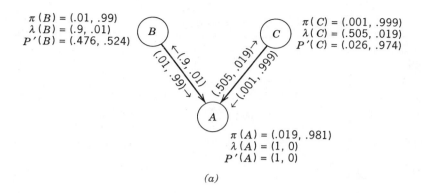

$\pi(B) = (.01, .99)$
$\lambda(B) = (.9, .01)$
$P'(B) = (.476, .524)$

$\pi(C) = (.001, .999)$
$\lambda(C) = (.505, .019)$
$P'(C) = (.026, .974)$

$\pi(A) = (.019, .981)$
$\lambda(A) = (1, 0)$
$P'(A) = (1, 0)$

(a)

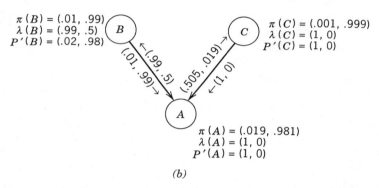

$\pi(B) = (.01, .99)$
$\lambda(B) = (.99, .5)$
$P'(B) = (.02, .98)$

$\pi(C) = (.001, .999)$
$\lambda(C) = (1, 0)$
$P'(C) = (1, 0)$

$\pi(A) = (.019, .981)$
$\lambda(A) = (1, 0)$
$P'(A) = (1, 0)$

(b)

FIGURE 6.19 (a) depicts the state of the causal network in Mr. Holmes' example after A is instantiated; (b) contains the state after both A and C arre instantiated.

When B receives a new λ message,

B.1. $\lambda(b_1) = \lambda_A(b_1) = .99$

$\lambda(b_2) = \lambda_A(b_2) = .5.$

B.2. $P'(b_1) = \alpha\lambda(b_1)\pi(b_1) = \alpha(.99)(.01) = .0099$

$P'(b_2) = \alpha\lambda(b_2)\pi(b_2) = \alpha(.5)(.99) = .495.$

Normalization yields $P'(B) = (.02, .98)$.

B.3. B has no parents to receive new λ messages.

B.4. B has no other children to receive new π messages.

Propagation ends.

So we see that the probability of Mr. Holmes' residence having been burglarized is now only .02, and indeed he can relax a bit. The state of the network after both A and C have been instantiated is shown in Figure 6.19(b).

Virtual evidence is handled in singly connected networks in exactly the same way it is handled in trees since in this case a variable containing virtual evidence is a leaf with only one parent, and therefore it does not affect the probabilities of other variables unless it is instantiated. The first problem below involves the propagation of virtual evidence.

PROBLEMS 6.3

1. Perform the initialization procedure on the causal network in Example 5.16. Suppose that it is then learned for certain that the patient is sneezing. Propagate to determine the probability of him having a cold and the probability of him having hay fever. Next suppose that the physician learns that the patient's family has a history of hay fever, and he judges that the probability of such a family history given that the patient has hay fever is .8, whereas the probability of such a history given that the patient does not have hay fever is .5. Propagate to determine how this virtual evidence affects the probability of a cold and the probability of hay fever. Notice that this is a case where the arc is not in the direction of perceived causation. The fact that a patient has hay fever would not cause him to have a family history of hay fever. However, the conditional independence assumptions for a causal network do appear to be satisfied. That is, the family history of hay fever is independent of the patient having a cold or sneezing given that the patient has hay fever.

2. Continue Problem 6.2.5 by including the variable for the canker sore in the original network and including another cause of a canker sore, namely a high fever. Suppose that the probability of a high fever is .02, the probability of such a sore given a high fever and syphilis is .9, the probability given no high fever and syphilis is .7, the probability given a high fever and no syphilis is .3, and the probability given no high fever and no syphilis is .01. Initialize the causal network and propagate to determine the probability of James having syphilis given that he has tested positive, has a canker sore, and has a high fever. (The values in this problem are entirely fictitious.)

3. Develop the theory which validates the algorithm for belief propagation in singly connected networks. This can either be done by following the development of the theory for propagation in trees or by consulting the original source [Pearl, 1986a].

COMPUTER PROBLEM 6.3

Write a program which performs probability propagation in singly connected networks. The specifications are exactly the same as those in Computer Problem 6.2. Test the program on the causal network created as in Problem 5.5.3 if that network is singly connected or on one of the singly connected causal networks we have discussed.

6.4. EXTENSIONS TO NONSINGLY CONNECTED NETWORKS

The method discussed in this chapter requires that the network be singly connected. That is, there can be at most one chain between any two variables. However, this is not always the case. In Example 5.13, metastatic cancer could indirectly cause a coma either by causing increased total serum calcium or by causing a brain tumor. Therefore there are two chains between metastatic cancer and a coma. Pearl [1986a, 1986c, 1987a, 1988] has presented several ways to apply his method to nonsingly connected networks. Although we discuss these methods briefly, for the most part the interested reader is referred to the original sources. In this text, we concentrate on a method recently developed by Lauritzen and Spiegelhalter [1988] (discussed in chapter 7) which fundamentally places no restrictions on the network (other than that it be a DAG).

The first method is called "conditioning." We will illustrate this method by considering the network in Figure 6.20, which is obviously not singly connected. Suppose that each variable has two alternatives, that G is instantiated for g_2, and we wish to calculate the conditional probabilities of the remaining variables in the network. Note that if we remove A from the network, the network becomes singly connected. Thus we instantiate A for each of its values, thereby locking A out of the network, and propagate the instantiation of G, using the method in this chapter, under each of these instantiations of A. For example, the conditional probability of Z given that G equals g_2 is given by

$$P(Z \mid g_2) = P(Z \mid g_2, a_1)P(a_1 \mid g_2) + P(Z \mid g_2, a_2)P(a_2 \mid g_2).$$

The values of $P(Z \mid g_2, a_1)$ and $P(Z \mid g_2, a_2)$, are obtained by propagating the effect of instantiating G for g_2, while holding A fixed first at a_1 and next at a_2. Owing to the definition of conditional probability, the value of $P(a_i \mid g_2)$ is given by

$$P(a_i \mid g_2) = \alpha P(g_2 \mid a_i)P(a_i),$$

where α is a normalizing constant which is equal to $1/P(g_2)$. The first term, in the expression on the right, $P(g_2 \mid a_i)$, can be obtained by propagating the instantiation of A for a_i to G through a singly connected network. The second term, $P(a_i)$, is the a priori probability of A and is stored in the network (since A is a root).

Suermondt and Cooper [1988] have investigated conditions which must be satisfied by the set of vertices, on which we condition, and present a heuristic algorithm for finding a set of vertices which satisfy these conditions.

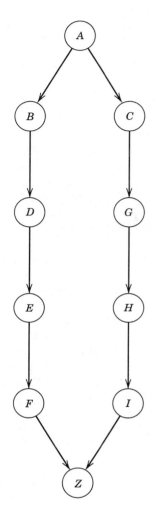

FIGURE 6.20 A nonsingly connected network.

The second method, suggested by Pearl, is stochastic simulation. This method uses Monte Carlo techniques to estimate probabilities by counting how frequently events occur in a series of simulation runs and is not an application of the method presented in this chapter. Stochastic simulation is discussed more in section 7.6.

The final method, called clustering, is similar to the method described in chapter 7. We will discuss clustering and this similarity in detail in chapter 8, when we cover abductive inference.

6.5. CONCLUDING REMARKS

We have already seen, in section 5.5, that the storage requirements for a causal network are modest as long as the number of parents of each variable is small. Furthermore, sparse, irregular networks are often appropriate. A sparse irregular network (i.e., one that contains small clusters of variables) always has a small number of parents for each variable. In particular, in trees the storage requirements will always be modest. In the case of a tree, if n is the maximum number of alternatives for a variable, we would need to store at most n^2 conditional probabilities and two n-dimensional vectors (the λ and π values) at each vertex, and at most two n-dimensional vectors (the λ and π messages) at each arc. Thus a tree would require $n^2 + 2n$ values stored at each vertex and $2n$ values stored at each arc.

As long as the network is sparse and irregular, it is also efficient to update a variable in terms of time. In the case of a tree, if n is the maximum number of alternatives and m is the maximum number of children for a given variable, at most the number of multiplications needed to update a variable would be n to compute the π message, n^2 to compute the λ message, n^2 to compute the π value, nm to compute the λ value, and n to compute the new probability. Hence the total number of multiplications needed to update a variable would be at most $2n^2 + nm + 2n$. It is easy to see that in the case of an arbitrary singly connected network, the time requirement for the update of all variables is linear with respect to the number of variables in the network. Thus, as long as the network is sparse and irregular, the methods outlined in this chapter are efficient in terms of both storage and time. We will discuss the computational considerations in nonsingly connected networks near the end of chapter 7.

Finally, we reference an existing expert system which uses the method covered in this chapter. Namely, Binford, Levitt, and Mann [1987] have implemented this method in expert systems for reasoning with geometry in machine vision (ADRIES and SUCCESSOR).

CHAPTER 7

PROBABILITY PROPAGATION IN TREES OF CLIQUES

By considering the way the human mind reasons, Pearl [1986a] developed the method for propagating probabilities in causal networks (i.e., determining the probabilities of remaining variables given that certain variables have been instantiated for specific values) discussed in chapter 6. Regardless of whether humans actually reason in the fashion described by Pearl, by conceiving the probability propagation problem in that light, he was able to devise a mathematically sound method. However, Pearl's method only works directly on networks which are singly connected. David Spiegelhalter and S. L. Lauritzen [Spiegelhalter, 1986b, 1987; Lauritzen & Spiegelhalter, 1988] approached the problem differently. Conceiving the problem strictly as a mathematical one, they explored how pure mathematical considerations might be used to propagate probabilities in causal networks. Using the graph theory developed in chapter 3 and research in the representation of conditional independencies by graphs [Darroch, Lauritzen, & Speed, 1980], they developed a method which works directly and efficiently in an arbitrary sparse causal network. It cannot be argued that the propagation in this method follows the lines of human reasoning. However, as mentioned at the beginning of chapter 6, in order to go beyond human capabilities, it may be necessary for computers to do things in a way different from the ways of humans. If we accept the conjecture that Pearl's propagation scheme does follow the lines of human reasoning, we might speculate that humans are not capable of reasoning rationally in networks which are not singly connected. Therefore, when reasoning in such networks, we must do things differently from humans.

Lauritzen and Spiegelhalter's [1988] method involves the extraction of an undirected triangulated graph from the DAG in the causal network, and the creation of a tree whose vertices are the cliques of this triangulated graph.

Such a tree is called a "join" tree. Probabilities in the original causal network are updated by passing messages among the vertices in this tree. Once the tree is built, we no longer refer to the original DAG. The method is therefore called "probability propagation in trees of cliques." The format of this chapter is similar to that of chapter 6. In section 7.1 we develop the mathematical theory necessary to the method (much of the mathematics has already been developed in chapter 3), in section 7.2 we illustrate the application of the theory with an example, and in section 7.3 we give a high-level algorithm for the method and an example of the application of the algorithm. Since section 7.1 contains a number of tedious theorems, a reader interested primarily in the application of the method should skip the proofs and read just for an understanding of the statements of the theorems.

7.1. THEORETICAL DEVELOPMENT OF PROBABILITY PROPAGATION IN TREES OF CLIQUES

We develop seven theorems in this section which will be applied in the following sections. As was done in chapter 6, we will develop the theory under the assumption that $P(V) > 0$ for all combinations of values of variables in V. The method is still valid when we remove this restriction; however, it would obscure the clarity of the theorems to address the special cases, which result from allowing probability values of 0. Accordingly, we do not allow such values in this initial development of the theory. First we need two definitions:

Definition 7.1. Let V be a finite set of propositional variables and P be a joint probability distribution of V. Suppose $\{W_i$ such that $1 \leq i \leq p\}$ is a collection of subsets of V, and ψ is a function which, for $1 \leq i \leq p$, assigns a unique real number to every combination of values of the propositional variables in W_i (i.e., ψ is a function from the cartesian product of the ranges of the propositional variables to the reals). Furthermore suppose, for some constant K, that

$$P(V) = K \prod_{i=1}^{p} \psi(W_i).$$

Then $(\{W_i$ such that $1 \leq i \leq p\}, \psi)$ is called a potential representation of P.

Recall, from section 5.1, that the event V, when used in a probability expression, is the event that each variable in V takes one of its values. Similarly, in the expression $\psi(W_i)$, the event W_i means each variable in W_i takes one of its values. As we have previously done in probability expressions, if, for example, $W = \{A, B, C\}$, we will usually denote $\psi(W)$ by $\psi(A, B, C)$ rather than by $\psi(\{A, B, C\})$.

Example 7.1. Let $V = \{A, B, C\}$, $W_1 = \{A, B\}$, and $W_2 = \{B, C\}$. Suppose each variable can take exactly two values, and for $1 \le i, j,\ k \le 2$,

$$P(a_i, b_j, c_k) = 1/8.$$

If we define for $1 \le i, j,\ k \le 2$,

$$\psi(a_i, b_j) = 1/4 \quad \text{and} \quad \psi(b_j, c_k) = 1/2,$$

then $(\{W_i \text{ such that } 1 \le i \le 2\}, \psi)$ is a potential representation of P.

Definition 7.2. Let V be a set and $\{W_i \text{ such that } 1 \le i \le p\}$ be an ordered set of subsets of V. Then we define for $1 \le i \le p$,

$$S_i = W_i \cap (W_1 \cup W_2 \cup \cdots \cup W_{i-1})$$
$$R_i = W_i - S_i.$$

Example 7.2. Let $V = \{A, B, C, D, E, F, G, H\}$, and

$$W_1 = \{A, B\} \qquad W_2 = \{B, C, E\} \qquad W_3 = \{C, E, G\}$$
$$W_4 = \{E, F, G\} \qquad W_5 = \{C, G, H\} \qquad W_6 = \{C, D\}.$$

Then

$$\begin{aligned} S_3 &= W_3 \cap (W_1 \cup W_2) \\ &= \{C, E, G\} \cap (\{A, B\} \cup \{B, C, E\}) \\ &= \{C, E\} \end{aligned}$$

and

$$\begin{aligned} R_3 &= W_3 - S_3 \\ &= \{C, E, G\} - \{C, E\} \\ &= \{G\}. \end{aligned}$$

Theorem 7.1. Let V be a finite set of propositional variables, P a joint probability distribution of V, and $\{W_i \text{ such that } 1 \le i \le p\}$ an ordered set of subsets of V. Then for $1 \le i \le p$,

$$P(W_i \mid S_i) = P(R_i \mid S_i).$$

Proof. By the definition of conditional probability, we have that

$$P(R_i \mid S_i) = \frac{P(R_i \cup S_i)}{P(S_i)}$$

$$= \frac{P(W_i)}{P(S_i)} \qquad \text{since} \quad W_i = R_i \cup S_i,$$

$$= \frac{P(W_i \cup S_i)}{P(S_i)} \qquad \text{since} \quad S_i \subseteq W_i,$$

$$= P(W_i \mid S_i). \quad \square$$

Theorem 7.2. Let V be a finite set of propositional variables, P a joint probability distribution of V, and $(\{W_i \text{ such that } 1 \leq i \leq p\}, \psi)$ a potential representation of P. Then

$$P(R_p \mid S_p) = \frac{\psi(W_p)}{\sum_{R_p} \psi(W_p)},$$

where again $\sum_{R_p}$ means all the variables in R_p are running through all their possible values.

Proof. Let

$$T_p = V - W_P.$$

Then clearly,

$$S_p = V - (T_p \cup R_p)$$

and therefore

$$P(R_p \mid S_p) = \frac{P(R_p \cup S_p)}{P(S_p)}$$

$$= \frac{P(W_p)}{P(S_p)}$$

$$= \frac{\sum_{T_p} (K \prod_{i=1}^{p} \psi(W_i))}{\sum_{T_p \cup R_p} (K \prod_{i=1}^{p} \psi(W_i))}$$

$$= \frac{\sum_{T_p} (\prod_{i=1}^{p} \psi(W_i))}{\sum_{R_p} (\sum_{T_p} (\prod_{i=1}^{p} \psi(W_i)))}.$$

The last equality is obtained by dividing out K and noticing that $T_p \cap R_p = \emptyset$. Noting further that $T_p \cap W_p = \emptyset$, we have that this last expression is equal to

$$\frac{\psi(W_p) \sum_{T_p} (\prod_{i=1}^{p-1} \psi(W_i))}{\sum_{R_p} (\psi(W_p) \sum_{T_p} (\prod_{i=1}^{p-1} \psi(W_i)))}.$$

Finally, since $W_i \cap R_p = \emptyset$ for $1 \le i \le p - 1$, this expression is equal to

$$\frac{\psi(W_p)\sum_{T_p}(\prod_{i=1}^{p-1}\psi(W_i))}{(\sum_{R_p}\psi(W_p))\sum_{T_p}(\prod_{i=1}^{p-1}\psi(W_i))},$$

which completes the proof. $\square$

Theorem 7.3. Let V be a finite set of propositional variables, P a joint probability distribution of V, and $(\{W_i \text{ such that } 1 \le i \le p\}, \psi)$ a potential representation of P. Suppose the ordering $[W_1, W_2, ..., W_p]$ has the running intersection property as defined for cliques in Definition 3.22. Let j, where $j < p$, be such that

$$S_p = W_p \cap (W_1 \cup W_2 \cup \cdots \cup W_{p-1}) \subseteq W_j. \tag{7.1}$$

If we define for $1 \le i \le p - 1$ and $i \ne j$

$$\tilde{\psi}(W_i) = \psi(W_i)$$

and

$$\tilde{\psi}(W_j) = \psi(W_j) \sum_{R_p} \psi(W_p),$$

then $(\{W_1, W_2, ..., W_{p-1}\}, \tilde{\psi})$ is a potential representation of the marginal distribution, relative to P, on $W_1 \cup W_2 \cup \cdots \cup W_{p-1}$.

Proof. Since the sets have the running intersection property, there is at least one j satisfying (7.1). Computing the marginal value, we have

$$P(W_1 \cup W_2 \cup \cdots \cup W_{p-1}) = \sum_{R_p} K \prod_{i=1}^{p} \psi(W_i)$$

$$= K \left(\sum_{R_p}\psi(W_p)\right)\left(\prod_{i=1}^{p-1}\psi(W_i)\right),$$

since $R_p \cap W_i = \emptyset$ for $1 \le i \le p - 1$. Now $\sum_{R_p}\psi(W_p)$ is a function only of the variables in S_p. Since $S_p \subseteq W_j$, we therefore have that $\psi(W_j)\sum_{R_p}\psi(W_p)$ is a function only of the variables in W_j, which completes the proof. $\square$

Lemma 7.1. Let V be a finite set of propositional variables, P a joint probability distribution of V, and $(\{W_i \text{ such that } 1 \le i \le p\}, \psi)$ a potential representation of P. Then

$$P(R_p \mid S_p) = P(R_p \mid W_1 \cup W_2 \cup \cdots \cup W_{p-1}).$$

Proof. Since $V - R_p = W_1 \cup W_2 \cup \cdots \cup W_{p-1}$, we have

$$P(R_p \mid W_1 \cup W_2 \cup \cdots \cup W_{p-1}) = \frac{P(R_p \cup W_1 \cup W_2 \cup \cdots \cup W_{p-1})}{P(W_1 \cup W_2 \cdots \cup W_{p-1})}$$

$$= \frac{P(V)}{P(W_1 \cup W_2 \cup \cdots \cup W_{p-1})}$$

$$= \frac{K \prod_{i=1}^{p} \psi(W_i)}{\sum_{R_p} (K \prod_{i=1}^{p} \psi(W_i))}$$

$$= \frac{\psi(W_p) \prod_{i=1}^{p-1} \psi(W_i)}{(\sum_{R_p} \psi(W_p))(\prod_{i=1}^{p-1} \psi(W_i))}.$$

The last equality is obtained by dividing out K and noticing that $W_i \cap R_p = \emptyset$ for $1 \le i \le p - 1$. Together with Theorem 7.2, this completes the proof. $\square$

Theorem 7.4. Let V be a finite set of propositional variables, P a joint probability distribution of V, and $(\{W_i \text{ such that } 1 \le i \le p\}, \psi)$ a potential representation of P. If the ordering $[W_1, W_2, \ldots, W_p]$ has the running intersection property, then

$$P(V) = P(W_1) \prod_{i=2}^{p} P(R_i \mid S_i).$$

Proof. Mimicking the first step in the proof of Lemma 7.1, we obtain

$$P(V) = P(R_p \mid W_1 \cup W_2 \cup \cdots \cup W_{p-1}) P(W_1 \cup W_2 \cup \cdots \cup W_{p-1})$$

$$= P(R_p \mid S_p) P(W_1 \cup W_2 \cup \cdots \cup W_{p-1}).$$

The last equality is due to Lemma 7.1. Now, owing to Theorem 7.3, there exists a potential representation, $(\{W_1, W_2, \ldots, W_{p-1}\}, \bar{\psi})$ of the marginal distribution, relative to P, on $W_1 \cup W_2 \cup \cdots \cup W_{p-1}$. Therefore, in exactly the same way, we can again mimic the first step in the proof of Lemma 7.1 and apply Lemma 7.1 to obtain

$$P(W_1 \cup W_2 \cup \cdots \cup W_{p-1}) = P(R_{p-1} \mid S_{p-1}) P(W_1 \cup W_2 \cup \cdots \cup W_{p-2}),$$

where we have used the fact that a marginal distribution and the original distribution are equal on the events on which the marginal distribution is defined. The proof is now completed by repeatedly applying these same steps. $\square$

To simplify the notation in the next theorem, we introduce new notation via the definition below:

Definition 7.3. Let V be a finite set of propositional variables, W a subset of V, φ a function which assigns a real number to every combination of values of variables in W, X a subset of V, and X^* a set of possible values of X. Then

$$\varphi(W \leftarrow X^*)$$

means the function φ evaluated on W with the variables in $W \cap X$ instantiated for their values in X^*. Note that $\varphi(W \leftarrow X^*)$ is a function that assigns a unique real number to every combination of values of variables in $W - X$.

Example 7.3. Let $V = \{A, B, C\}$, $W = \{A, B\}$, $X = \{B, C\}$. Suppose each variable has exactly two possible values, and $\varphi(W)$ is defined as follows:

$$\varphi(a_1, b_1) = 1, \qquad \varphi(a_1, b_2) = 2, \qquad \varphi(a_2, b_1) = 3, \qquad \varphi(a_2, b_2) = 4.$$

If $X^* = \{b_2, c_1\}$, then

$$\varphi(W \leftarrow X^*) = \varphi(A, b_2).$$

If we denote $\hat{W} = W - X = \{A\}$ and define $\hat{\varphi}(\hat{W}) = \varphi(W \leftarrow X^*)$, then

$$\hat{\varphi}(a_1) = \varphi(a_1, b_2) = 2.$$

Theorem 7.5. Let V be a finite set of propositional variables, P a joint probability distribution of V, $(\{W_i \text{ such that } 1 \leq i \leq p\}, \psi)$ be a potential representation of P, X a subset V, and X^* a set of values of variables in X. If we define

$$\hat{W}_i = W_i - X \qquad \text{and} \qquad \hat{\psi}(\hat{W}_i) = \psi(W_i \leftarrow X^*),$$

then $(\{\hat{W}_i \text{ such that } 1 \leq i \leq p\}, \hat{\psi})$ is a potential representation of

$$P'(V - X) = P(V - X \mid X^*).$$

That is, P' is defined to be the joint distribution of $V - X$ which is the conditional probability, relative to P, of $V - X$ given X^*.

Proof. We have

$$P(V - X \mid X^*) = \frac{P((V - X) \cup X^*)}{P(X^*)}$$

$$= \frac{P(V \leftarrow X^*)}{P(X^*)}$$

$$= \frac{K \prod_{i=1}^{p} \psi(W_i \leftarrow X^*)}{P(X^*)},$$

which completes the proof if we let $\hat{K} = K / P(X^*)$. $\square$

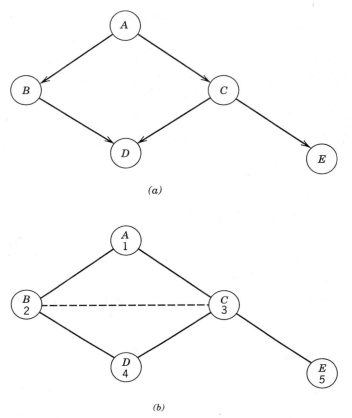

FIGURE 7.1   The undirected graph in (b) is the moral graph relative to the DAG in (a). The new arc is dashed. The vertices in Figure 7.1(b) have been ordered according to maximum cardinality search.

The next theorem has to do specifically with causal networks. First we need the following definition:

Definition 7.4. Let $G = (V, E)$ be a DAG. If $v \in V$, and u and w are parents of v, create an undirected arc between u and w (i.e., "marry" the parents). Do this for every $v \in V$ and every pair of parents of v, and call the set of all these arcs F. If $G_m = (V, E')$ is the undirected graph formed by including all the arcs in E (with their directions dropped) and all the arcs in F, then G_m is called the moral graph relative to G.

The undirected graph in Figure 7.1(b) is the moral graph relative to the DAG in Figure 7.1(a), and the undirected graph in Figure 7.2(b) is the moral graph relative to the DAG in Figure 7.2(a).

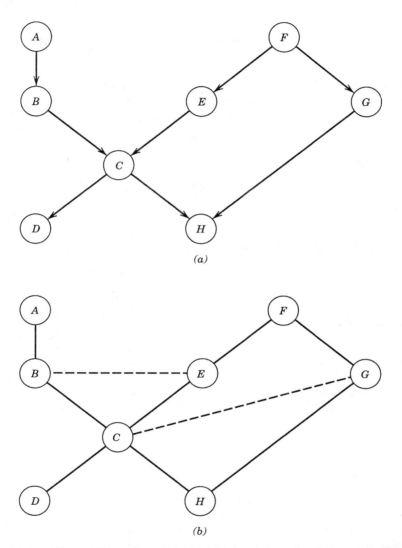

(a)

(b)

FIGURE 7.2   The undirected graph in (b) is the moral graph relative to the DAG in (a). The new arcs are dashed.

Theorem 7.6. Let $C = (V, E, P)$ be a causal network, $G = (V, E)$ the DAG in C, G_m the moral graph relative to G, and G_u a graph formed by triangulating G_m (as discussed in section 3.2). Furthermore, let $\{\text{Clq}_i \text{ such that } 1 \leq i \leq p\}$ be the cliques of G_u (here, we will denote a clique by Clq, since C or C is often used to represent a propositional variable). For each $v \in V$, assign a unique clique Clq_i such that

$$\{v\} \cup c(v) \subseteq \mathrm{Clq}_i.$$

This is always possible, since parents in the original graph are married and therefore $\{v\} \cup c(v)$ is a complete set in G_m and thus in G_u. Recall that every complete set is a subset of at least one clique. It is possible that a complete set could be a subset of more than one clique. In this case the choice is arbitrary; however, v must be assigned to only one clique. The clique assigned to v will be denoted as $f(v)$. Next define for $1 \le i \le p$

$$\psi(\mathrm{Clq}_i) = \prod_{f(v)=\mathrm{Clq}_i} P(v \mid c(v)), \qquad (7.2)$$

where the product means that $P(v \mid c(v))$ is included in the product if $f(v) = \mathrm{Clq}_i$ (i.e., v is assigned to Clq_i). The product is meant to represent the value 1 if there is no v assigned to Clq_i. Then

$$(\{\mathrm{Clq}_i \text{ such that } 1 \le i \le p\}, \psi)$$

is a potential representation of P.

Proof. Every v is assigned to exactly one clique. Therefore, for every v, $P(v \mid c(v))$ is included in the product in (7.2) for exactly one i. Clearly, no other terms are in that product. The proof therefore follows immediately from Theorem 5.1, which states that

$$P(V) = \prod_{v \in V} P(v \mid c(v)). \quad \square$$

Example 7.4. Looking back at Examples 5.13 and 5.14, we see that the undirected graph in Figure 7.1(b) is the moral graph relative to the DAG in the causal network in those examples. The moral graph is already triangulated; consequently, there is no need to triangulate. The cliques of this undirected graph are

$$\{A, B, C\}, \{B, C, D\}, \{C, E\}.$$

Due to Theorem 7.6, if we define

$$\psi(A, B, C) = P(C \mid A)P(B \mid A)P(A)$$

$$\psi(B, C, D) = P(D \mid B, C)$$

$$\psi(C, E) = P(E \mid C),$$

then $(\{\{A,B,C\},\{B,C,D\},\{C,E\}\},\psi)$ is a potential representation of P. That is,

$$P(V) = \psi(A,B,C)\psi(B,C,D)\psi(C,E).$$

We looked back at Examples 5.13 and 5.14 to determine the clique in which each variable and its parents are contained.

Example 7.5. This example is a continuation of Problem 5.5.2. Based on the verbal description in that problem, let

$$a_1 = \text{person visited Asia}$$

$$b_1 = \text{tuberculosis present}$$

$$c_1 = \text{lung cancer or tuberculosis present}$$

$$d_1 = \text{positive X ray}$$

$$e_1 = \text{lung cancer present}$$

$$f_1 = \text{person is a smoker}$$

$$g_1 = \text{bronchitis present}$$

$$h_1 = \text{dyspnea present.}$$

Since each variable has exactly two alternatives, only one alternative has been listed. For example, $a_2 = $ no visit to Asia. The DAG in Figure 7.2(a) then represents the causal relationships described in Problem 5.5.2. Notice that the variable C represents the presence of lung cancer or tuberculosis. Based on the description of the problem, a positive X ray is equally likely given either lung cancer or tuberculosis. The same is true for dyspnea. Hence the use of the variable C is a shorthand way of having arcs from B to both D and H and arcs from E to both D and H. When we assign probabilities, $P(c_1 \mid b_1,e_1)$, $P(c_1 \mid b_1,e_2)$, $P(c_1 \mid b_2,e_1)$ will all be 1, while the $P(c_1 \mid b_2,e_2)$ will be 0.

As indicated earlier, the undirected graph in Figure 7.2(b) is the moral graph relative to the DAG in Figure 7.2(a). Note that this graph is not triangulated, since the simple cycle $[C,E,F,G]$ does not possess a chord. If we triangulate the graph by adding the arc (E,G), we obtain the triangulated graph in Figure 7.3. We could have added the arc (C,F); the choice is arbitrary. The cliques of this undirected graph are

$$
\begin{array}{ccc}
\{A,B\} & \{B,E,C\} & \{E,G,F\} \\
\{C,D\} & \{E,C,G\} & \{C,G,H\}.
\end{array}
$$

Owing to Theorem 7.6, if we define

$$\psi(A,B) = P(B \mid A)P(A) \qquad\qquad \psi(B,E,C) = P(C \mid B,E)$$

$$\psi(E,G,F) = P(E \mid F)P(G \mid F)P(F) \qquad \psi(C,D) = P(D \mid C)$$

$$\psi(E,C,G) = 1 \qquad\qquad \psi(C,G,H) = P(H \mid C,G),$$

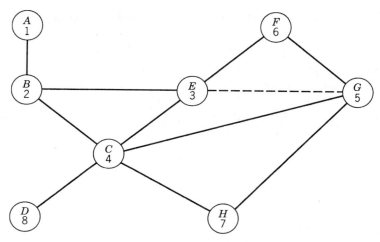

FIGURE 7.3 The graph obtained by triangulating the graph in Figure 7.2(b). The new arc is dashed. The vertices have been ordered according to maximum cardinality search.

then $(\{\{A,B\},\{B,E,C\},\{E,G,F\},\{C,D\},\{E,C,G\},\{C,G,H\}\},\psi)$ is a potential representation of P. That is,

$$P(V) = \psi(A,B)\psi(B,E,C)\psi(E,G,F)\psi(C,D)\psi(E,C,G)\psi(C,G,H).$$

We looked at Figure 7.2(a) to determine the clique in which each variable and its parents are contained. Notice that the clique $\{E,C,G\}$ does not contain any variable along with all its parents, and hence $\psi(E,C,G)$ is equal to 1.

The final theorem is for probability spaces in general. It is included here because this is where we have occasion to apply it.

Theorem 7.7. Let $(\Omega,\mathcal{F},P)$ be a probability space, E_1 an event in $\mathcal{F}$ such that $P(E_1) > 0$, and P' the conditional probability measure defined on $\mathcal{F}$ as follows:

$$P'(E) = P(E \mid E_1).$$

Then, for every two events E_2 and E_3 in $\mathcal{F}$, we have

$$P'(E_3 \mid E_2) = P(E_3 \mid E_1 \wedge E_2).$$

That is, the conditional probability of E_3 given both E_1 and E_2 is equal to the conditional probability, in the space that E_1 has already occurred, of E_3 given E_2.

Proof. By the definition of conditional probability, we have

$$P'(E_3 \mid E_2) = \frac{P'(E_2 \wedge E_3)}{P'(E_2)}$$

$$= \frac{P(E_2 \wedge E_3 \mid E_1)}{P(E_2 \mid E_1)}$$

$$= \frac{P(E_2 \wedge E_3 \wedge E_1)/P(E_1)}{P(E_2 \wedge E_1)/P(E_1)}$$

$$= \frac{P(E_2 \wedge E_3 \wedge E_1)}{P(E_2 \wedge E_1)}$$

$$= P(E_3 \mid E_2 \wedge E_1). \quad \square$$

7.2. ILLUSTRATING THE METHOD OF PROBABILITY PROPAGATION IN TREES OF CLIQUES

By continuing Example 7.4, we illustrate how the theorems in the previous section enable us to compute the probabilities of remaining variables given that certain variables have realized instantiated values. As in chapter 6, we will call this probability P'. Initially, P' is the a priori probability P. We determine the a priori probability as follows:

In Example 7.4, we identified the cliques, $\{Clq_1, Clq_2, Clq_3\}$, of G_u and used these cliques in a potential representation, $(\{Clq_1, Clq_2, Clq_3\}, \psi)$, of P. Next we obtain an ordering of these cliques which has the running intersection property. An ordering of the vertices according to maximum cardinality search (see section 3.2) is

$$[A, B, C, D, E].$$

This ordering is depicted in Figure 7.1(b). If we order the cliques according to their highest labeled vertex, we obtain the following clique ordering:

$$Clq_1 = \{A, B, C\}, \qquad Clq_2 = \{B, C, D\}, \qquad Clq_3 = \{C, E\}.$$

Due to Theorem 3.1, this clique ordering has the running intersection property. Therefore, for each i such that $2 \le i \le 3$, there exists at least one $j < i$ such that

$$S_i = Clq_i \cap (Clq_1 \cup Clq_2 \cup \cdots \cup Clq_{i-1}) \subseteq Clq_j.$$

S_i and R_i are defined as in Definition 7.2. In this example, identifying such a j is a simple matter:

$$S_2 \subseteq Clq_1 \qquad \text{and} \qquad S_3 \subseteq Clq_2.$$

After these *parent* cliques are identified, we build a tree of cliques according to the parents identified. In a more complex example, there can be more than

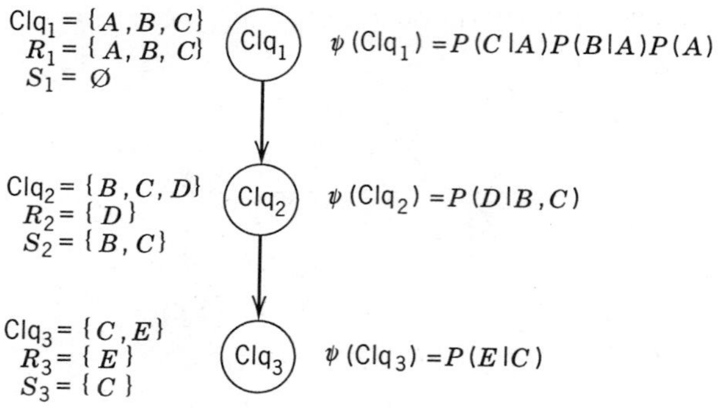

$Clq_1 = \{A, B, C\}$
$R_1 = \{A, B, C\}$ $\quad \boxed{Clq_1}$ $\quad \psi(Clq_1) = P(C|A)P(B|A)P(A)$
$S_1 = \emptyset$

$Clq_2 = \{B, C, D\}$
$R_2 = \{D\}$ $\quad \boxed{Clq_2}$ $\quad \psi(Clq_2) = P(D|B, C)$
$S_2 = \{B, C\}$

$Clq_3 = \{C, E\}$
$R_3 = \{E\}$ $\quad \boxed{Clq_3}$ $\quad \psi(Clq_3) = P(E|C)$
$S_3 = \{C\}$

FIGURE 7.4   The permanent tree of cliques for the cancer example (Example 7.4).

one possible parent. In that case the choice is arbitrary. Stored at each vertex in the tree are Clq_i, R_i, S_i, and $\psi(Clq_i)$. It is a trivial exercise to determine R_i and S_i. In Example 7.4, we determined the function $\psi(Clq_i)$ for $1 \leq i \leq 3$. Using the conditional probabilities from Example 5.14, we determine the values of $\psi(Clq_i)$ as follows:

$$\psi(Clq_1) = P(C \mid A)P(B \mid A)P(A).$$

For example

$$\psi(a_1, b_1, c_1) = P(b_1 \mid a_1)P(c_1 \mid a_1)P(a_1)$$
$$= (.8)(.2)(.2) = .032.$$

It is left as an exercise to compute the remaining values. Figure 7.4 shows the structure of the tree, and Table 7.1 contains all the values stored in the tree. Owing to the large amount of data in these examples, we will always illustrate the structure of the tree with a figure and show the actual data in a table. In the tables we also identify the parent cliques so that the structure can be determined from the table as well.

Once this tree is built, it is stored permanently until the causal network is modified. The tree is therefore a permanent part of the expert system. All probabilities are determined from the tree rather than from the DAG in the causal network. During a particular use of the expert system, a copy of the permanent tree is made. In addition to the information in the permanent tree, $P'(Clq_i')$, the current probability based on the variables thus far instantiated, is stored at each vertex in the copy. We will call the information stored at each vertex in the copy Clq_i', R_i', S_i', $\psi'(Clq_i')$, and $P'(Clq_i')$. Initially, Clq_i', R_i', S_i', $\psi'(Clq_i')$ all have the same values as Clq_i, R_i, S_i, and $\psi(Clq_i)$, respectively, while $P'(Clq_i')$ is not yet determined. We determine probabilities by changing the state of this tree. This is accomplished by passing messages among the vertices in the tree.

TABLE 7.1 The Values in the Permanent Tree of Cliques in the Cancer Example (Example 7.4)

Clq No.	Parent	Clq_i	R_i	S_i	Configuration	$\psi(\mathrm{Clq}_i)$
1	—	$\{A,B,C\}$	$\{A,B,C\}$	$\varnothing$	$\{a_1,b_1,c_1\}$	.032
					$\{a_1,b_1,c_2\}$	.128
					$\{a_1,b_2,c_1\}$	.008
					$\{a_1,b_2,c_2\}$	.032
					$\{a_2,b_1,c_1\}$	.008
					$\{a_2,b_1,c_2\}$	.152
					$\{a_2,b_2,c_1\}$	.032
					$\{a_2,b_2,c_2\}$	.608
2	1	$\{B,C,D\}$	$\{D\}$	$\{B,C\}$	$\{b_1,c_1,d_1\}$	.8
					$\{b_1,c_1,d_2\}$	.2
					$\{b_1,c_2,d_1\}$	.9
					$\{b_1,c_2,d_2\}$	.1
					$\{b_2,c_1,d_1\}$	.7
					$\{b_2,c_1,d_2\}$	.3
					$\{b_2,c_2,d_1\}$	.05
					$\{b_2,c_2,d_2\}$	.95
3	2	$\{C,E\}$	$\{E\}$	$\{C\}$	$\{c_1,e_1\}$	.8
					$\{c_1,e_2\}$	.2
					$\{c_2,e_1\}$	.6
					$\{c_2,e_2\}$	.4

Recall that $P'(\mathrm{Clq}_i')$ is the current probability of the clique Clq_i', conditional on the variables thus far instantiated. Initially, P' is the a priori probability, which is the probability that we currently wish to ascertain. It is trivial to detemine the variable probabilities from the clique probabilities. We obtain the clique probabilities as follows:

Owing to Theorem 7.1,

$$P'(\mathrm{Clq}_i' \mid S_i') = P'(R_i' \mid S_i')$$

and therefore, since $S_i' \subseteq \mathrm{Clq}_i'$,

$$P'(\mathrm{Clq}_i') = P'(R_i' \mid S_i')P'(S_i').$$

Thus, if for each i we can determine $P'(R_i' \mid S_i')$ and $P'(S_i')$, we will be done.

First we determine $P'(R_3' \mid S_3')$. At this point, $P' = P$ and, for $1 \le i \le 3$, $\mathrm{Clq}_i' = \mathrm{Clq}_i$, $R_i' = R_i$, $S_i' = S_i$ and $\psi'(\mathrm{Clq}_i') = \psi(\mathrm{Clq}_i)$. Therefore $(\{\mathrm{Clq}_1', \mathrm{Clq}_2', \mathrm{Clq}_3'\}, \psi')$ is a potential representation of P' and we have, owing to Theorem 7.2, that

$$P'(R_3' \mid S_3') = \frac{\psi'(\mathrm{Clq}_3')}{\sum_{R_3'} \psi'(\mathrm{Clq}_3')}.$$

For example,

$$P'(e_1 \mid c_1) = \frac{\psi'(c_1, e_1)}{\psi'(c_1, e_1) + \psi'(c_1, e_2)}$$

$$= \frac{.8}{.8 + .2} = .8.$$

It turns out, in this example, the denominator is 1 for all the calculations of $P'(e_i \mid c_j)$. This is not always the case.

Next, to obtain $P'(R_2' \mid S_2')$, we proceed as follows. Owing to Theorem 7.3, if we redefine $\psi'(\text{Clq}_2')$ as follows:

$$\psi'(\text{Clq}_2') = \psi'(\text{Clq}_2') \sum_{R_3'} \psi'(\text{Clq}_3'),$$

then $(\{\text{Clq}_1', \text{Clq}_2'\}, \psi')$ is now a potential representation of the marginal probability distribution of $\text{Clq}_1' \cup \text{Clq}_2'$. Since all values of $\sum_{R_3'} \psi'(\text{Clq}_3')$ are 1 in this example, $\psi'(\text{Clq}_2')$ is unchanged in this example. We see that $\sum_{R_3'} \psi'(\text{Clq}_3')$ is a message which Clq_3' sends to Clq_2'.

Before leaving Clq_3', we redefine $\psi'(\text{Clq}_3')$ as follows (in this example it will be unchanged):

$$\psi'(\text{Clq}_3') = P'(R_3' \mid S_3').$$

We will see shortly why we are redefining $\psi'(\text{Clq}_3')$.

Proceeding now to Clq_2', again using Theorem 7.2, we have

$$P'(R_2' \mid S_2') = \frac{\psi'(\text{Clq}_2')}{\sum_{R_2'} \psi'(\text{Clq}_2')}.$$

For instance,

$$P'(d_2 \mid b_1, c_1) = \frac{\psi'(b_1, c_1, d_2)}{\psi'(b_1, c_1, d_1) + \psi'(b_1, c_1, d_2)}$$

$$= \frac{.2}{.8 + .2} = .2.$$

It turns out again that the denominators are all 1 in this example.

We proceed in the same fashion to obtain $P'(R_1' \mid S_1')$. Again applying Theorem 7.3, if we redefine

$$\psi'(\text{Clq}_1') = \psi'(\text{Clq}_1') \sum_{R_2'} \psi'(\text{Clq}_2'),$$

then $(\{\text{Clq}_1'\}, \psi')$ is a potential representation of the marginal distribution of Clq_1'. Since all values of $\sum_{R_2'} \psi'(\text{Clq}_2')$ are 1 in this example, $\psi'(\text{Clq}_1')$ remains unchanged in this example. Again we see that $\sum_{R_2'} \psi'(\text{Clq}_2')$ is a message which Clq_2' sends to Clq_1'.

TABLE 7.2   The State of the Copy Tree of Cliques in the Cancer Example (Example 7.4) After the A Priori Probabilities Are Computed

Clq$'$ No.	Parent	Clq$'_i$	R'_i	S'_i	Configuration	$\psi'(\text{Clq}'_i)$	$P'(\text{Clq}'_i)$
1	—	$\{A,B,C\}$	$\{A,B,C\}$	$\emptyset$	$\{a_1,b_1,c_1\}$	.032	.032
					$\{a_1,b_1,c_2\}$	.128	.128
					$\{a_1,b_2,c_1\}$	.008	.008
					$\{a_1,b_2,c_2\}$	.032	.032
					$\{a_2,b_1,c_1\}$	.008	.008
					$\{a_2,b_1,c_2\}$	.152	.152
					$\{a_2,b_2,c_1\}$	.032	.032
					$\{a_2,b_2,c_2\}$	.608	.608
2	1	$\{B,C,D\}$	$\{D\}$	$\{B,C\}$	$\{b_1,c_1,d_1\}$	.8	.032
					$\{b_1,c_1,d_2\}$	.2	.008
					$\{b_1,c_2,d_1\}$	.9	.252
					$\{b_1,c_2,d_2\}$	.1	.028
					$\{b_2,c_1,d_1\}$	.7	.028
					$\{b_2,c_1,d_2\}$	.3	.012
					$\{b_2,c_2,d_1\}$	.05	.032
					$\{b_2,c_2,d_2\}$	.95	.608
3	2	$\{C,E\}$	$\{E\}$	$\{C\}$	$\{c_1,e_1\}$	.8	.064
					$\{c_1,e_2\}$	.2	.016
					$\{c_2,e_1\}$	.6	.552
					$\{c_2,e_2\}$	.4	.368

Before leaving Clq'_2, we redefine $\psi'(\text{Clq}'_2)$ as follows:

$$\psi'(\text{Clq}'_2) = P'(R'_2 \mid S'_2).$$

Proceeding now to Clq'_1, again applying Theorem 7.2, we have

$$P'(\text{Clq}'_1 \mid S'_1) = P'(R'_1 \mid S'_1) = \frac{\psi'\text{Clq}'_1)}{\sum_{R'_1} \psi'(\text{Clq}'_1)}.$$

But since S'_1 is always equal to $\emptyset$, this value is just $P'(\text{Clq}'_1)$. For example,

$$P'(a_1,b_1,c_1) = \frac{\psi'(a_1,b_1,c_1)}{\sum_{i=1}^{2} \sum_{j=1}^{2} \sum_{k=1}^{2} \psi'(a_i,b_j,c_k)}$$

$$= \frac{.032}{.032 + .128 + .008 + .032 + .008 + .152 + .032 + .608} = .032.$$

It turns out again that all the denominators are 1. Again we redefine $\psi'(\text{Clq}'_1)$ as follows:

$$\psi'(\text{Clq}'_1) = P'(\text{Clq}'_1).$$

The new values of $\psi'(\text{Clq}'_i)$ (which in this example have not been changed), along with the values of $P'(\text{Clq}'_1)$, appear in Table 7.2.

Now that we have $P'(\mathrm{Clq}_1')$, we can obtain $P'(\mathrm{Clq}_2')$ as follows. Owing to Theorem 7.1,

$$P'(\mathrm{Clq}_2') = P'(R_2' \mid S_2')P'(S_2')$$
$$= \psi'(\mathrm{Clq}_2')P'(S_2').$$

Recall that $\psi'(\mathrm{Clq}_2')$ was redefined to be $P'(R_2' \mid S_2')$. Now since $S_2' \subseteq \mathrm{Clq}_1'$, $P'(S_2')$ can be computed from $P'(\mathrm{Clq}_1')$. That is,

$$P'(S_2') = \sum_{\mathrm{Clq}_1' - S_2'} P'(\mathrm{Clq}_1').$$

or

$$P'(B,C) = \sum_{A} P'(A,B,C).$$

For example,

$$P'(b_1, c_1) = P'(a_1, b_1, c_1) + P'(a_2, b_1, c_1)$$
$$= .032 + .008 = .040.$$

We then have

$$P'(b_1, c_1, d_1) = \psi'(b_1, c_1, d_1)P'(b_1, c_1)$$
$$= (.8)(.04) = .032.$$

We see that $P'(S_2')$ is a message which Clq_1' sends to Clq_2'. After computing all the values of $P'(\mathrm{Clq}_2')$, in the same way we can use those values to obtain $P'(\mathrm{Clq}_3')$. This is due to the fact that Clq_3' is a child of Clq_2' and therefore $S_3' \subseteq \mathrm{Clq}_2'$. In this fashion, regardless of the size of the tree, we can propagate information from parents to children until leaves are reached and the values of $P'(\mathrm{Clq}_i')$ are determined for every i. The values of $P'(\mathrm{Clq}_i')$ for $1 \leq i \leq 3$ are in Table 7.2, which contains the state of the copy tree after the a priori probabilities are computed. In practice, $P(\mathrm{Clq}_i)$ could be stored at each vertex in the permanent tree, and the previous determination of the a priori probabilities would only need to be performed once, when the permanent tree was created. The values of $P(\mathrm{Clq}_i)$ would then be copied into $P'(\mathrm{Clq}_i')$ when the copy tree was created. We have not included $P(\mathrm{Clq}_i)$ in the permanent tree to emphasize that it is not part of the data necessary to propagation. Once $P'(\mathrm{Clq}_i')$ is determined, it is a trivial calculation to obtain the variable probabilities from the clique probabilities. For example,

$$P'(A) = \sum_{B,C} P'(A,B,C).$$

It is left as an exercise to actually compute these values.

Suppose next that D is instantiated for d_1 and we wish to calculate the conditional probabilities of the remaining variables based on this instantiation. We obtain them in the same manner as we obtained the a priori probabilities. That is, after performing one new operation, we perform the exact same operations on the copy tree as when the a priori probabilities were computed. However, we start with the tree in the state in Table 7.2.

First, since for each i the current value of $\psi'(\mathrm{Clq}_i')$ is $P'(R_i' \mid S_i')$, we have, owing to Theorem 7.4, that $(\{\mathrm{Clq}_1', \mathrm{Clq}_2', \mathrm{Clq}_3'\}, \psi')$ is a potential representation of our current P'. We see now why we redefine $\psi'(\mathrm{Clq}_i')$ to be $P'(R_i' \mid S_i')$ during the process of computing the probabilities. It is so that $(\{\mathrm{Clq}_1', \mathrm{Clq}_2', \mathrm{Clq}_3'\}, \psi')$ is always a potential representation of our current P'. We use this fact as follows. Owing to $(\{\mathrm{Clq}_1', \mathrm{Clq}_2', \mathrm{Clq}_3'\}, \psi')$ being a potential representation of P' and to Theorem 7.5, if we set

$$\hat{\mathrm{Clq}}_1 = \mathrm{Clq}_1' - \{D\} = \{A, B, C\}$$

$$\hat{\mathrm{Clq}}_2 = \mathrm{Clq}_2' - \{D\} = \{B, C\}$$

$$\hat{\mathrm{Clq}}_3 = \mathrm{Clq}_3' - \{D\} = \{C, E\}$$

and for $1 \leq i \leq 3$ define $\hat{\psi}$ as follows:

$$\hat{\psi}(\hat{\mathrm{Clq}}_i) = \psi'(\mathrm{Clq}_i' \leftarrow \{d_1\}),$$

then $(\{\hat{\mathrm{Clq}}_1, \hat{\mathrm{Clq}}_2, \hat{\mathrm{Clq}}_3\}, \hat{\psi})$ is a potential representation of $P'(V - \{D\} \mid d_1)$, where P' still represents our old P', namely the a priori probability. This distribution will be our new P' (recall P' is always defined to be the conditional probability of remaining variables given the variables thus far instantiated). Our goal is to determine the values of this new P'. Since we have a potential representation of this new P', we can proceed in exactly the same way as when we computed the a priori probabilities. First, however, we must actually compute $\hat{\psi}(\hat{\mathrm{Clq}}_i)$ for $1 \leq i \leq 3$. To that end, we have

$$\hat{\psi}(\hat{\mathrm{Clq}}_1) = \hat{\psi}(A, B, C) = \psi'(\{A, B, C\} \leftarrow \{d_1\}) = \psi'(A, B, C).$$

Hence all of $\hat{\mathrm{Clq}}_1$'s potential values are the same as those of Clq_1'. The same relationship will hold between $\hat{\mathrm{Clq}}_3$ and Clq_3', since D is also not in Clq_3'. For Clq_2', we have

$$\hat{\psi}(\hat{\mathrm{Clq}}_2) = \hat{\psi}(B, C) = \psi'(\{B, C, D\} \leftarrow \{d_1\}) = \psi'(B, C, d_1).$$

For example,

$$\hat{\psi}(b_1, c_1) = \psi'(b_1, c_1, d_1) = .8.$$

It is left as an exercise to compute the remaining values. After all these values are computed, for $1 \leq i \leq 3$, Clq_i' and $\psi'(\mathrm{Clq}_i')$ are replaced in the tree by $\hat{\mathrm{Clq}}_i$ and $\hat{\psi}(\hat{\mathrm{Clq}}_i)$, respectively. The new cliques are always subsets of the previous cliques. In a particular application, after instantiation, we have no further need for the original cliques, so we can replace them by the ones with the instantiated values. Thus the structure of the tree remains intact, but the value of the cliques stored at each vertex changes. Table 7.3 contains this new state of the

TABLE 7.3   The State of the Copy Tree of Cliques in the Cancer Example (Example 7.4) After D Is Instantiated but Before Propagation[a]

Clq' No.	Parent	Clq'_i	R'_i	S'_i	Configuration	$\psi'(\text{Clq}'_i)$	$P'(\text{Clq}'_i)$
1	—	$\{A,B,C\}$	$\{A,B,C\}$	$\varnothing$	$\{a_1,b_1,c_1\}$	.032	
					$\{a_1,b_1,c_2\}$	.128	
					$\{a_1,b_2,c_1\}$	.008	
					$\{a_1,b_2,c_2\}$	.032	
					$\{a_2,b_1,c_1\}$	.008	
					$\{a_2,b_1,c_2\}$	.152	
					$\{a_2,b_2,c_1\}$	.032	
					$\{a_2,b_2,c_2\}$	.608	
2	1	$\{B,C\}$	$\varnothing$	$\{B,C\}$	$\{b_1,c_1\}$	.8	
					$\{b_1,c_2\}$	.9	
					$\{b_2,c_1\}$	.7	
					$\{b_2,c_2\}$	.05	
3	2	$\{C,E\}$	$\{E\}$	$\{C\}$	$\{c_1,e_1\}$	.8	
					$\{c_1,e_2\}$	.2	
					$\{c_2,e_1\}$	.6	
					$\{c_2,e_2\}$	.4	

[a] $P'(\text{Clq}_i)$ is not yet determined.

copy tree. It is the state after D is instantiated, but before the values of P' are computed. Notice also in Table 7.3 that for $1 \le i \le 3$,

$$R'_i = R'_i - \{D\}$$

and

$$S'_i = S'_i - \{D\}.$$

This is trivially true. Consequently, there is no need to to recompute R'_i and S'_i based on their definitions when Clq'_i changes. We need only subtract the set of instantiated variables.

We now proceed in exactly the same way as when we computed the a priori probabilities. Remember, by P' we now mean the conditional probability based on the instantiation of D for d_1. Again owing to Theorem 7.2,

$$P'(R'_3 \mid S'_3) = \frac{\psi'(\text{Clq}'_3)}{\sum_{R'_3} \psi'(\text{Clq}'_3)}.$$

Since, in this example, the values of $\psi'(\text{Clq}'_3)$ have not changed since we computed the a priori probabilities, the values obtained will be the same as those obtained before.

Next we determine $P'(R'_2 \mid S'_2)$. Owing to Theorem 7.3, if we redefine

$$\psi'(\text{Clq}'_2) = \psi'(\text{Clq}'_2) \sum_{R'_3} \psi(\text{Clq}'_3),$$

then $(\{\text{Clq}'_1, \text{Clq}'_2\}, \psi')$ is a potential representation of the marginal probability distribution of $\text{Clq}'_1 \cup \text{Clq}'_2$. Again, since all values of the $\sum_{R'_3} \psi'(\text{Clq}'_3)$ are 1, $\psi'(\text{Clq}'_2)$ is unchanged in this example.

Before leaving Clq'_3, we again redefine the values of $\psi'(\text{Clq}'_3)$ as follows (in this example it will be unchanged):

$$\psi'(\text{Clq}'_3) = P'(R'_3 \mid S'_3).$$

Proceeding now with Clq'_2, we again use Theorem 7.2 to obtain that

$$P'(R'_2 \mid S'_2) = \frac{\psi'(\text{Clq}'_2)}{\sum_{R'_2} \psi'(\text{Clq}'_2)}.$$

Since $R'_2 = \emptyset$ and $\sum_{\emptyset} \psi'(\text{Clq}'_2) = \psi'(\text{Clq}'_2)$, we have that

$$P'(R'_2 \mid S'_2) = \frac{\psi'(\text{Clq}'_2)}{\sum_{\emptyset} \psi'(\text{Clq}'_2)} = \frac{\psi'(\text{Clq}'_2)}{\psi'(\text{Clq}'_2)} = 1$$

for all combinations of values of variables in Clq'_2. This can be seen immediately, since the probability that the variables in the empty set take specific values is always 1.

Next we determine $P'(R'_1 \mid S'_1)$. Again applying Theorem 7.3, if we set

$$\psi'(\text{Clq}'_1) = \psi'(\text{Clq}'_1) \sum_{R'_2} \psi'(\text{Clq}'_2),$$

then $(\{\text{Clq}'_1\}, \psi')$ is a potential representation of the marginal distribution of Clq'_1. In this case

$$\sum_{R'_2} \psi'(\text{Clq}'_2) = \psi'(\text{Clq}'_2).$$

Thus we have

$$\psi'(\text{Clq}'_1) = \psi'(\text{Clq}'_1)\psi'(\text{Clq}'_2).$$

For example,

$$\psi'(a_1, b_1, c_1) = \psi'(a_1, b_1, c_1)\psi'(b_1, c_1)$$
$$= (.032)(.8) = .0256.$$

Computing the remaining values of $\psi'(\text{Clq}_1')$ yields

$$\psi'(a_1,b_1,c_1) = .0256 \qquad \psi'(a_1,b_1,c_2) = .1152$$
$$\psi'(a_1,b_2,c_1) = .0056 \qquad \psi'(a_1,b_2,c_2) = .0016$$
$$\psi'(a_2,b_1,c_1) = .0064 \qquad \psi'(a_2,b_1,c_2) = .1368$$
$$\psi'(a_2,b_2,c_1) = .0224 \qquad \psi'(a_2,b_2,c_2) = .0304.$$

Before leaving Clq_2', we set

$$\psi'(\text{Clq}_2') = P'(R_2' \mid S_2') = 1.$$

Proceeding now with Clq_1', we apply Theorem 7.2 again to obtain

$$P'(\text{Clq}_1' \mid S_1') = P'(R_1' \mid S_1') = \frac{\psi'(\text{Clq}_1')}{\sum_{R_1'} \psi'(\text{Clq}_1')}.$$

Since S_1' is always equal to $\varnothing$, this value is just $P'(\text{Clq}_1')$. For example,

$$P'(a_1,b_1,c_1) = \frac{\psi'(a_1,b_1,c_1)}{\sum_{i=1}^{2}\sum_{j=1}^{2}\sum_{k=1}^{2}\psi'(a_i,b_j,c_k)}$$

$$= \frac{.0256}{.0256 + .1152 + .0056 + .0016 + .0064 + .1368 + .0224 + .0304}$$

$$= .074.$$

The remaining values have been computed and are listed in Table 7.4. Again we redefine $\psi'(\text{Clq}_1')$ as follows:

$$\psi'(\text{Clq}_1') = P'(\text{Clq}_1').$$

The new values of $\psi'(\text{Clq}_i')$ for $1 \le i \le 3$ are also listed in Table 7.4. Now that we have $P'(\text{Clq}_1')$, we can obtain $P'(\text{Clq}_2')$ as follows. Owing to Theorem 7.1,

$$P'(\text{Clq}_2') = P'(R_2' \mid S_2')P'(S_2')$$
$$= \psi'(\text{Clq}_2')P'(S_2')$$

or

$$P'(B,C) = \psi'(B,C)P'(B,C).$$

However, since $S_2' \subseteq \text{Clq}_1'$, $P'(S_2')$ can be computed from $P'(\text{Clq}_1')$. That is,

$$P'(S_2') = \sum_{\text{Clq}_1' - S_2'} P'(\text{Clq}_1')$$

or

$$P'(B,C) = \sum_{A} P'(A,B,C).$$

TABLE 7.4   The State of the Copy Tree of Cliques in the Cancer Example (Example 7.4) After D Is Instantiated and the New Values of $P'(\mathrm{Clq}_i)$ Are Determined

Clq' No.	Parent	Clq'_i	R'_i	S'_i	Configuration	$\psi'(\mathrm{Clq}'_i)$	$P'(\mathrm{Clq}'_i)$
1	—	$\{A,B,C\}$	$\{A,B,C\}$	Ø	$\{a_1,b_1,c_1\}$	.074	.074
					$\{a_1,b_1,c_2\}$	.335	.335
					$\{a_1,b_2,c_1\}$	.016	.016
					$\{a_1,b_2,c_2\}$	.005	.005
					$\{a_2,b_1,c_1\}$	.019	.019
					$\{a_2,b_1,c_2\}$	.398	.398
					$\{a_2,b_2,c_1\}$	.065	.065
					$\{a_2,b_2,c_2\}$	.088	.088
2	1	$\{B,C\}$	Ø	$\{B,C\}$	$\{b_1,c_1\}$	1	.093
					$\{b_1,c_2\}$	1	.733
					$\{b_2,c_1\}$	1	.081
					$\{b_2,c_2\}$	1	.093
3	2	$\{C,E\}$	$\{E\}$	$\{C\}$	$\{c_1,e_1\}$	.8	.139
					$\{c_1,e_2\}$	.2	.035
					$\{c_2,e_1\}$	.6	.496
					$\{c_2,e_2\}$	.4	.330

For example,

$$P'(b_1,c_1) = P'(a_1,b_1,c_1) + P'(a_2,b_1,c_1)$$
$$= .074 + .019 = .093.$$

We then have

$$P'(b_1,c_1) = \psi'(b_1,c_1)P'(b_1,c_1)$$
$$= (1)(.093) = .093.$$

After computing all the values of $P'(\mathrm{Clq}'_2)$, in the same way we can use those values to obtain $P'(\mathrm{Clq}'_3)$. These values are all listed in Table 7.4. The table contains the entire state of the copy tree after the conditional probabilities given the instantiation of D for d_1 have been determined.

Therefore we see that the procedure for computing the conditional probabilities based on the instantiation of a variable is identical to that used to compute the a priori probabilities. Suppose next that we wish to compute the conditional probabilities of remaining variables given that D has already been instantiated for d_1 and C is now instantiated for c_2. This is the conditional probability given both C and D. Owing to Theorem 7.7, we can compute it from the conditional distribution given D. We proceed in exactly the same manner. In the present state of the copy tree (i.e., the state in Table 7.4),

$(\{Clq'_1, Clq'_2, Clq'_3\}, \psi')$ is a potential representation of the conditional distribution given that D is instantiated for d_1. We can thus reapply the exact same procedure using these cliques and potentials. That is, we start out by setting

$$\hat{Clq}_1 = Clq'_1 - \{C\} = \{A, B, C\} - \{C\} = \{A, B\}$$
$$\hat{Clq}_2 = Clq'_2 - \{C\} = \{B, C\} - \{C\} = \{B\}$$
$$\hat{Clq}_3 = Clq'_3 - \{C\} = \{C, E\} - \{C\} = \{E\}.$$

Next we proceed in exactly the same manner as when D was instantiated. It is left as an exercise to actually perform these steps.

Note that when we calculated the conditional distribution given the instantiation of a variable, there was nothing in the procedure to limit us to the instantiation of only one variable. Thus D and C do not need to be instantiated in sequence. We can instantiate them simultaneously and compute the conditional distribution based on their instantiations in only one pass. That is, starting with the network in the state shown in Table 7.2, we set

$$\hat{Clq}_1 = Clq'_1 - \{C, D\} = \{A, B, C\} - \{C, D\} = \{A, B\}.$$
$$\hat{Clq}_2 = Clq'_2 - \{C, D\} = \{B, C, D\} - \{C, D\} = \{B\}$$
$$\hat{Clq}_3 = Clq'_3 - \{C, D\} = \{C, E\} - \{C, D\} = \{E\}.$$

For $1 \le i \le 3$ we must also set

$$R'_i = R'_i - \{C, D\}$$

and

$$S'_i = S'_i - \{C, D\}.$$

We then proceed in exactly the same manner as when one variable was instantiated. It is left as an exercise to complete these steps and show that the same values are obtained as when the variables are instantiated in sequence. Therefore we have the option of accumulating evidence and computing the combined effect quickly in one pass, or computing the effect as the evidence arrives. The former option would often be taken when a large amount of evidence arrives simultaneously.

In the next section we give a high-level algorithm for this procedure.

PROBLEMS 7.2

1. Perform all the calculations in the cancer example (Example 7.4) which were left as exercises.

2. Starting with the copy tree in the initialized state in the cancer example (i.e., the state in Table 7.2), instantiate D for d_1 and C for c_2 simultaneously. Substantiate that the results are the same as when they are instantiated in sequence.

7.3. AN ALGORITHM FOR PROBABILITY PROPAGATION IN TREES OF CLIQUES

Below is a high-level algorithm for propagating probabilities in an arbitrary causal network (i.e., the only restriction on the digraph is that it be a DAG). This algorithm follows directly from the seven theorems in section 7.1 in the same way that the examples in section 7.2 followed from those theorems. Following the algorithm is an example of its application.

PROBABILITY PROPAGATION IN TREES OF CLIQUES

Creating the Permanent Tree of Cliques

A. Create the moral graph G_m relative to the DAG in the causal network.
B. Triangulate G_m:
   begin
       1. Create an order α of the vertices in V using maximum cardinality search {algorithm in section 3.2};
       2. Compute the elimination graph G_u of G_m with respect to that order by computing the fill-in of G_m with respect to that order {algorithm in section 3.2}
          {the original order α is a perfect order relative to G_u.}
   end.
C. Determine the cliques in G_u and order them according to their highest labeled vertex according to order α. Let $[\text{Clq}_1, \text{Clq}_2, \ldots, \text{Clq}_p]$ be that order. {Golumbic [1980] contains an algorithm, for determining the cliques of a triangulated graph, which runs in $O(n + e)$ time where n is the number of vertices and e is the number of arcs.}
D. For $i = 1$ to p, do
   begin
       1. $S_i := \text{Clq}_i \cap (\text{Clq}_1 \cup \text{Clq}_2 \cup \cdots \cup \text{Clq}_{i-1})$;
       2. $R_i := \text{Clq}_i - S_i$;
       3. If $i > 1$, then identify a $j < i$ such that Clq_j is a parent of Clq_i. That is, $S_i \subseteq \text{Clq}_j$. If there is more than one such j, the choice is arbitrary.
   end.
E. Create a tree of cliques according to the parents identified in step D.
F. For each $v \in V$, do
   assign v to a unique clique Clq_i such that $v \cup c(v) \subseteq \text{Clq}_i$ {the clique assigned to v is denoted as $f(v)$};

For $i := 1$ to p, do
    begin
        1. $\psi(\text{Clq}_i) := \prod_{f(v)=\text{Clq}_i} P(v \mid c(v))$;
          $\{v$ is included in the product if Clq_i is assigned to $v.\}$
          $\{$If no v is assigned to Clq_i, the product represents 1.$\}$
        2. Store Clq_i, R_i, S_i, and $\psi(\text{Clq}_i)$ at the vertex determined in step F
    end.

The tree becomes a permanent part of the expert system and is changed only when the causal network changes. All the information necessary to determine probabilities is stored in the permanent tree of cliques. Stored at each vertex in the tree are the following:

1. A clique Clq_i where $1 \le i \le p$.
2. S_i.
3. R_i.
4. $\psi(\text{Clq}_i)$.

Once the tree is built there is no longer a need to index the cliques. However, we will continue to refer to them as indexed in order to distinguish them. When the expert system is used, a copy of the permanent tree is made. This is the initialization process:

Initialization

Create a copy of the permanent tree of cliques. At each vertex in this copy tree, store the following:

1. A clique Clq_i' where $1 \le i \le p$.
2. S_i'.
3. R_i'.
4. $\psi'(\text{Clq}_i')$.
5. $P'(\text{Clq}_i')$.

For $1 \le i \le p$, set the initial values of Clq_i', S_i', R_i', and $\psi'(\text{Clq}_i')$ to their respective values in the permanent tree. For $1 \le i \le p$, leave $P'(\text{Clq}_i')$, the current probability of Clq_i' based on variables thus far instantiated, unassigned.

The values of $P'(\text{Clq}_i)$ are not yet determined after initialization. Their initial values, namely the a priori probablities, are obtained using the same updating algorithm as the one used to determine the conditional probabilities based on instantiated values. The updating is all done on the copy tree of cliques; that is, we no longer refer to the original causal network. If we let X be a set of instantiated variables and X^* be a set of instantiated values, the updating algorithm is as follows (when computing the a priori probabilities, $X = \emptyset$):

Updating

{The procedures Lambda_Prop and Pi_Prop, called by this algorithm, are given after the algorithm.}
{In this algorithm, we refer to a vertex by the clique stored at that vertex.}

A. For every vertex Clq_i', in the tree, do
    begin
       1. $\hat{\text{Clq}} := \text{Clq}_i' - X$;
       2. $\hat{\psi}(\hat{\text{Clq}}) := \psi'(\text{Clq}_i' \leftarrow X^*)$;
       3. $\text{Clq}_i' := \hat{\text{Clq}}$;
       4. $\psi'(\text{Clq}_i') := \hat{\psi}(\hat{\text{Clq}})$;
       5. $S_i' := S_i' - X$;
       6. $R_i' := R_i' - X$
    end.
B. For all leaves Clq_i', do Lambda_Prop(i).
C. When a vertex Clq_j', receives a λ message from a child Clq_i', do
    begin
       1. $\psi'(\text{Clq}_j') := \lambda_{\text{Clq}_i'}(S_i')\psi'(\text{Clq}_j')$;
       2. If Clq_j' has received λ messages from all its children, then
          begin
             do Lambda_Prop(j);
             if Clq_j' is the root, then
                begin
                   $P'(\text{Clq}_j') := \psi'(\text{Clq}_j')$;
                   if $P'(\text{Clq}_j') = 0$ for all combinations of values of
                   variables in Clq_j', then Error-Exit;
                   do Pi_Prop(j)
                end
          end
    end.
D. When a vertex Clq_i' receives a π message from its parent Clq_j', do
    begin
       1. $P'(\text{Clq}_i') := \pi_{\text{Clq}_j'}(S_i')\psi'(\text{Clq}_i')$;
       2. If Clq_i' is not a leaf, then do Pi_Prop(i)
    end.

E. When the probabilities of all cliques are determined, for each vertex Clq_i and each variable $v \in \mathrm{Clq}_i$, do

$$P'(v) := \sum_{\substack{w \in \mathrm{Clq}_i' \\ w \neq v}} P'(\mathrm{Clq}_i').$$

Procedure Lambda_Prop (i: integer);
   begin
      $\lambda_{\mathrm{Clq}_i'}(S_i') := \sum_{R_i'} \psi'(\mathrm{Clq}_i');$ {If $R_i' = \emptyset$, the sum is $\psi'(\mathrm{Clq}_i')$.}
                                     {If $S_i' = \emptyset$, the message is a constant.}
      for all values of S_i' for which $\lambda_{\mathrm{Clq}_i'}(S_i') \neq 0$, do

$$\psi'(\mathrm{Clq}_i') := \frac{\psi'(\mathrm{Clq}_i')}{\lambda_{\mathrm{Clq}_i'}(S_i')};$$

      if Clq_i' is not the root, then
         send Clq_i''s parent the message $\lambda_{\mathrm{Clq}_i'}(S_i')$
   end.

Procedure Pi_Prop (j: integer);
   begin
      for all children Clq_i' of Clq_j', do
         begin
            $\pi_{\mathrm{Clq}_j'}(S_i') := \sum_{\mathrm{Clq}_j' - S_i'} P'(\mathrm{Clq}_j');$ {If $\mathrm{Clq}_j' - S_i' = \emptyset$, the sum
                                                        is $P'(\mathrm{Clq}_j')$.}
                                                        {If $\mathrm{Clq}_j' = \emptyset$, $P'(\mathrm{Clq}_j') = 1$;
                                                        this implies that $S_i' = \emptyset$.}
                                                        {If $S_i' = \emptyset$, the message is 1.}
            send Clq_i' the message $\pi_{\mathrm{Clq}_j'}(S_i')$
         end
   end.

Procedure Error-Exit;
   begin
      write ("The joint probability of the set of instantiated values is equal
              to 0");
      halt
   end;

Notice that the λ and π messages sent during one update are not needed in the next update. Thus they are not stored in the tree as was done in Pearl's method. Note further that we exit the algorithm with an error if for the root

Clq'_j, $P'(Clq'_j)$ is equal to 0 for all combinations of values of the variables in Clq'_j. This is a situation which would arise only if we attempted to instantiate a set of values whose current joint probability is equal to 0. It is left as an exercise to show that this is the case. We need the error check because there is no way to check if the joint probability of the set of instantiated values is 0 beforehand without using the chain rule and propagating individual instantiations, the very act which we are trying to avoid by instantiating the variables simultaneously (in chapter 8 we will see how to use the chain rule and individual instantiations to compute the joint probability of several variables).

We will now illustrate this algorithm by continuing Example 7.5. First we must assign conditional probabilities to the causal network. Suppose we have the following fictitious values:

$$P(a_1) = .01 \qquad P(c_1 \mid b_1, e_1) = 1$$
$$P(c_1 \mid b_1, e_2) = 1$$
$$P(b_1 \mid a_1) = .05 \qquad P(c_1 \mid b_2, e_1) = 1$$
$$P(b_1 \mid a_2) = .01 \qquad P(c_1 \mid b_2, e_2) = 0$$

$$P(f_1) = .5 \qquad P(d_1 \mid c_1) = .98$$
$$P(d_1 \mid c_2) = .05$$

$$P(e_1 \mid f_1) = .1$$
$$P(e_1 \mid f_2) = .01 \qquad P(h_1 \mid c_1, g_1) = .9$$
$$P(h_1 \mid c_1, g_2) = .7$$
$$P(g_1 \mid f_1) = .6 \qquad P(h_1 \mid c_2, g_1) = .8$$
$$P(g_1 \mid f_2) = .3 \qquad P(h_1 \mid e_2, g_2) = .1.$$

Again only the probability of one of two alternatives has been listed. Next we perform the routine to create the permanent tree of cliques:

A. We have already done this step in Example 7.5. The moral graph is in Figure 7.2(b).
B. We already triangulated the graph in Example 7.5 by inspection (i.e., without computing the fill-in). The triangulated graph is in Figure 7.3. By Theorem 3.2 we can now do a maximum cardinality search to obtain a perfect ordering of the vertices. The following is such an ordering:

$$[A, B, E, C, G, F, H, D]$$

This ordering is depicted in Figure 7.3.
C. An ordering of the cliques according to their highest labeled vertex is the following:

$$Clq_1 = \{A, B\} \qquad Clq_2 = \{B, E, C\} \qquad Clq_3 = \{E, C, G\}$$
$$Clq_4 = \{E, G, F\} \qquad Clq_5 = \{C, G, H\} \qquad Clq_6 = \{C, D\}.$$

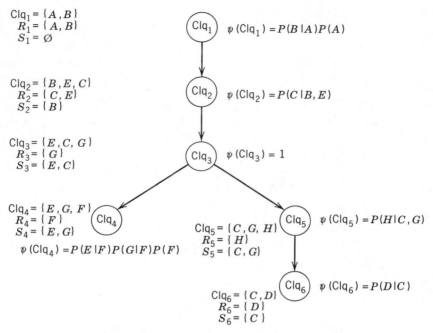

FIGURE 7.5 The permanent tree of cliques for the dyspnea example (Example 7.5).

$D.\quad S_2 = Clq_2 \cap (Clq_1) = \{B,E,C\} \cap \{A,B\} = \{B\}$

$\quad\;\; R_2 = Clq_2 - S_2 = \{B,E,C\} - \{B\} = \{C,E\}.$

It is left as an exercise to determine the remaining values of S_i and R_i for $1 \leq i \leq 6$ and to identify the parents.

E. A tree has been created and is depicted in Figure 7.5. Notice that Clq_5 is the parent of Clq_6. Since $S_6 \subseteq Clq_2$ and $S_6 \subseteq Clq_3$, Clq_2 or Clq_3 also could have been made the parent. The choice is arbitrary.

F. For $1 \leq i \leq 6$, the function $\psi(Clq_i)$ was determined in Example 7.5. Next we must compute the values:

$$\psi(Clq_1) = P(B \mid A)P(A)$$

$$\psi(a_1, b_1) = P(b_1 \mid a_1)P(a_1) = (.05)(.01) = .0005.$$

It is left as an exercise to compute the remaining values. The structure of the tree is depicted in Figure 7.5, while all the values are in Table 7.5.

The tree is now a permanent part of the expert system. Suppose we next wish to use the system. First we perform the initialization procedure which simply creates a copy of the permanent tree with Clq_i', R_i', S_i', $\psi'(Clq_i')$, and $P'(Clq_i')$ stored at each vertex. $P'(Clq_i')$ is given no value, whereas Clq_i', R_i', S_i',

TABLE 7.5 The Values in the Permanent Tree of Cliques in the Dyspnea Example (Example 7.5)

Clq' No.	Parent	Clq_i	R_i	S_i	Configuration	$\psi(Clq_i)$
1	—	$\{A,B\}$	$\{A,B\}$	$\emptyset$	$\{a_1,b_1\}$	.0005
					$\{a_1,b_2\}$	.0095
					$\{a_2,b_1\}$	.0099
					$\{a_2,b_2\}$	.9801
2	1	$\{B,E,C\}$	$\{C,E\}$	$\{B\}$	$\{b_1,e_1,c_1\}$	1
					$\{b_1,e_1,c_2\}$	0
					$\{b_1,e_2,c_1\}$	1
					$\{b_1,e_2,c_2\}$	0
					$\{b_2,e_1,c_1\}$	1
					$\{b_2,e_1,c_2\}$	0
					$\{b_2,e_2,c_1\}$	0
					$\{b_2,e_2,c_2\}$	1
3	2	$\{E,C,G\}$	$\{G\}$	$\{E,C\}$	$\{e_1,c_1,g_1\}$	1
					$\{e_1,c_1,g_2\}$	1
					$\{e_1,c_2,g_1\}$	1
					$\{e_1,c_2,g_2\}$	1
					$\{e_2,c_1,g_1\}$	1
					$\{e_2,c_1,g_2\}$	1
					$\{e_2,c_2,g_1\}$	1
					$\{e_2,c_2,g_2\}$	1
4	3	$\{E,G,F\}$	$\{F\}$	$\{E,G\}$	$\{e_1,g_1,f_1\}$	.03
					$\{e_1,g_1,f_2\}$	.0015
					$\{e_1,g_2,f_1\}$	.02
					$\{e_1,g_2,f_2\}$	.0035
					$\{e_2,g_1,f_1\}$	.27
					$\{e_2,g_1,f_2\}$	.1485
					$\{e_2,g_2,f_1\}$	.18
					$\{e_2,g_2,f_2\}$	.3465
5	3	$\{C,G,H\}$	$\{H\}$	$\{C,G\}$	$\{c_1,g_1,h_1\}$	.9
					$\{c_1,g_1,h_2\}$	.1
					$\{c_1,g_2,h_1\}$	.7
					$\{c_1,g_2,h_2\}$	.3
					$\{c_2,g_1,h_1\}$	.8
					$\{c_2,g_1,h_2\}$	.2
					$\{c_2,g_2,h_1\}$	.1
					$\{c_2,g_2,h_2\}$	.9
6	5	$\{C,D\}$	$\{D\}$	$\{C\}$	$\{c_1,d_1\}$	.98
					$\{c_1,d_2\}$	.02
					$\{c_2,d_1\}$	.05
					$\{c_2,d_2\}$	.95

and $\psi'(\text{Clq}'_i)$ are assigned the values of Clq_i, R_i, S_i, and $\psi(\text{Clq}_i)$, respectively. Next we compute the a priori probabilities using the update algorithm:

A. Since $X = \emptyset$, when computing a priori distributions, all values will be unchanged by this step.

B. Procedure Lambda_Prop executes on the leaf Clq'_4 (the order in which this procedure executes on the leaves is immaterial):

$$\lambda_{\text{Clq}'_4}(S'_4) = \sum_{R'_4} \psi'(\text{Clq}'_4)$$

$$\lambda_{\text{Clq}'_4}(e_1, g_1) = \psi'(e_1, g_1, f_1) + \psi'(e_1, g_1, f_2)$$

$$= .03 + .0015 = .0315$$

$$\lambda_{\text{Clq}'_4}(e_1, g_2) = \psi'(e_1, g_2, f_1) + \psi'(e_1, g_2, f_2)$$

$$= .02 + .0035 = .0235.$$

It is left as an exercise to compute $\lambda_{\text{Clq}'_4}(e_2, g_1)$ and $\lambda_{\text{Clq}'_4}(e_2, g_2)$. We will follow the propagation of only a few values. After all the values are computed, Clq'_4 sends the λ message to Clq'_3. Finishing procedure Lambda_Prop's execution on Clq'_4, we have

$$\psi'(\text{Clq}'_4) = \frac{\psi'(\text{Clq}'_4)}{\lambda_{\text{Clq}'_4}(S'_4)}$$

$$\psi'(e_1, f_1, g_1) = \frac{.03}{.0315} = .9524.$$

This value, along with the other new values of $\psi'(\text{Clq}'_4)$, is listed in Table 7.6.

B. Procedure Lambda_Prop executes on the leaf Clq'_6:

$$\lambda_{\text{Clq}'_6}(S'_6) = \sum_{R'_6} \psi'(\text{Clq}'_6)$$

$$\lambda_{\text{Clq}'_6}(c_1) = \psi'(c_1, d_1) + \psi'(c_1, d_2)$$

$$= .98 + .02 = 1.$$

The other λ value also evaluates to 1. Clq'_6 then sends this λ message to Clq'_5. Finally, in this execution of procedure Lambda_Prop, we have

$$\psi'(\text{Clq}'_6) = \frac{\psi'(\text{Clq}'_6)}{\lambda_{\text{Clq}'_6}(S'_6)}$$

$$\psi'(c_1, d_1) = \frac{.98}{1} = .98.$$

This value, along with the other new values of $\psi'(\text{Clq}'_6)$, is listed in Table 7.6.

TABLE 7.6 The State of the Copy Tree of Cliques in the Dyspnea Example (Example 7.5) After the A Priori Probabilities Are Computed

Clq' No.	Parent	Clq'_i	R'_i	S'_i	Configuration	$\psi'(\text{Clq}'_i)$	$P'(\text{Clq}'_i)$
1	—	$\{A,B\}$	$\{A,B\}$	$\emptyset$	$\{a_1,b_1\}$	.0005	.0005
					$\{a_1,b_2\}$	.0095	.0095
					$\{a_2,b_1\}$	.0099	.0099
					$\{a_2,b_2\}$	.9801	.9801
2	1	$\{B,E,C\}$	$\{C,E\}$	$\{B\}$	$\{b_1,e_1,c_1\}$	.055	.00057
					$\{b_1,e_1,c_2\}$	0	0
					$\{b_1,e_2,c_1\}$	.9450	.00983
					$\{b_1,e_2,c_2\}$	0	0
					$\{b_2,e_1,c_1\}$	.055	.05443
					$\{b_2,e_1,c_2\}$	0	0
					$\{b_2,e_2,c_1\}$	0	0
					$\{b_2,e_2,c_2\}$	.945	.9352
3	2	$\{E,C,G\}$	$\{G\}$	$\{E,C\}$	$\{e_1,c_1,g_1\}$	.5727	.0315
					$\{e_1,c_1,g_2\}$	.4273	.0235
					$\{e_1,c_2,g_1\}$	.5727	0
					$\{e_1,c_2,g_2\}$	.4273	0
					$\{e_2,c_1,g_1\}$	.4429	.00435
					$\{e_2,c_1,g_2\}$	.5571	.00548
					$\{e_2,c_2,g_1\}$	.4429	.4142
					$\{e_2,c_2,g_2\}$	.5571	.521
4	3	$\{E,G,F\}$	$\{F\}$	$\{E,G\}$	$\{e_1,g_1,f_1\}$	.9524	.03
					$\{e_1,g_1,f_2\}$	.0476	.0015
					$\{e_1,g_2,f_1\}$	.8511	.02
					$\{e_1,g_2,f_2\}$	.1489	.0035
					$\{e_2,g_1,f_1\}$	.6452	.27
					$\{e_2,g_1,f_2\}$	.3548	.1486
					$\{e_2,g_2,f_1\}$	.3419	.18
					$\{e_2,g_2,f_2\}$	.6581	.3464
5	3	$\{C,G,H\}$	$\{H\}$	$\{C,G\}$	$\{c_1,g_1,h_1\}$	.9	.03227
					$\{c_1,g_1,h_2\}$	.1	.00359
					$\{c_1,g_2,h_1\}$	.7	.02029
					$\{c_1,g_2,h_2\}$	.3	.00869
					$\{c_2,g_1,h_1\}$	.8	.3314
					$\{c_2,g_1,h_2\}$	.2	.08284
					$\{c_2,g_2,h_1\}$	.1	.05210
					$\{c_2,g_2,h_2\}$	.9	.4689
6	5	$\{C,D\}$	$\{D\}$	$\{C\}$	$\{c_1,d_1\}$	.98	.06354
					$\{c_1,d_2\}$	.02	.00130
					$\{c_2,d_1\}$	.05	.04676
					$\{c_2,d_2\}$	.95	.8884

C.1. Clq_5' receives a λ message from Clq_6':

$$\psi'(\text{Clq}_5') = \lambda_{\text{Clq}_6'}(S_6')\psi'(\text{Clq}_5') = \psi'(\text{Clq}_5'),$$

since this particular λ message is a vector of 1's.

C.2. Since Clq_5' has now received λ messages from all its children, procedure Lambda_Prop executes on Clq_5':

$$\lambda_{\text{Clq}_5'}(S_5') = \sum_{R_5'} \psi'(\text{Clq}_5')$$

$$\lambda_{\text{Clq}_5'}(c_1, g_1) = \psi'(c_1, g_1, h_1) + \psi'(c_1, g_1, h_2)$$

$$= .9 + .1 = 1.$$

The other λ values also all evaluate to 1. Clq_5' then sends this λ message to Clq_3'. Finally, in this execution of procedure Lambda_Prop, we have

$$\psi'(\text{Clq}_5') = \frac{\psi'(\text{Clq}_5')}{\lambda_{\text{Clq}_5'}(S_5')}$$

$$\psi'(c_1, g_1, h_1) = \frac{.9}{1} = .9.$$

This value, along with the other new values of $\psi'(\text{Clq}_5')$, is listed in Table 7.6.

C.1. Clq_3' receives λ messages from both Clq_4' and Clq_5'. Since all values of the message from Clq_5' are 1, we need only concern ourselves with the message from Clq_4'. We have

$$\psi'(\text{Clq}_3') = \lambda_{\text{Clq}_4'}(S_4')\psi'(\text{Clq}_3')$$

$$\psi'(e_1, c_1, g_1) = \lambda_{\text{Clq}_4'}(e_1, g_1)\psi'(e_1, c_1, g_1)$$

$$= (.0315)(1) = .0315$$

$$\psi'(e_1, c_1, g_2) = \lambda_{\text{Clq}_4'}(e_1, g_2)\psi'(e_1, c_1, g_2)$$

$$= (.0235)(1) = .0235.$$

It is left as an exercise to compute the other values of $\psi'(\text{Clq}_3')$.

C.2. Since Clq_3' has now received messages from all its children, procedure Lambda_Prop executes on Clq_3':

$$\lambda_{\text{Clq}_3'}(S_3') = \sum_{R_3'} \psi'(\text{Clq}_3')$$

$$\lambda_{\text{Clq}_3'}(e_1, c_1) = \psi'(e_1, c_1, g_1) + \psi'(e_1, c_1, g_2)$$

$$= .0315 + .0235 = .055.$$

It is left as an exercise to compute the other λ values. Clq_3' then sends this λ message to Clq_2'. Finally, in this execution of procedure Lambda_Prop, we have

$$\psi'(Clq_3') = \frac{\psi'(Clq_3')}{\lambda_{Clq_3'}(S_3')}$$

$$\psi'(e_1, c_1, g_1) = \frac{.0315}{.055} = .5727.$$

This value, along with the other new values of $\psi'(Clq_3')$, is listed in Table 7.6.

The λ message propagation continues up the tree until the root is reached. The new values of $\psi'(Clq_i')$ for $1 \leq i \leq 6$ are listed in Table 7.6. Before discussing the propagation of the π messages, we make an observation. The validity of the steps in the Lambda_Prop routine are based on Theorems 7.2 and 7.3, which are true for the last clique in a set of cliques ordered with the running intersection property. However, we arbitrarily apply the steps starting at the leaves. In the example in the previous section this was immaterial, as the tree was simply an ordered linked list. Thus we automatically processed the cliques in decreasing order starting with the last one, and by the time a clique was processed it became the last clique in a set of cliques ordered with the running intersection property. Therefore the theorems were always applicable. By looking at the present example, we can see why it is not really necessary to process the cliques in decreasing order starting with the last one. It is easy to see that if we process the cliques in this example in decreasing order starting with the sixth one, the message Clq_3' receives from Clq_4' is the same as when we processed the message from Clq_4' first. Accordingly, it does not matter if we process the message from Clq_4' first. This line of reasoning easily generalizes to all vertices in a tree. That is to say, the message coming into a vertex from one branch is unaffected by those coming in from other branches. Consequently, the order in which the messages are passed is immaterial as long as we do not allow a vertex to send a message until it receives its messages from all its children.

We now continue the example. When Clq_1' receives its λ message, since Clq_1' is the root, the remainder of step C.2 executes:
First,

$$P'(Clq_1') = \psi'(Clq_1').$$

Next Pi_Prop executes on Clq_1':

$$\pi_{Clq_1'}(S_2') = \sum_{Clq_1' - S_2'} P'(Clq_1').$$

That is,

$$\pi_{\text{Clq}_1'}(B) = \sum_{\{A,B\}-\{B\}} P'(A,B)$$

$$= \sum_A P'(A,B)$$

$$\pi_{\text{Clq}_1'}(b_1) = P'(a_1,b_1) + P'(a_2,b_1) = .0005 + .0099 = .0104$$

$$\pi_{\text{Clq}_1'}(b_2) = P'(a_1,b_2) + P'(a_2,b_2) = .0095 + .9801 = .9896.$$

Clq_1' sends this π message to Clq_2'.

D.1. Clq_2' receives a π message from Clq_1':

$$P'(\text{Clq}_2') = \pi_{\text{Clq}_1'}(S_2')\psi'(\text{Clq}_2')$$

$$P'(b_1,e_1,c_1) = \pi_{\text{Clq}_1'}(b_1)\psi'(b_1,e_1,c_1)$$

$$= (.0104)(.055) = .00057$$

$$P'(b_2,e_1,c_1) = \pi_{\text{Clq}_1'}(b_2)\psi'(b_2,e_1,c_1)$$

$$= (.9896)(.055) = .05443.$$

It is left as an exercise to compute the remaining values.

D.2. Since Clq_2' is not a leaf, procedure Pi_Prop executes on Clq_2' and therefore Clq_2' sends a π message to Clq_3'.

The propagation of π messages thus continues down to the leaves, thereby determining the probabilities of all the cliques. Table 7.6 contains the state of the copy tree after this propagation is completed.

E. The final step, to compute the probabilities of variables from the probabilities of cliques, is left as an exercise.

Suppose next that a patient arrives and it is known for certain that he has recently visited Asia and has dyspnea. We wish to compute the impact that this has on the probabilities of the other variables in the network. We use the same update procedure employed to compute the a priori probabilities. In this case, X, the set of instantiated variables, is equal to $\{A,H\}$, and X^*, the set of instantiated values, is equal to $\{a_1,h_1\}$. The updating proceeds as follows:

A.1. $\hat{\text{Clq}} = \text{Clq}_1' - X = \{A,B\} - \{A,H\} = \{B\}.$

A.2. $\hat{\psi}(\hat{\text{Clq}}) = \psi'(\text{Clq}'_1 \leftarrow \{a_1, h_1\})$

$\qquad \hat{\psi}(B) = \psi'(\{A, B\} \leftarrow \{a_1, h_1\})$

$\qquad \hat{\psi}(b_1) = \psi'(a_1, b_1) = .0005$

$\qquad \hat{\psi}(b_2) = \psi'(a_1, b_2) = .0095.$

A.3. $\text{Clq}'_1 = \hat{\text{Clq}} = \{B\}.$

A.4. $\psi'(\text{Clq}'_1) = \hat{\psi}(\hat{\text{Clq}})$

$\qquad \psi'(B) = \hat{\psi}(B)$

$\qquad \psi'(b_1) = .0005$

$\qquad \psi'(b_2) = .0095.$

A.5. $S'_1 = S'_1 - X = \varnothing - \{A, H\} = \varnothing.$

A.6. $R'_1 = R'_1 - X = \{A, B\} - \{A, H\} = \{B\}.$

Steps A.1 through A.5 have been performed for the remaining five cliques and the results are in Table 7.7. This table contains the state of the causal network after A and H are instantiated but before propagation. The propagation (i.e., steps B through E of the update algorithm) is identical to that for computing the a priori probabilities and is left as an exercise. The state of the causal network after this propagation is in Table 7.8.

PROBLEMS 7.3

1. Perform all the calculations in the dyspnea example (Example 7.5) which were left as exercises.

2. Suppose, in the dyspnea example, after A is instantiated for a_1 and H for h_1, it is learned that the patient is a nonsmoker and has a negative X ray. Determine the new probabilities of the remaining variables. Does it appear more likely that the patient has tuberculosis, lung cancer, or bronchitis?

3. In the dyspnea example, construct the tree of cliques by making Clq_2 the parent of Clq_6. Compute the a priori probabilities and the conditional probabilities based on the instantiation of A for a_1 and H for h_1. Substantiate that the results are the same as when Clq_5 is the parent of Clq_6.

4. Show that if we attempt to instantiate a set of values whose current joint probability is equal to 0, then, if Clq'_j is the root, $P'(\text{Clq}'_j)$ will end up being equal to 0 for all combinations of values of the variables in Clq'_j.

TABLE 7.7 The State of the Copy Tree of Cliques in the Dyspnea Example (Example 7.5) After A and H Are Instantiated but Before Propagation[a]

Clq' No.	Parent	Clq'_i	R'_i	S'_i	Configuration	$\psi'(\text{Clq}'_i)$	$P'(\text{Clq}'_i)$
1	—	$\{B\}$	$\{B\}$	$\emptyset$	$\{b_1\}$	.0005	
					$\{b_2\}$	.0095	
2	1	$\{B,E,C\}$	$\{C,E\}$	$\{B\}$	$\{b_1,e_1,c_1\}$	.055	
					$\{b_1,e_1,c_2\}$	0	
					$\{b_1,e_2,c_1\}$	.9450	
					$\{b_1,e_2,c_2\}$	0	
					$\{b_2,e_1,c_1\}$	.055	
					$\{b_2,e_1,c_2\}$	0	
					$\{b_2,e_2,c_1\}$	0	
					$\{b_2,e_2,c_2\}$	.945	
3	2	$\{E,C,G\}$	$\{G\}$	$\{E,C\}$	$\{e_1,c_1,g_1\}$	.5727	
					$\{e_1,c_1,g_2\}$	.4273	
					$\{e_1,c_2,g_1\}$	.5727	
					$\{e_1,c_2,g_2\}$	.4273	
					$\{e_2,c_1,g_1\}$	.4429	
					$\{e_2,c_1,g_2\}$	.5571	
					$\{e_2,c_2,g_1\}$	.4429	
					$\{e_2,c_2,g_2\}$	.5571	
4	3	$\{E,G,F\}$	$\{F\}$	$\{E,G\}$	$\{e_1,g_1,f_1\}$	.9524	
					$\{e_1,g_1,f_2\}$	.0476	
					$\{e_1,g_2,f_1\}$	.8511	
					$\{e_1,g_2,f_2\}$	.1489	
					$\{e_2,g_1,f_1\}$	.6452	
					$\{e_2,g_1,f_2\}$	.3548	
					$\{e_2,g_2,f_1\}$	.3419	
					$\{e_2,g_2,f_2\}$	.6581	
5	3	$\{C,G\}$	$\emptyset$	$\{C,G\}$	$\{c_1,g_1\}$	.9	
					$\{c_1,g_2\}$	.7	
					$\{c_2,g_1\}$	.8	
					$\{c_2,g_2\}$	.1	
6	5	$\{C,D\}$	$\{D\}$	$\{C\}$	$\{c_1,d_1\}$	.98	
					$\{c_1,d_2\}$	.02	
					$\{c_2,d_1\}$	.05	
					$\{c_2,d_2\}$	.95	

[a]$P'(\text{Clq}'_i)$ is not yet determined.

TABLE 7.8   The State of the Copy Tree of Cliques in the Dyspnea Example (Example 7.5) After A and H Are Instantiated and the New Values of $P'(\mathrm{Clq}_i')$ Have Been Determined

Clq' No.	Parent	Clq_i'	R_i'	S_i'	Configuration	$\psi'(\mathrm{Clq}_i')$	$P'(\mathrm{Clq}_i')$
1	—	$\{B\}$	$\{B\}$	$\varnothing$	$\{b_1\}$	.08788	.08788
					$\{b_2\}$	.91212	.91212
2	1	$\{B,E,C\}$	$\{C,E\}$	$\{B\}$	$\{b_1,e_1,c_1\}$	.0567	.00498
					$\{b_1,e_1,c_2\}$	0	0
					$\{b_1,e_2,c_1\}$	.9433	.0828
					$\{b_1,e_2,c_2\}$	0	0
					$\{b_2,e_1,c_1\}$	.1036	.0945
					$\{b_2,e_1,c_2\}$	0	0
					$\{b_2,e_2,c_1\}$	0	0
					$\{b_2,e_2,c_2\}$	.8964	.81762
3	2	$\{E,C,G\}$	$\{G\}$	$\{E,C\}$	$\{e_1,c_1,g_1\}$	.6328	.063
					$\{e_1,c_1,g_2\}$	.3672	.0366
					$\{e_1,c_2,g_1\}$	.9148	0
					$\{e_1,c_2,g_2\}$	.0852	0
					$\{e_2,c_1,g_1\}$	.5055	.0419
					$\{e_2,c_1,g_2\}$	.4945	.041
					$\{e_2,c_2,g_1\}$	.8641	.7065
					$\{e_2,c_2,g_2\}$	.1369	.111
4	3	$\{E,G,F\}$	$\{F\}$	$\{E,G\}$	$\{e_1,g_1,f_1\}$	.9524	.06
					$\{e_1,g_1,f_2\}$	.0476	.003
					$\{e_1,g_2,f_1\}$	.8511	.0312
					$\{e_1,g_2,f_2\}$	.1489	.0055
					$\{e_2,g_1,f_1\}$	.6452	.4827
					$\{e_2,g_1,f_2\}$	.3548	.2655
					$\{e_2,g_2,f_1\}$	.3419	.0522
					$\{e_2,g_2,f_2\}$	.6581	.1005
5	3	$\{C,G\}$	$\varnothing$	$\{C,G\}$	$\{c_1,g_1\}$	1	.1049
					$\{c_1,g_2\}$	1	.0776
					$\{c_2,g_1\}$	1	.7065
					$\{c_2,g_2\}$	1	.111
6	5	$\{C,D\}$	$\{D\}$	$\{C\}$	$\{c_1,d_1\}$	.98	.1789
					$\{c_1,d_2\}$	.02	.0036
					$\{c_2,d_1\}$	.05	.0409
					$\{c_2,d_2\}$	.95	.7766

5. In the dyspnea example, directly after determining the a priori probabilities, instantiate E, C, and G for e_1, c_1, and g_1, respectively, and use the algorithm to compute the new probabilities of the other variables in the network. This problem illustrates the case in which a clique becomes empty when variables are instantiated.

7.4. AN IMPROVEMENT OF THE ALGORITHM

We can improve the efficiency of the algorithm in the previous section in the following way. Suppose P' is either the a priori probability or the conditional probability based on the instantiation of some variables, and $(\{\mathrm{Clq}_1', \mathrm{Clq}_2', \ldots, \mathrm{Clq}_p'\}, \psi')$ is the potential representation of P' which is currently stored in the copy tree of cliques. Suppose further that we next instantiate some more variables. Let P'' be the conditional probability based on these new instantiations and $(\{\mathrm{Clq}_1'', \mathrm{Clq}_2'', \ldots, \mathrm{Clq}_p''\}, \psi'')$ be the potential representation for P'' obtained from $(\{\mathrm{Clq}_1', \mathrm{Clq}_2', \ldots, \mathrm{Clq}_p'\}, \psi')$ using the updating algorithm. Suppose Clq_i' is a vertex which contains no newly instantiated variables, and additionally that Clq_i' is either a leaf or is a vertex which has received all its λ messages from its children and they are all vectors of 1's. Then Clq_i'', R_i'', S_i'', and $\psi''(\mathrm{Clq}_i'')$ are equal to Clq_i', R_i', S_i', and $\psi'(\mathrm{Clq}_i')$, respectively. Therefore, when procedure Lambda_Prop computes the λ message for Clq_i'', we have

$$\lambda_{\mathrm{Clq}_i''}(S_i'') = \sum_{R_i''} \psi''(\mathrm{Clq}_i'')$$

$$= \sum_{R_i'} \psi'(\mathrm{Clq}_i')$$

$$= \sum_{R_i'} P'(R_i' \mid S_i') = 1.$$

We see then that the λ message computed for such a vertex will always be a vector of 1's. Notice that this is the case only after the a priori probabilities have been computed, since it is only then that $\psi'(\mathrm{Clq}_i') = P'(R_i' \mid S_i')$. Therefore the execution of procedure Lambda_Prop on such a vertex will change nothing at the vertex, and the λ message to the parent of the vertex will be a vector of 1's. Thus we need only execute procedure Lambda_Prop bottom up starting with vertices which contain instantiated variables rather than with leaves. However, if the vertices are to remain autonomous processors, a vertex must always receive λ messages from all its children in order to know when to send its parent a λ message. Hence we must still begin propagation from all the leaves. We therefore modify the algorithm, for updating probabilities *after* the a priori probabilities have been computed, as follows: If a leaf contains no newly instantiated variables, we only send a dummy message to its parent. No calculations are performed when a dummy message is received. Such a message is only a trigger, indicating that the child has sent a message. If a vertex receives dummy messages from all its children and the vertex contains no newly instantiated variables, then it sends its parent a dummy message.

PROBLEM 7.4

In the dyspnea example, starting with the copy tree in the state in Table 7.6, propagate the instantiation of A for a_1 and H for h_1 using the new updating algorithm.

7.5. VIRTUAL EVIDENCE

Virtual evidence is handled in much the same way as in Pearl's method. We will illustrate the technique with an example. Suppose the causal network is in the state depicted in Table 7.8 and a family history of cancer is revealed. Suppose further that the physician estimates that the probability of this particular history given that the patient has lung cancer is .5, while the probability of this history given that the patient does not have lung cancer is .1. We wish to determine the impact of this evidence on the remaining four variables in the causal network.

As in Pearl's method, we can create a virtual vertex J in the causal network. J takes exactly one value, j_1, where

$$j_1 = \text{the family has this particular history.}$$

We add an arc in the causal network from E to J with

$$P(j_1 \mid e_1) = .5 \qquad \text{and} \qquad P(j_1 \mid e_2) = .1.$$

Now, if J had been in the network at the start, (J, E) would have been a seventh clique Clq_7, and we would have determined that

$$R_7 = \{J\}, \qquad S_7 = \{E\}, \qquad \text{and} \qquad \psi(\text{Clq}_7) = P(J \mid E).$$

The values of Clq_7', R_7', S_7', and $\psi'(\text{Clq}_7')$ would have been initialized to these values, respectively. Clq_7 could have been the child of any clique which contains E. Now, if this clique had been in the network from the start, any subsequent updating would not have changed $\psi'(\text{Clq}_7')$ from its initial values, since for a given clique Clq_i', $\psi'(\text{Clq}_i')$ can only be changed by the instantiation of vertices in Clq_i' or by messages coming from Clq_i''s children. (Note that the λ message computed for Clq_7' would initially have been a vector of 1's.) Thus we can conceive of this clique as having always been in the network, and its present state is still its initial state. We can therefore instantiate J for j_1 and use the update algorithm in the usual fashion. We will begin that task and leave the remainder of it as an exercise.

First, we must add the virtual clique Clq_7' to the copy tree by making it the child of some vertex containing E. Arbitrarily, we pick Clq_3'. Figure 7.6 depicts

Clq₁ = {B}

Clq₂ = {B, E, C}

Clq₃ = {E, C, G}

Clq₇ = {E, J}   Clq₄ = {E, G, F}   Clq₅ = {C, G}

Clq₆ = {C, D}

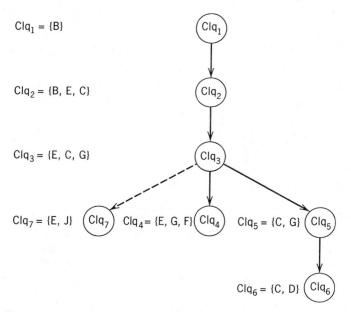

FIGURE 7.6 The copy tree of cliques with the addition of the virtual clique Clq_7 in the dyspnea example (Example 7.5).

the state of the copy tree with the addition of this clique. Next we instantiate J for j_1 and execute procedure Update:

A. The first six cliques will of course be unchanged. For Clq_7, we have

1. $\hat{Clq} = \{E,J\} - \{J\} = \{E\}$.

2. $\hat{\psi}(\hat{Clq}) = \psi'(Clq_7' \leftarrow \{j_1\})$
   $\hat{\psi}(e_1) = \psi'(\{e_1,J\} \leftarrow \{j_1\}) = \psi'(e_1,j_1) = .5$
   $\hat{\psi}(e_2) = \psi'(\{e_2,J\} \leftarrow \{j_1\}) = \psi'(e_2,j_1) = .1.$

3. $Clq_7' = \hat{Clq} = \{E\}$.

4. $\psi'(Clq_7') = \hat{\psi}(\hat{Clq})$
   $\psi'(e_1) = .5$ and $\psi'(e_2) = .1.$

5. $S_7' = \{E\} - \{J\} = \{E\}$.

6. $R_7' = \{J\} - \{J\} = \emptyset.$

Next, procedure Lambda_Prop executes on Clq_7:

$$\lambda_{\text{Clq}_7'}(S_7') = \sum_{R_7'} \psi'(\text{Clq}_7')$$

$$\lambda_{\text{Clq}_7'}(E) = \sum_{\varnothing} \psi'(E) = \psi'(E) = P(j_1 \mid E).$$

Thus we have shown that when virtual evidence for a vertex E is received, all we need do is start propagation by sending any clique which contains E a λ message which is equal to the probability of the evidence given E. There is no need to actually create the virtual clique and go through all of the above steps. It is left as an exercise to complete the propagation of the virtual evidence in the above example.

PROBLEMS 7.5

1. Complete the problem started in the text. That is, finish the updating which was started by the virtual evidence for E.

2. Suppose the copy tree of cliques is in the state determined by the four instantiations given in Problem 7.3.2 and that the same virtual evidence is received for E. Update the copy tree based on this evidence. How are the probabilities of tuberculosis, lung cancer, and bronchitis, respectively, affected?

COMPUTER PROBLEM 7.5

Write a program which performs probability propagation in an arbitrary causal network. The specifications are exactly the same as those in Computer Problem 6.2. The permanent tree of cliques is built directly after the causal network is constructed. Allow for virtual evidence. Test the program on the expert system created in Problem 5.5.3 or on one of the nonsingly connected causal network discussed in this text.

7.6. COMPUTATIONAL CONSIDERATIONS

Lauritzen and Spiegelhalter [1988] have accomplished what researchers have been attempting since the pioneer efforts of Duda, Hart, and Nilsson [1976]. That is, they have devised a method for propagating probabilities in an arbitrary network which makes no unwarranted assumptions of conditional independence and no ad hoc adjustments. The question remains as to the circumstances in which this method is computationally feasible. We investigate that issue next.

7.6.1. An Analysis of the Necessary Computations

In order to discuss the computations in this method, we introduce the following quantities:

n = number of vertices (variables) in the causal network

e = number of arcs in the undirected graph G_u

p = number of vertices (cliques) in the tree of cliques

m = maximum number of variables in a clique

r = maximum number of alternatives for a variable in the network

$k(\text{Clq}_i)$ = number of combinations of values of variables in Clq_i

$k(R_i)$ = number of combinations of values of variables in R_i

$k(S_i)$ = number of combinations of values of variables in S_i

$\overline{k}$ = $\max(k(\text{Clq}_i)$ such that $1 \leq i \leq p)$.

The following example illustrates the new quantities:

Example 7.6. Suppose $V = \{A, B, C, D\}$, the cliques are

$$\text{Clq}_1 = \{A, B, C\} \quad \text{and} \quad \text{Clq}_2 = \{C, D\},$$

and A has three possible values, B has four possible values, and C and D each have two possible values. Then

$$m = 3 \quad \text{and} \quad r = 4$$

$$k(\text{Clq}_1) = (3)(4)(2) = 24$$

$$k(\text{Clq}_2) = (2)(2) = 4$$

$$\overline{k} = \max(24, 4) = 24.$$

Next we determine how many elementary computations are needed in the algorithm for propagation in trees of cliques. First we analyze the portion which creates the permanent tree of cliques, and second we analyze the algorithm for updating. Since there are no computations in the initialization algorithm, we need not analyze that portion. Remember that the algorithm which creates the permanent tree of cliques does not reexecute each time the expert system is used.

1. Computations in the Algorithm Which Creates the Permanent Tree of Cliques: The graph manipulation steps (namely steps A–F) have already been discussed in section 3.2. We know that the time involved is $O(n + e)$. Clearly, this time is feasible, regardless of the size of n and e. As to step G, for $1 \leq i \leq p$ we perform the following calculation:

$$\psi(\text{Clq}_i) := \prod_{f(v)=\text{Clq}_i} P(v \mid c(v)).$$

For a given clique Clq_i, the number of terms in this product is the number of variables such that the variable and its parents are entirely contained in Clq_i, and the variable has been assigned to Clq_i. Call that number x_i. Remember for a given clique Clq_i', x_i may be equal to 0. Then the total number of multiplications necessary to determine $\psi(\text{Clq}_i)$ for all combinations of values of variables in Clq_i is bounded above by

$$\max(0, (x_i - 1)k(\text{Clq}_i))$$

and the total number required for all the cliques is bounded above by

$$\sum_{i=1}^{p} \max(0, (x_i - 1)k(\text{Clq}_i)) \leq \overline{k} \sum_{i=1}^{p}(x_i).$$

Finally, since each variable, along with all its parents, is assigned to exactly one Clq_i, this last expression is equal to

$$\overline{k}n,$$

which is therefore an upper bound for the number of computations needed in this part of the algorithm. Trivially, the upper bound on $\overline{k}$ is

$$\overline{k} \leq r^m.$$

We see then that the critical quantity in the upper bound is m, the maximum number of variables in a clique. We will return to this matter after discussing the computations in the update algorithm.

2. Computations in the Update Algorithm: We will discuss the computations needed to compute the a priori probabilities. When used to compute the probabilities based on the instantiation of variables, the algorithm will always execute somewhat faster owing to there being less variables in some of the cliques and to the improvement in the algorithm introduced in section 7.4. When computing the a priori probabilities, every clique except the root sends one λ message, and every clique except the leaves sends one π message. Since we do not know how many leaves there are, we will use p, the number of cliques, as the upper bound for the number of λ and π messages sent. Next we analyze the computations needed to send and receive each of these messages.

When a clique Clq_i sends a λ message, procedure Lamda_Prop executes on that clique. First the instruction

$$\lambda_{\text{Clq}_i'}(S_i') := \sum_{R_i'} \psi'(\text{Clq}_i')$$

executes. For each combination of values of variables in S_i', this instruction requires $k(R_i')$ additions. Therefore, to determine the value for all combinations of values of variables in S_i, the number of additions required is

$$k(R_i)k(S_i') = k(\text{Clq}_i').$$

Next the instruction

$$\psi'(\text{Clq}_i') := \frac{\psi'(\text{Clq}_i)}{\lambda_{\text{Clq}_i'}(S_i')}$$

executes. Plainly, this instruction requires $k(\text{Clq}_i')$ divisions. Thus the total number of computations needed for Clq_i' to send a λ message is $2k(\text{Clq}_i')$, and the total number of computations needed to send all λ messages is bounded above by

$$\sum_{i=1}^{p} 2k(\text{Clq}_i) \leq 2p\overline{k}.$$

When a clique Clq_j' receives a λ message from Clq_i', the following instruction executes:

$$\psi'(\text{Clq}_j') := \lambda_{\text{Clq}_i'}(S_i)\psi'(\text{Clq}_j').$$

Clearly, this instruction requires $k(\text{Clq}_j')$ multiplications, and the total number of multiplications required to receive all λ messages is bounded above by

$$p\overline{k}.$$

When a clique Clq_j' sends a π message to Clq_i', the following instruction executes:

$$\pi_{\text{Clq}_j'}(S_i') := \sum_{\text{Clq}_j' - S_i'} P'(\text{Clq}_j').$$

The number of additions needed to compute each of these values is equal to $k(\text{Clq}_j' - S_i')$. Since there are $k(S_i')$ values, the total number of additions required is

$$k(\text{Clq}_j' - S_i')k(S_i') = k(\text{Clq}_j')$$

and the total number of computations needed to send all the π messages is bounded above by

$$p\overline{k}.$$

When a clique Clq_i' receives a π message from Clq_j', the following instruction executes:

$$P'(\text{Clq}_i') := \pi_{\text{Clq}_j'}(S_i')\psi'(\text{Clq}_j').$$

Clearly, to compute all values requires $k(\mathrm{Clq}_i')$ multiplications, and an upper bound on the number of computations needed to receive all π messages is therefore

$$p\overline{k}.$$

The total number of computations needed to compute the a priori probabilities is therefore bounded above by

$$2p\overline{k} + p\overline{k} + p\overline{k} + p\overline{k} = 5p\overline{k}.$$

It is a simple matter to show that in the case of a triangulated graph, the number of cliques is less than or equal to the number of vertices (use Theorem 3.2). Thus an upper bound on p, the number of cliques, is n, where n is the number of variables in the causal network. Therefore, just as in the case of the algorithm to create the permanent tree, since $\overline{k}$ is bounded above by r^m, the critical value in the upper bound for the number of calculations in the update algorithm is m, the maximum number of variables in a clique. We have only shown that an upper bound on the number of computations is exponential with regard to m. In other words, we have shown that the algorithm has computing time $O(pr^m)$. Since, for the largest clique Clq_i' we must compute $\psi'(\mathrm{Clq}_i')$ for every combination of values of variables in Clq_i', plainly the number of computations is bounded below by 2^m; that is, the algorithm has computing time $\Omega(2^m)$. It may seem then that we should search for a better general-purpose algorithm to perform probability propagation. We will see, in subsection 7.6.3, that this may be a fruitless endeavor. Notice, however, that for the most part, a large value of m would occur if the network were highly connected. As mentioned in chapter 5, people usually conceptualize causal relationships by forming hierarchies of small clusters of variables. Thus causal networks created by humans should often contain small clusters of variables, and therefore a small value of m.

The method for propagating probabilities in a tree of cliques has been implemented in MUNIN [Andreassen et al., 1987], a system for diagnosis and test planning in electromyography. In that system, m, the maximum number of variables in a clique, is only 4, p is equal to 25, and $\overline{k}$ is equal to 495. Therefore an upper bound for the number of computations in the update algorithm is

$$5p\overline{k} = (5)(25)(495) = 61,875,$$

a quite manageable number.

7.6.2. A Brief Comparison to the Method of Pearl

At the end of chapter 6, we mentioned some extensions of Pearl's method to nonsingly connected networks. Shachter [1989] has compared Pearl's method, Lauritzen and Spiegelhalter's method, and a third exact method called arc reversal/node reduction [Shachter, 1986, 1988]. In particular, he shows that all three methods are identical whenever the DAG is a forest. We briefly offer some direct comparisons here, many of which are obtained from Suermondt and Cooper [1989].

First we compare Lauritzen and Spiegelhalter's method to Pearl's [1988] method of clustering. In this latter method, which we will discuss in detail in chapter 8, variables are also collapsed into single vertices to create a tree. Updating then takes place by the passing of λ and π messages in this tree, as discussed in chapter 6. A major difference between this and Lauritzen and Spiegelhalter's method is that, no matter how many variables are instantiated simultaneously, there are only two passes of messages in the tree of cliques: one pass going up the tree and another pass going down the tree. Consequently, in general, Lauritzen and Spiegelhalter's method should be more efficient.

On the other hand, there are situations in which Pearl's method of conditioning is more efficient. Consider the network in Figure 7.7. Assume that each variable has two alternatives. The total number of vertices in the network is $\theta(n + \sqrt{n})$ (in the figure, n is equal to 16). If we condition on C, we can compute $P(B \mid A)$, using Pearl's propagation method, swiftly in one pass. Hence the time required by Pearl's method is also $\theta(n + \sqrt{n})$. However, if we marry parents and triangulate, we end up with a relatively inefficient computation of $P(B \mid A)$. It is left as an exercise to actually determine the order of the running time in terms of n.

However, there are other situations in which Pearl's method of conditioning will be exponentially difficult, while Lauritzen and Spiegelhalter's method will be very efficient. Consider the causal network in Figure 7.8. Again assume each variable only has two alternatives. Clearly, Lauritzen and Spiegelhalter's method will require $\theta(n)$ time to compute $P(B \mid A)$, where n is the number of vertices in the depth of the graph. In the figure, n is equal to 5. Using Pearl's method of conditioning, we must condition on $\theta(n/2)$ vertices to render the network singly connected. That is, we must condition on all the vertices on the far left or on the far right side of the graph. Since each variable only has two alternatives, we must therefore propagate probabilities in $\theta(2^{n/2})$ singly connected networks in order to compute $P(B \mid A)$. Thus Pearl's method of conditioning requires $\theta(2^{n/2})$ time to compute that probability.

Unfortunately, there are causal networks in which neither method is computationally efficient. Consider the causal network in Figure 7.9. The total number of vertices in the network is $\theta(n + \sqrt{n})$ (in the figure, n is equal to 16). Again assume that each variable has two alternatives. Even though each variable has at most two parents and the total number of probabilities in the network is $\theta(n + \sqrt{n})$, both Pearl's method of conditioning and Lauritzen and Spiegelhalter's method require at least $\theta(2^{\sqrt{n}})$ time to compute $P(B \mid A)$. It is left as an exercise to show that this is the case. It may be thought that after the parents are married, the graph will be triangulated (except for one path near the bottom) and that therefore the maximum number of vertices in a clique is equal to 3 regardless of the value of n, contradicting the claim that the time in Lauritzen and Spiegelhalter's method is at least $\theta(2^{\sqrt{n}})$. However, by looking

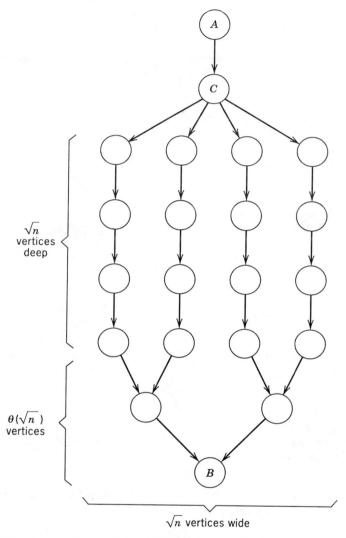

FIGURE 7.7 A causal network in which Pearl's method is more efficient than Lauritzen and Spiegelhalter's method.

at a portion of the undirected graph which results from marrying the parents, we can see that this is not the case. Consider the portion in Figure 7.10. Even though the graph appears to be composed of triangles, it is not triangulated. Any cycle of length 6 on the border of the graph has no chord.

In light of this last example, it may appear that we should try to develop a better general-purpose method than either of these two methods. We will now

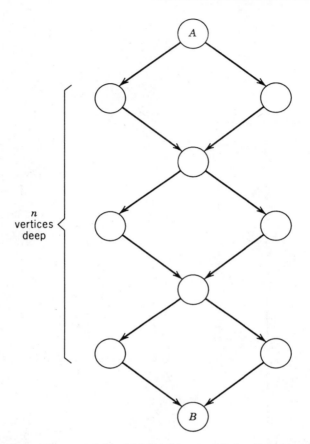

FIGURE 7.8 A causal network in which Lauitzen and Spiegelhalter's method is more efficient than Pearl's method.

see that the development of a method which is computationally efficient for an arbitrary network appears unlikely.

7.6.3. Probability Propagation Is NP-Hard

Cooper [1988] has shown that, for the set of causal networks which are restricted to having no more than three parents per variable and no more than five sons per variable, the problem of determining the probabilities of remaining variables given that certain variables have been instantiated, in nonsingly connected causal networks, is NP-hard in terms of the number of variables in the network. Briefly, NP-hard problems are problems which are at least as computationally complex as NP-complete problems, while NP-complete problems are decision problems which appear to be computationally very difficult. No algorithm has ever been found which can solve an NP-complete in

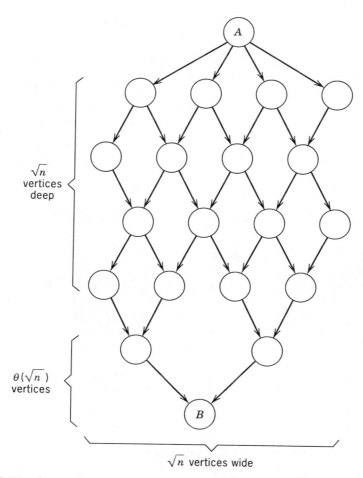

FIGURE 7.9 A causal network in which any known probability propagation method is computationally inefficient.

polynomial time, although it has never been proved that it is not possible to find such an algorithm. Many believe that indeed it is not possible, and a great deal of effort goes into attempts to prove whether a polynomial time algorithm exists (i.e., whether $P = NP$). In any case, any known algorithm, in the worst case, requires an exponential number of calculations to solve an NP-complete problem. See Garey and Johnson [1979] or Balcazar and Dias [1988] for a complete introduction to NP-complete and NP-hard problems. Thus there exists causal networks, in which each variable has at most three parents, for which it can take an exponentially large number of calculations (in terms of the number of variables in the network) to determine the probabilities of remaining variables given certain variables are instantiated.

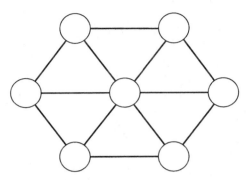

FIGURE 7.10 A portion of the undirected graph which results from the graph in Figure 7.9 after the parents are married.

Cooper [1988] has also obtained the stronger result that, for the set of causal networks which are restricted to having no more than two parents per variable with no restriction on the number of children per variable, the problem of determining the probabilities of remaining variables given that certain variables have been instantiated, in nonsingly connected causal networks, is $\#P$-complete. $\#P$-complete problems are a special class of NP-hard problems. Namely, the answer to a $\#P$-complete problem is the number of solutions to some decision problem. Thus the only way that we will ever find an algorithm, which can determine probabilities for arbitrary causal networks in polynomial time (in terms of the number of variables in the network) is if someone proves that $P = \#P$.

Both Cooper's NP-hard and $\#P$-complete results hold even if we restrict the number of alternatives for each variable to two. Furthermore, by adding dummy parents it is always possible to obtain a new causal network, containing the same distribution as an existing one, which has an arbitrary number of parents per variable. Clearly, probability propagation in the new network is at least as hard as probability propagation in the original one. Therefore both these results hold in causal networks in which we require any number of parents per variable.

We saw, at the end of chapter 6, that probability propagation in singly connected networks can be performed in linear time relative to the the number of variables in the network. However, in nonsingly connected networks, the number of computations can become exponentially large even for some networks in which the maximum number of parents of each variable is two. This is not an attribute of any particular method, but an attribute of the problem itself (unless, of course, $P = \#P$). Accordingly, it is unlikely that we could develop an efficient method for propagating probabilities in an arbitrary nonsingly connected causal network.

In practice, we must determine whether a particular method is applicable in a given situation. For example, we have shown above that the number of calculations needed in probability propagation in trees of cliques is bounded above by $5p\bar{k}$, where p is the number of cliques and $\bar{k}$ is bounded above by r^m, where r is the maximum number of alternatives for a variable and m is the maximum number of variables in a clique. Often, as in the case of MUNIN, if there exist small clusters of variables, then m is small. But, in the worst case, if the network is highly connected, all the variables could end up in one clique, in which case m equals the number of variables. In practice, we must determine the tree of cliques and compute $5p\bar{k}$ to assess whether it is feasible to use the method described in this chapter.

Some efforts can be made to optimize the structure of the tree of cliques. Maximum cardinality search, along with the fill-in algorithm in chapter 3, yields a fill-in which contains no optimization properties. Furthermore, one maximum cardinality search can yield a fill-in which contains substantially more arcs than a fill-in obtained from another maximum cardinality search. As noted in chapter 3, the problem of determining a fill-in which adds a minimum number of arcs is NP-hard. Lexicographic search, as described in Golumbic [1980], yields a fill-in which is minimal. A fill-in is minimal if no renumbering of the vertices gives a fill-in which is a subset of the given fill-in. Lexicographic search runs in $O(ne)$ time where n is the number of vertices and e is the number of arcs. However, neither a minimum nor a minimal fill-in necessarily optimizes the algorithm in section 7.3 for propagating probabilities in trees of cliques. We need the following considerations in order to optimize that algorithm. For each clique, take the product of the number of alternatives for each variable over all variables in the clique. Take the sum of these products over all cliques. It is not hard to see that the fill-in which minimizes this sum optimizes the algorithm in section 7.3. Let us call the problem of determining the fill-in which minimizes this sum MINSUM and the problem of determining the fill-in which minimizes the size of the largest clique MINCLQ. The results in Arnborg, Corneil, and Proskurowski [1987] show that MINCLQ is NP-hard. We can "reduce" MINCLQ to MINSUM by letting the number of alternatives at each node be big enough that the largest clique dominates the sum. Thus MINSUM is also NP-hard. The possibility remains that MINSUM may not be NP-hard in the case where a fixed bound is placed on the number of alternatives. However, this possibility seems remote.

Thus it is unlikely that we could develop an efficient general-purpose algorithm which would yield an optimal fill-in. The development of special case and heuristic algorithms is a fertile area for research.

Similarly, we would need to investigate whether Pearl's method of conditioning is feasible for a particular network before using it. As mentioned at the end of chapter 6, Suermondt and Cooper [1988] have developed some conditions for using this method.

Pople [1982] has observed that many problems in medicine require large networks which are not sparsely connected. In such a case, both Pearl's method and Lauritzen and Spiegelhalter's method would rarely be feasible even if we searched for an optimal structure. Since probability propagation is NP-hard, the development of an efficient general purpose algorithm which would perform probability propagation in all causal networks appears unlikely. This suggests that research should be directed towards obtaining alternative methods which work in different cases.

One alternative would be to use Monte Carlo techniques, termed stochastic simulation, as discussed at the end of the last chapter. Pearl [1987a] has developed a Monte Carlo technique which produces a unique point-valued mean probability, plus a variance of that mean. Chin and Cooper [1987], however, noted that "although this method appears promising, current stochastic simulation algorithms have slow convergence properties in some cases, and thus are not generally applicable." Chavez [1989] has developed a version of Pearl's stochastic simulation algorithm which, although it does not necessarily run faster than the original algorithm, is able to give precise a priori bounds for its running time as a function of the relative or interval error in the probabilities.

A second alternative is the use of approximation algorithms. Approximation algorithms produce an inexact, bounded solution; however, it is guaranteed that the exact solution is within those bounds. Cooper [1984], Peng and Reggia [1987a, 1987b], and Horvitz, Suermondt, and Cooper [1989] obtain such approximation algorithms.

A third alternative would be to use heuristic algorithms. Heuristic algorithms are not guaranteed to yield a correct solution. However, they may often yield answers which are acceptably close. Their accuracy is tested by measuring how close their solutions are to the correct ones in networks for which exact determinations are possible. One heuristic approach involves the use of a connectionist architecture [Wald et al., 1989]. Another heuristic approach involves simplifying the network by deleting arcs according to some preassigned criteria of importance [Horvitz, 1987]. We remove such arcs until the problem becomes computationally feasible.

A final alternative is to create special case algorithms. Special case algorithms determine exact solutions for special classes of networks. The algorithm in section 7.3 is such an algorithm for the case where there are small clusters of variables. Heckerman [1989] has obtained an algorithm for the case of the classical diagnostic problem with the probabilistic causal assumptions (to be discussed in the next section), which is exponential only in the numbers of positive findings.

Further investigation of all these alternatives is an important area of current research.

Cooper [1988] has also shown that abductive inference (to be discussed in chapter 8) in causal networks is NP-hard. Therefore the above discussion pertains to abductive inference, as well as to probability propagation.

PROBLEMS 7.6

1. For the network in Figure 7.7, determine the order of the running time, in terms of n, of Lauritzen and Spiegelhalter's method in the computation of $P(B \mid A)$. Hint: The problem here is not that the maximum number of vertices in a clique increases with n (if we choose the correct triangulation). That is, if we triangulate by connecting to C every vertex which is not already connected to C, then the maximum number of vertices in a clique is four regardless of n.

2. For the network in Figure 7.9, show that both Pearl's method and Lauritzen and Spiegelhalter's method require at least $\theta(2^{\sqrt{n}})$ time to compute $P(B \mid A)$. For Lauritzen and Spiegelhalter's method it is necessary to show that the optimal triangulation requires at least $\theta(2^{\sqrt{n}})$ time.

7.7. THE CLASSICAL DIAGNOSTIC PROBLEM

Before ending this chapter, we discuss a special case of the application of the methods developed in chapters 5 through 7—the application to the classical diagnostic problem.

Up to this point we have assumed that we can somehow obtain the probabilities needed in a causal network either from an expert or from a data base. We have not discussed whether it is really feasible to obtain them for a particular problem. Consider now a causal network, which is to be used for medical diagnosis, in which a particular symptom can be caused by four different diseases. We would need to ascertain the probability of that symptom given 16 different combinations of the diseases. For instance, we would need the probability of the symptom given that disease 1 and disease 2 are present and disease 3 and disease 4 are absent. It would be quite difficult to obtain such a value, particularly from a physician. The physician should know something about the individual probabilities of the symptom given each disease, but would have little idea as to the probability of the symptom given arbitrary combinations of the diseases.

This difficulty has been well-studied for an important class of problems, namely those in which we have a possible set of symptoms (features, manifestations) and set of direct causes (e.g., diseases) of the symptoms. This is the classical diagnostic problem. Although traditionally the set of cause/effect relationships in this problem is not called a causal network, this set can be represented by a two-level causal network in which each variable can take one of exactly two possible values. Plainly, as Example 7.5 illustrates, there exist diagnostic problems for which the set of cause/effect relationships cannot be represented by a two-level network; however, for some diagnostic problems it can. We shall call such problems classical diagnostic problems (this concept will soon be defined rigorously). In the case of such two-level causal networks, Peng and Reggia [1987a] have devised a method for obtaining the conditional

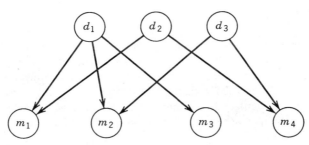

FIGURE 7.11 The DAG which represents the same causal relationships as relation C in Example 7.7.

probability of a symptom given combinations of its causes from conditional probabilities of the symptom based on individual causes. These latter probabilities are much easier to ascertain. Although their efforts were actually aimed at abductive inference (to be discussed in chapter 8), some of their results are applicable here. We will discuss these results in subsection 7.7.2. First, in subsection 7.7.1, we discuss early efforts to solve this problem. Pearl [1988] has obtained results equivalent to those of Peng and Reggia by considering the problem as one involving a noisy OR gate.

Since, in the case of the classical diagnostic problem each variable has only two possible values, we will alter our notation to be consistent with that used traditionally for such problems. The set of all possible causes (e.g., diseases) will be represented by

$$D = \{d_1, d_2, \ldots, d_n\}.$$

The set of all possible manifestations (symptoms) will be represented by

$$M = \{m_1, m_2, \ldots, m_k\}.$$

The DAG in the causal network is then a relation C, consisting of ordered pairs of causes and manifestations, where

$$\langle d_i, m_j \rangle \in C$$

if there is an arc from d_i to m_j. The possible values of propositional variables will be denoted by $+d_i$ and $\neg d_i$ and by $+m_j$ and $\neg m_j$, while the the propositional variables themselves will be represented by d_i and m_j.

Example 7.7. Suppose

$$D = \{d_1, d_2, d_3\}$$

$$M = \{m_1, m_2, m_3, m_4\}$$

$$C = \{\langle d_1, m_1 \rangle, \langle d_1, m_2 \rangle, \langle d_1, m_3 \rangle, \langle d_2, m_1 \rangle, \langle d_2, m_4 \rangle, \langle d_3, m_2 \rangle, \langle d_3, m_4 \rangle\}.$$

Then the DAG in a causal network which corresponds to C is the one in Figure 7.11.

We now define the classical diagnostic problem:

Definition 7.5. Let $P = \langle D, M, C, M^+, M^- \rangle$ where D, M, and C are as defined above and M^+ and M^- are disjoint subsets of M. Then P is called the classical diagnostic problem.

The variables in M^+ are variables which are known to be instantiated for their positive values (i.e., the manifestations are known to be present), whereas those in M^- are variables which are known to be instantiated for their negative values (i.e., the manifestations are known to be absent). We will denote the set of all these instantiations by M^*. That is,

$$M^* = M^+ \cup M^-.$$

There are actually two problems in the classical diagnostic problem. The first is to find the probability of each $d_i \in D$ given M^*. The solution to this problem is just probability propagation. The technique covered in this chapter can solve this problem if we can ascertain the a priori probabilities of each disease and the probability of each symptom given its causes. In subsection 7.7.2, we develop a method for obtaining these latter probabilities. The second problem is to determine the overall most probable set of diseases. The process which solves this problem is abductive inference and is the focus of chapter 8.

7.7.1. Early Efforts to Solve the Classical Diagnostic Problem

Traditionally, this problem has been solved by appealing directly to Bayes' theorem in a fashion similar to the following:

$$P(+d_i \mid M^*) = \frac{P(M^* \mid +d_i)P(+d_i)}{P(M^* \mid +d_i)P(+d_i) + P(M^* \mid \neg d_i)P(\neg d_i)}.$$

The values which we could not possibly obtain directly are $P(M^* \mid +d_i)$ and $P(M^* \mid \neg d_i)$. Since M^* includes any manifestations (not only those caused by d_i), as noted by Charniak [1983], we would need an astronomical number of values. If we *assume*, however, that the manifestations in M^* are independent given that d_i occurs and given that d_i does not occur, then we have

$$P(M^* \mid +d_i) = \prod_{m_j \in M^+} P(+m_j \mid +d_i) \prod_{m_j \in M^-} P(\neg m_j \mid +d_i)$$

$$P(M^* \mid \neg d_i) = \prod_{m_j \in M^+} P(+m_j \mid \neg d_i) \prod_{m_j \in M^-} P(\neg m_j \mid \neg d_i).$$

With this assumption we need only be able to ascertain the probability of each manifestation given a disease and given its complement, a far more manageable task. As mentioned in chapter 4, these independence assumptions have become notorious. Also as mentioned in that chapter, Charniak [1983] has shown that they often are justified in two-level networks and has suggested

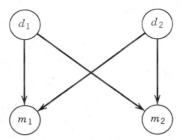

FIGURE 7.12 A situation in which the conditional independence assumptions, made in early expert systems, are not justified.

ways of handling situations in which they are not applicable. However, as can be seen by the simple DAG in Figure 7.12, there are many situations in which they definitely are not applicable. If two diseases both cause the same two manifestations as in that DAG, given that d_1 has occurred, m_1 and m_2 are, in general, not independent, since knowledge of m_1 tells us something about d_2, which in turns tells us something about m_2.

There are successful expert systems which apply Bayes' theorem directly and make the above independence assumptions. In particular, we mention MEDAS [Ben-Bassat et al., 1980] and de Dombel's system for diagnosing acute abdominal pain [de Dombel et al., 1972]. However, as we have illustrated, in many cases the assumptions are not justified. When making these assumptions, Ben-Bassat [Ben-Bassat et al., 1980] noted that

> preliminary experience reveals that correct decisions are not sensitive to non-drastic violations of this assumption. Therefore, for practical purposes, this assumption can usually be made without significantly affecting the performance of the system. However, further research should be made to explore the effect of this assumption and possible techniques to avoid or correct it.

The possible techniques to which Ben-Bassat referred in 1980 now exist. Namely, we have the propagation techniques given in this and the previous chapter, the method for obtaining the necessary probability values given in the next subsection, and the abductive inference techniques given in chapter 8.

7.7.2. The Probabilistic Causal Method

As mentioned previously, the results obtained here are part of Peng and Reggia's probabilistic causal model [1987a, 1987b], which was created to handle abductive inference. We therefore call this method for obtaining probability values the probabilistic causal method. We cover the probabilistic causal method here because it is applicable regardless of whether we perform probability propagation or abductive inference. Before stating the assumptions necessary to this method, we need some definitions:

Definition 7.6. Causation Event. For any $d_i \in D$ and $m_j \in M$, $d_i \to m_j$ denotes the event that d_i actually causes m_j. This causation event is true if and only if both d_i and m_j are present and m_j is actually being caused by d_i. Better notation for this event would be $+d_i \to +m_j$, since the event means the presence of d_i is causing the presence of m_j. However, we will stay with the simpler notation to avoid unnecessary symbols.

The concept of a causation event is innovative. It is not equivalent to m_j and d_i both being present, since this latter situation could happen while some other disease is causing m_j.

Definition 7.7. $P(d_i \to m_j \mid +d_i)$ is the conditional probability of d_i causing m_j given that d_i is present.

The $P(d_i \to m_j \mid +d_i)$ is not equivalent to the $P(+m_j \mid +d_i)$, since

$$P(d_i \to m_j \mid +d_i) = \frac{P(d_i \to m_j \wedge +d_i)}{P(+d_i)} = \frac{P(d_i \to m_j)}{P(+d_i)}.$$

The latter equality is due to the fact that $d_i \to m_j$ entails $+d_i$. We therefore have that if d_i cannot cause m_j (i.e., $P(d_i \to m_j) = 0$), then $P(d_i \to m_j \mid +d_i)$ is equal to 0. However, clearly $P(+m_j \mid +d_i)$ need not be 0 in this case, since another disease could cause m_j while d_i is present. It is quite easy to show that

$$P(d_i \to m_j \mid +d_i) \leq P(+m_j \mid +d_i).$$

Definition 7.8. Let X be a conjunction of any cause and causation events or their negations other than $d_i \to m_j$ and $\neg(d_i \to m_j)$. Then X is said to be a context of $d_i \to m_j$.

Example 7.8. Let

$$X = d_1 \to m_2 \wedge \neg(d_2 \to m_3) \wedge +d_3 \wedge \neg d_4.$$

Then X is a context of $d_1 \to m_3$.

The concept of a context is explained best in the original words of Peng and Reggia [1987a]:

Intuitively, when the occurrence of a causation event $d_i \to m_j$ in a given case is being considered, it is possible that some other disorder and causation events may actually be occurring simultaneously, while still others may not be simultaneously present in that particular case. Note that these other disorders and causation events have no causal relation to $d_i \to m_j$; their occurrence/nonoccurrence only provides a context for the occurrence/nonoccurrence of $d_i \to m_j$. Thus such a context can be expressed by a conjunction of these events. Note that a causation event cannot be part of its own context, and thus itself and its negation should not be included in any of its contexts.

We can now state the assumptions in the probabilistic causal model:

Assumption 7.1. For $d_i \in D$,

$$0 < P(+d_i) < 1.$$

This assumption simply says that every disease has a possibility of occurring and that no disease is certain to occur.

Assumption 7.2. For $d_i \in D$ and $m_j \in M$,

$$P(d_i \to m_j \mid +d_i) > 0$$

if and only if $\langle d_i, m_j \rangle \in C$; that is, there is an arc from d_i to m_j.

   This assumption says that the probability of d_i causing m_j is positive if and only if d_i is a cause of m_j.

Assumption 7.3. The set of all members of D are independent. That is, for $d_i \in D$ and $D' \subseteq D - \{d_i\}$,

$$P\left(d_i \wedge \left(\bigwedge_{d_j \in D'} d_j\right)\right) = P(d_i)P\left(\bigwedge_{d_j \in D'} d_j\right).$$

   This assumption is already made in the causal network representation, since all the members of D are roots.

Assumption 7.4. Causation event $d_i \to m_j$ is independent of any of its contexts X given that d_i has occurred. That is,

$$P(d_i \to m_j \mid +d_i \wedge X) = P(d_i \to m_j \mid +d_i).$$

   This assumption says that the tendency of d_i to cause m_j does not change in the context of occurrence of other cause and causation events. In other words, d_i always causes m_j with a fixed strength.

Assumption 7.5. No effect event (manifestation) occurs without being caused by some disorder through some causation event. This implies that for $m_j \in M$,

$$+m_j = \bigvee_{d_i \in D} d_i \to m_j.$$

   This assumption simply says that every effect must be caused by something.

   These assumptions are not arbitrarily made just to make things works; rather they are, like the general assumptions made in a causal network, relationships which we feel often hold among causes and effects. For a particular application, however, we must take care to observe whether indeed they do hold. For example, let

$$d_1 = \text{arteriosclerosis}$$

$$m_1 = \text{lung symptom 1}$$

$$m_2 = \text{lung symptom 2}.$$

As noted by Charniak [1983], given that a person has d_1, the presence of one lung symptom $(+m_1)$ still makes the presence of another lung symptom $(+m_2)$ more likely. Thus the event $d_1 \to m_1$ is not independent of $d_1 \to m_2$ given $+d_1$, a violation of Assumption 7.4. The problem here is that arteriosclerosis can cause left heart syndrome, which in turn can cause cause either of the two lung symptoms. Therefore, if we let

$$d_2 = \text{left heart syndrome,}$$

the DAG which properly represents these causal relationships is the one in Figure 7.13(b), not the one in Figure 7.13(a). Thus it takes a three-level causal network to represent the relationships. We see then that we cannot simply group a set of symptoms on one level and a set of diseases on the other and expect the above assumptions to hold. We must take care to notice whether they really do hold.

Our main result will be a formula which gives the conditional probability of a manifestation, m_j, given its causes from the individual conditional probabilities $P(d_i \to m_j \mid +d_i)$. Therefore it is these latter probabilities which we must ascertain. A physician would be familiar with the strength with which a disease causes a particular symptom (i.e., $P(d_i \to m_j \mid +d_i)$); he would have less familiarity with the purely statistical quantity $P(m_j \mid d_i)$ needed in the standard application of Bayes' formula. Therefore he should have an easier time judging $P(d_i \to m_j \mid +d_1)$ than $P(m_j \mid d_i)$. Moreover, in this model he need only judge one value rather than the two values needed in Bayes' formula. That is, he need only judge $P(d_i \to m_j \mid +d_i)$ rather than both $P(+m_j \mid +d_i)$ and $P(+m_j \mid \neg d_i)$. On the other hand, if we are obtaining the probabilities statistically, then it is straightforward to obtain $P(m_j \mid d_i)$, while $P(d_i \to m_j \mid +d_i)$ appears to be a rather elusive quantity. The following theorem eliminates this elusiveness:

Theorem 7.8. For any $d_i \in D$ and $m_j \in M$,

$$P(d_i \to m_j \mid +d_i) = P(+m_j \mid [+d_i]),$$

where $[+d_i]$ stands for the event that d_i is present and all the other causes of m_j are definitely not present.

Proof. Let

$$X = \bigwedge_{d_r \in c(m_j) - \{d_i\}} \neg d_r.$$

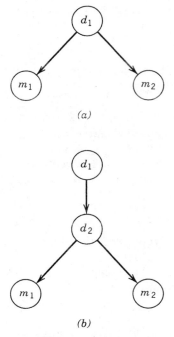

(a)

(b)

FIGURE 7.13 (b) represents the proper causal relationships among arteriosclerosis and lung syndromes, whereas (a) does not.

(Recall that $c(m_j)$ is the set of all causes of m_j.) Then

$$[+d_i] = X \land +d_i. \tag{7.3}$$

Therefore $[+d_i]$ entails $\neg d_r$ for $d_r \in c(m_j) - \{d_i\}$. We have then, by Definition 7.6, that for any such d_r,

$$P(d_r \to m_j \mid [+d_i]) = 0. \tag{7.4}$$

Now if $d_r \notin c(m_j)$, then, by Assumption 7.2,

$$P(d_r \to m_j \mid +d_r) = 0$$

and hence, as was shown earlier in this section,

$$P(d_r \to m_j) = 0.$$

Thus equality (7.4) holds for any $r \neq i$. Now, by Assumption 7.5,

$$+m_j = \bigvee_{d_r \in D} d_r \to m_j.$$

and therefore, by equalities (7.3) and (7.4),

$$P(+m_j \mid [+d_i]) = P\left(\bigvee_{d_r \in D} d_r \rightarrow m_j \mid [+d_i]\right)$$

$$= P(d_i \rightarrow m_j \mid [+d_i]).$$
$$= P(d_i \rightarrow m_j \mid X \wedge +d_i)$$
$$= P(d_i \rightarrow m_j \mid +d_i).$$

The last equality is due to Assumption 7.4 and the fact that X is a context for $d_i \rightarrow m_j$. $\square$

By Theorem 7.8, we can obtain the value of $P(d_i \rightarrow m_j \mid +d_i)$ statistically from patients who have d_i but do not have any other disease which is a cause of m_j. We now state the main result of this section:

Theorem 7.9. If we let

$$c_{ij} = P(d_i \rightarrow m_j \mid +d_i)$$

and D' be a subset of $c(m_j)$, then

$$P\left(\neg m_j \mid \left(\bigwedge_{d_i \in D'} +d_i\right) \wedge \left(\bigwedge_{d_i \in c(m_j) - D'} \neg d_i\right)\right) = \prod_{d_i \in D'} (1 - c_{ij}).$$

Proof. The proof, which is quite lengthy, relies on an induction argument on the number of diseases and manipulations similar to those in Theorem 7.8. The reader is referred to Peng and Reggia [1987a] for the complete proof. $\square$

The following example gives an application of Theorem 7.9:

Example 7.9. In Figure 7.14(a) the values of c_{ij} have been drawn on the arcs in the DAG from Figure 7.11. By Theorem 7.9 we have

$$P(\neg m_1 \mid +d_1, +d_2) = (1 - c_{11})(1 - c_{21}) = (1 - .2)(1 - .9) = .08$$
$$P(\neg m_1 \mid +d_1, \neg d_2) = (1 - c_{11}) = (1 - .2) = .8$$
$$P(\neg m_1 \mid \neg d_1, +d_2) = (1 - c_{21}) = (1 - .9) = .1$$
$$P(\neg m_1 \mid \neg d_1, \neg d_2) = 1.$$

The corresponding conditional probabilities of $+m_1$ (each of which is simply equal to 1 minus the conditional probability of $\neg m_1$) are listed in Figure 7.14(b). It is left as an exercise to compute the conditional probabilities of the other manifestations.

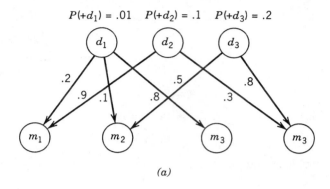

$P(+d_1) = .01$ $P(+d_2) = .1$ $P(+d_3) = .2$

(a)

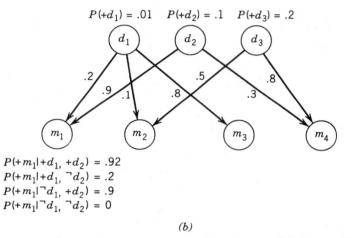

$P(+d_1) = .01$ $P(+d_2) = .1$ $P(+d_3) = .2$

$P(+m_1|+d_1, +d_2) = .92$
$P(+m_1|+d_1, \neg d_2) = .2$
$P(+m_1|\neg d_1, +d_2) = .9$
$P(+m_1|\neg d_1, \neg d_2) = 0$

(b)

FIGURE 7.14   In (a) the values of c_{ij} are stored on each arc. In (b) the conditional probabilities of symptom m_1 given its causes have been computed.

After obtaining the needed conditional probabilities using Theorem 7.9, we can propagate probabilities upon the instantiation of variables using either the technique from chapter 6 or the one in this chapter.

It is clear that if we use the probabilistic causal method to obtain probabilities, then we can assign probabilities even if the number of parents of a variable is large. Furthermore, Pearl [1988] has streamlined his method for probability propagation in this case by removing the need to actually compute the required conditional probabilities from the values of c_{ij}, and also by removing the need to perform the additions in operative formulas 1 and 4 (see section 6.3). If m_j is a manifestation of d_i and we define

$$q_{ij} = 1 - c_{ij},$$

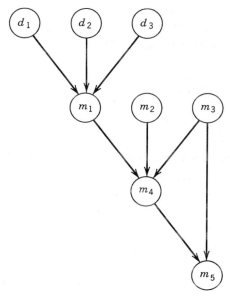

FIGURE 7.15 The probabilistic causal method is appropriate for m_1 and m_4, but not for m_5.

then we can replace operative formula 1 by

$$\lambda_{m_j}(+d_i) = \lambda(\neg m_j)q_{ij}P_{ij} + \lambda(+m_j)(1 - q_{ij}P_{ij})$$
$$\lambda_{m_j}(\neg d_i) = \lambda(\neg m_j)P_{ij} + \lambda(+m_j)(1 - P_{ij}),$$

where

$$P_{ij} = \prod_{d_r \in c(m_j) - \{d_i\}} (1 - c_{rj}\pi_{m_j}(+d_r)).$$

We can replace operative formula 4 by

$$\pi(+m_j) = \left(1 - \prod_{d_i \in c(m_j)} (1 - c_{ij}\pi_{m_j}(+d_i))\right)$$
$$\pi(\neg m_j) = \prod_{d_i \in c(m_j)} (1 - c_{ij}\pi_{m_j}(+d_i)).$$

Thus we see that if we use the probabilistic causal method to obtain probabilities, we can propagate probabilities using Pearl's method even if a variable has a large number of parents. It is straightforward to obtain the above formulas for this special case from the general formulas in section 6.3, and it is left as an exercise to do so.

The probabilistic causal method can also be used in networks which have more than two levels as long as the assumptions are valid. Indeed, since most relationships cannot really be represented by a two-level network, this is the way in which the method would often be applied. For example, one assumption is that all the causes of a variable are independent. Hence we could use this method (if the other assumptions are also valid) to determine the $P(m_1 \mid d_1, d_2, d_3)$ and the $P(m_4 \mid m_1, m_2, m_3)$ in Figure 7.15, but we could not use it to determine the $P(m_5 \mid m_3, m_4)$.

PROBLEMS 7.7.2

1. Compute the remaining conditional probabilities in Example 7.9.

2. Using probability propagation in trees of cliques compute the probabilities of d_1, d_2, d_3, and m_4 given that the patient has m_1, m_2, and m_3 in Example 7.9.

3. Derive the streamlined versions of operative formulas 1 and 4 in probability propagation in singly connected networks from the original formulas (see section 6.3).

4. Investigate ways to streamline the algorithm for probability propagation in trees of cliques in the case of the classical diagnostic problem and the probabilistic causal method. Try to simplify the calculations as Pearl was able to do for his method.

CHAPTER 8

ABDUCTIVE INFERENCE

The previous two chapters derive methods for computing the conditional probabilities of all the variables in a causal network given that certain variables are instantiated or virtual evidence is obtained. In many cases these probabilities are all we need. For example, in the PROSPECTOR expert system, if we find out that it is highly probable that there are massive sulfide deposits and we are interested in excavating for sulfide minerals, this individual probability of one variable is sufficient. In many cases, however, we are interested in more than just the individual conditional probabilities of each variable. In particular, given that certain symptoms are present, the physician is not only interested in the individual conditional probability of each disease, but also in the most likely overall hypothesis or explanation for the occurring symptoms. Reggia, Nau, and Wang [1983] state:

> As an example, medical diagnosis has been relatively well studied by cognitive psychologists. This empirical work has shown that diagnostic reasoning involves a sequential hypothesize-and-test process during which the physician conceptually constructs a model of the patient. This "hypothesis" postulates the presence of one or more diseases that could explain the patient's manifestations.

For instance, suppose in Example 7.5 that a physician learns that a particular patient is a smoker and has a positive chest X ray. The physician not only determines the individual probabilities of lung cancer, tuberculosis, and bronchitis, each being present, but also the overall most likely explanation. He might decide that it is most probable that the patient has lung cancer and bronchitis and does not have tuberculosis. If, after querying the patient, he

317

learns that the patient recently visited Asia, he might change the most probable explanation to be that the patient has tuberculosis and that the other diseases are not present. The physician might also determine a second, third, etc., most likely explanation. The actual diagnostic process involves cycles. After the physician formulates a hypothesis, he generates questions based on this hypothesis and runs additional tests to obtain answers to these questions. He then formulates a (possibly) new most probable explanation. This process continues until a satisfactory diagnosis is reached. Before proceeding, we formally define an "explanation" and "abductive inference."

Definition 8.1. Let $C = (V, E, P)$ be a causal network. Suppose some evidence has arrived in the form of instantiated variables and/or virtual evidence. A set of values of some specifed set of variables in the network is called an *explanation* for the evidence. This specified set of variables is called the *explanation set*.

   This definition of an explanation is a bit loose, since it does not tell us what the explanation set should be. There is no hard and fast rule for deciding on this set in general. It is always the set of variables in whose values we are interested. In some specific cases what this set should be is straightforward. For example, in the classical diagnostic problem, the instantiated variables will always be a subset of M, while the set of variables in the explanation set will always be all of D. In some problems the explanation set will include all the roots, whereas in others it will include all the uninstantiated variables. But, if we look again at Example 7.5, we see that we have neither of these cases. In that example we are interested in which diseases the patient may have. The explanation set would therefore be $\{B, C, E, G\}$. None of these variables are roots; indeed, we have eliminated both the roots from the explanation set. In reality, a positive chest X ray may have been caused indirectly by a patient smoking because smoking caused him to get lung cancer. Therefore the fact that the patient smokes might be part of the primordial explanation for this evidence. However, we don't care if he smokes; we only care about which diseases he has. Thus we are only interested in a set of values of the diseases. Note also in this example that evidence (smoking) can be a cause of some members (lung cancer and bronchitis) of the explanation set.

Definition 8.2. (Peng & Reggia, 1987a). *Abductive inference* is any reasoning process which derives the best explanation(s) for a given set of problem features (evidence).

   This chapter derives methods which perform abductive inference. One explanation will be considered better than another if the joint probability (conditional on the evidence) of the values in the first explanation is greater than the joint probability of the values in the second. Clearly, probability propagation, as described in chapters 6 and 7, is just a special case of abductive inference in which the explanation set contains exactly one variable.

In general, the determination of the most probable explanation is not equivalent to simply finding the most probable values of each individual variable in the explanation set. For example, in the classical diagnostic problem, if the diseases are independent given the manifestations, then trivially the most probable explanation is indeed the set of most probable values of the individual diseases. However, ordinarily this would only happen if each item of evidence was indicative of only one disease. For instance, if two diseases both caused a certain symptom, we would have the type of causal relationship depicted in Figure 5.8(a). Given that the symptom occurred, the presence of one disease should make the other disease less probable. The following simple example illustrates that, in general, the most probable explanation is not the set of most probable values of the individual variables.

Example 8.1. In the classical diagnostic problem, let d_1 and d_2 both be causes of m_1 and m_2, as illustrated in Figure 7.12, and suppose we have the probabilities below:

$$P(+d_1) = .2 \qquad\qquad P(+d_2) = .1$$
$$P(+m_1 \mid +d_1, +d_2) = .6 \qquad P(+m_1 \mid +d_1, \neg d_2) = .3$$
$$P(+m_1 \mid \neg d_1, +d_2) = .5 \qquad P(+m_1 \mid \neg d_1, \neg d_2) = .1$$
$$P(+m_2 \mid +d_1, +d_2) = .4 \qquad P(+m_2 \mid +d_1, \neg d_2) = .2$$
$$P(+m_2 \mid \neg d_1, +d_2) = .3 \qquad P(+m_2 \mid \neg d_1, \neg d_2) = .1.$$

We can compute the probability of each disease combination given that both m_1 and m_2 occur directly from the definition of conditional probability. For instance,

$$
\begin{aligned}
&P(+d_1, +d_2 \mid +m_1, +m_2) \\[4pt]
&= \frac{P(+d_1, +d_2, +m_1, +m_2)}{P(+m_1, +m_2)} \\[4pt]
&= \frac{P(+d_1, +d_2, +m_1, +m_2)}{\sum_{d_1, d_2} P(d_1, d_2, +m_1, +m_2)} \\[4pt]
&= \frac{(.6)(.4)(.2)(.1)}{(.6)(.4)(.2)(.1) + (.3)(.2)(.2)(.9) + (.5)(.3)(.8)(.1) + (.1)(.1)(.8)(.9)} \\[4pt]
&= .139.
\end{aligned}
$$

Similarly,

$$P(+d_1, \neg d_2 \mid +m_1, +m_2) = .310$$
$$P(\neg d_1, +d_2 \mid +m_1, +m_2) = .344$$
$$P(\neg d_1, \neg d_2 \mid +m_1, +m_2) = .207.$$

We see that the most probable explanation is $\{\neg d_1, +d_2\}$. That is, it is most probable that the patient has d_2 but not d_1. On the other hand, simple summations yield

$$P(\neg d_1 \mid +m_1, +m_2) = .551 \qquad \text{and} \qquad P(\neg d_2 \mid +m_1, +m_2) = .517.$$

Hence the most probable explanation is not obtained by simply taking the most probable values of the individual variables.

In Example 8.1 we determined the most probable explanation by direct computation. As in the case of probability propagation, this is not possible in many meaningful problems because the number of possible explanations is exponentially large with regard to the number of variables in the explanation set. Pearl [1987] has obtained a method for generating the top two most probable explanations in a singly connected causal network. That method is discussed in section 8.1. An extension of the method to arbitrary causal networks is covered in section 8.2. Since we are often interested in more than just the top two explanations, in section 8.3 we derive a method for obtaining the other most probable explanations for a special class of problems.

8.1. ABDUCTIVE INFERENCE IN SINGLY CONNECTED CAUSAL NETWORKS

Initially, in subsection 8.1.1, we assume that the explanation set includes all the uninstantiated variables. In subsection 8.1.2, we show how to modify the method for the case in which it does not.

8.1.1. The Method

The algorithm for this method is quite similar to that for propagating probabilities in singly connected causal networks (section 6.3). The main difference is that summation is replaced by maximization. For a proof of the correctness of the algorithm, the reader is referred to Pearl [1987b, 1988]. Briefly, we explain why maximization replaces summation as follows: Let V be the set of all variables in the network and let X' be the set of instantiated values of the instantiated variables X. Since we are now assuming that the explanation set includes all the uninstantiated variables, our aim is to find the set of values of $V - X$ which maximizes the quantity

$$P(V - X \mid X'),$$

which is clearly equivalent to finding the set of values of V which maximizes the quantity

$$P(V \mid X').$$

Let B be an arbitrary variable, b_i a possible value of B, and $W = V - \{B\}$. If we then define

$$P^*(b_i) = \max_W P(W \cup \{b_i\} \mid X'),$$

then of all the sets of values of all variables in the network containing B equal to b_i, $P^*(b_i)$ will be the probability of that set with maximum probability. First we determine, for each variable B, the value of $P^*(b_i)$, for each i, by propagating messages as done in chapter 7, with maximization replacing summation. Then the value of B, which we shall call b^*, in the most probable explanation, is equal to the b_i for which $P^*(b_i)$ is a maximum.

We now give the algorithm. For the sake of notational simplicity, the algorithm is given for the case where there are exactly two parents. The case where there are more than two parents is a straightforward generalization. Given the current instantiations, for a given variable, say B, $P^*(b_i)$ is not equal to the conditional probability of the most probable explanation which has B instantiated for b_i, but rather is only proportional to that conditional probability. However, the constant of proportionality does not depend on i. Therefore the value of B in the most probable explanation is still the b_i for which $P^*(b_i)$ is maximal.

ABDUCTIVE INFERENCE IN SINGLY CONNECTED NETWORKS

Operative Formulas

1. If B is a child of A, B has k possible values, A has m possible values, and B has one other parent, D, with n possible values, then for $1 \leq j \leq m$ the λ message from B to A is given by

$$\lambda_B(a_j) = \max_{\substack{1 \leq i \leq k \\ 1 \leq p \leq n}} (\pi_B(d_p) P(b_i \mid a_j, d_p) \lambda(b_i)).$$

2. If B is a child of A and A has m possible values, then for $1 \leq j \leq m$ the π message from A to B is given by

$$\pi_B(a_j) = \begin{cases} 1 & \text{if } A \text{ is instantiated for } a_j \\ 0 & \text{if } A \text{ is instantiated, but not for } a_j \\ \dfrac{P^*(a_j)}{\lambda_B(a_j)} & \text{if } A \text{ is not instantiated,} \end{cases}$$

where $P^*(a_j)$ is obtained in operative formula 5.

3. If B is a variable with k possible values, $s(B)$ is the set of B's children, then for $1 \leq i \leq k$ the λ value of B is given by

$$\lambda(b_i) = \begin{cases} \displaystyle\prod_{C \in s(B)} \lambda_C(b_i) & \text{if } B \text{ is not instantiated} \\ 1 & \text{if } B \text{ is instantiated for } b_i \\ 0 & \text{if } B \text{ is instantiated, but not for } b_i. \end{cases}$$

4. If B is a variable with k possible values and exactly two parents, A and D, A has m possible values, and D has n possible values, then for $1 \le i \le k$ the π value of B is given by

$$\pi(b_i) = \max_{\substack{1 \le j \le m \\ 1 \le p \le n}} (P(b_i \mid a_j, d_p)\pi_B(a_j)\pi_B(d_p)).$$

5. If B is a variable with k possible values, then for $1 \le i \le k$, $P^*(b_i)$ is given by

$$P^*(b_i) = \lambda(b_i)\pi(b_i).$$

6. If B is a variable with k possible values, then b^*, the value of B in the most probable explanation, is the value of b_j which maximizes $P^*(b_i)$. That is, b^* is given by

$$b^* = b_j \qquad \text{such that} \quad P^*(b_j) \ge P^*(b_i) \qquad \text{for} \quad 1 \le i \le k.$$

The causal network is first initialized to compute the a priori most probable explanation (i.e., the most probable initial combination of values of all variables in the network) as follows: ·

Initialization

A. Set all λ values, λ messages, and π messages to 1.

B. For all roots A, do

      begin

          1. If A has m possible values, then for $1 \le j \le m$ set

$$\pi(a_j) = P(a_j);$$

$$P^*(a_j) = P(a_j);$$

          2. Compute a^* using operative formula 6

      end.

C. For all roots A for all children B of A, do

      Post a new π message to B using operative formula 2.

      {A propagation flow will then begin owing to updating procedure C.}

D. For all leaves B for all parents A of B, do

      Post a new λ message to A using operative formula 1.

      {A propagation flow will then begin due to updating procedure B.}

When a variable is instantiated, or a λ or π message is received by a variable, one of the following updating procedures is used:

Updating

A. If a variable B is instantiated for b_j, then
> begin
>> 1. Set $P^*(b_j) = 1$ and for $i \neq j$, set $P^*(b_j) = 0$;
>> 2. Compute $\lambda(B)$ using operative formula 3;
>> 3. Post new λ messages to all B's parents using operative formula 1;
>> 4. Post new π messages to all B's children using operative formula 2;
>> 5. Compute b^* using operative formula 6
>
> end.

B. If a variable B receives a new λ message from one of its children, then if B is not already instantiated,
> begin
>> 1. Compute the new value of $\lambda(B)$ using operative formula 3;
>> 2. Compute the new value of $P^*(B)$ using operative formula 5;
>> 3. Post λ messages to all B's parents using operative formula 1;
>> 4. Post new π messages to B's other children using operative formula 2;
>> 5. Compute the new value of b^* using operative formula 6
>
> end.

C. If a variable B receives a new π message from a parent, then
> begin
>> If B is not already instantiated, then
>> begin
>>> 1. Compute the new value of $\pi(B)$ using operative formula 4;
>>> 2. Compute the new value of $P^*(B)$ using operative formula 5;
>>> 3. Post new π messages to all B's children using operative formula 2;
>>> 4. Compute the new value of b^* using operative formula 6
>> end;
>> 5. For each other parent of B, do
>> begin
>>> Compute a new λ message to that parent using operative formula 1;
>>> If the new λ message is different from the old one, then post it
>> end
>
> end.

The formulas in the above algorithm can easily be generalized to the case in which a vertex has more than two parents. For example, if B has another parent, E, with r possible values, then operative formula 1 would be

$$\lambda_B(a_j) \max_{\substack{1 \le i \le k \\ 1 \le p \le n \\ 1 \le s \le r}} \left(\pi_B(e_s)\pi_B(d_p)P(b_i \mid a_j, d_p, e_s)\lambda(b_i)\right).$$

Notice that in the initialization procedure we propagate initial λ messages from the leaves. Recall that in the case of probability propagation, we simply initialized all λ messages to $(1, 1, \ldots, 1)$ and performed no initial propagation of λ messages. Here we also initialize all λ messages to $(1, 1, \ldots, 1)$; however, in addition we propagate initial λ messages. It is easy to see from operative formula 1 that, since the λ values at the leaves are all initially equal to 1, all the λ messages from the leaves will end up being equal to $\alpha(1, 1, \ldots, 1)$, where α is a constant which depends on the leaf sending the message. Therefore, inductively all λ messages and all λ values will each end up being equal to a constant times $(1, 1, \ldots, 1)$. Consequently, as can easily be seen from operative formulas 5 and 6, the values obtained for the initial most probable explanation will be no different than if we left all λ messages with a value of $(1, 1, \ldots, 1)$. The reason we initially propagate λ messages is to set up an initial pointer structure. The need for this pointer structure and the method for setting it up as we pass the messages will be discussed in subsection 8.1.3 on breaking ties.

Note further that in operative formula 1, even if $\lambda(b_i)$ has the same value for all i, the new λ messages to other parents may be changed when B receives a new π message from one parent. Remember that in the case of probability propagation this was not the case. This is due to the fact that the summation in probability propagation is replaced by maximization in abductive inference. It thus seems that our abductive inference method may draw conclusions which are different from those of a human. In the case of probability propagation, we obtained results which are consistent with deductions which we feel would made by a human. For example, when Mr. Holmes learns that there has been an earthquake, he does not think that it is more or less probable that his residence has been burglarized. However, in the case of abductive inference, the fact of an earthquake could change the value of having been burglarized in the most probable set of values of the other two variables. Recall that our fundamental goal is to obtain the best possible results, not necessarily to obtain the same results as a human. We will investigate this more in the exercises. We note here that this phenomenon results in many more message passes and thereby more computation in abductive inference. For that reason, in updating procedure C we make certain that the new λ message is indeed different from the old one before posting it. Remember that λ messages, which are the same except for a multiplicative constant, are not different. After initialization, we could also make certain that the λ message computed in updating procedure B is different before posting it. We must post it at initialization to set up the initial pointer structure.

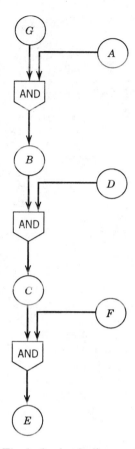

FIGURE 8.1   The logic circuit diagram for Example 8.2.

As noted at the beginning of this subsection, the above algorithm determines the most probable set of values of all the remaining variables in the network. However, we want the most probable set of values of only the variables in the explanation set, and this set often does not include all the remaining variables in the network. Just as finding the most probable value of each individual variable does not yield the most probable set of values of variables in the explanation set, finding the most probable set of values of a larger set of variables does not necessarily yield this set either. You will be asked to verify this in the exercises. We will show in the next subsection how to modify this algorithm for the case in which the explanation set does not include all the uninstantiated variables. By limiting the explanation set to a subset of all the variables, we also help solve the problem involving the large number of messages discussed above.

We now give an example of the application of the above algorithm:

Example 8.2. [Pearl, 1987b]. Suppose we have three binary inputs, A, D, and F, joined by AND gates, and their total output is represented by E. Then

$$a_1 = \text{input } A \text{ is on}$$

$$a_2 = \text{input } A \text{ is off}$$

$$e_1 = \text{the total output is on}$$

$$e_2 = \text{the total output is off.}$$

The values of D and F are defined analogously. This situation can be represented by the logic circuit diagram in Figure 8.1, where G, B, and C are unseen variables. G is a variable which is always on, while B is on if A is on, and C is on only if both B and D are on. Clearly, E is on only if both C and F are on. If

$$P(a_1) = .55, \qquad P(d_1) = .6, \qquad P(f_1) = .6,$$

then this situation can be represented by the causal network in Figure 8.2. Suppose we find a failure at E (i.e., $E = e_2$) and we know D to be on (i.e., $D = d_1$). The problem is to find the most probable explanation for the failure. That is, to determine whether it is most probable that A and F are both off, or that A is on and F is off, or that A is off and F is on. The explanation set is hence $\{A,F\}$. As mentioned previously, the most probable set of values of the remaining variables in general does not contain the most probable set of values in the explanation set. However, it is easy to show that in this particular problem it does. We will therefore find the most probable set of values of the remaining variables.

First we perform the initialization:

A. Set all λ values, λ messages, and π messages to 1.

B.1. $\pi(A) = P(A) = (.55, .45) \qquad P^*(A) = P(A) = (.55, .45)$

$\quad\;\; \pi(D) = P(D) = (.6, .4) \qquad\;\; P^*(D) = P(D) = (.6, .4)$

$\quad\;\; \pi(F) = P(F) = (.6, .4) \qquad\;\; P^*(F) = P(F) = (.6, .4).$

B.2. $a^* = a_1, \qquad$ since $\quad P^*(a_1) > P^*(a_2)$

$\quad\;\; d^* = d_1 \qquad$ since $\quad P^*(d_1) > P^*(d_2)$

$\quad\;\; f^* = f_1 \qquad$ since $\quad P^*(f_1) > P^*(f_2).$

C. $\pi_B(a_1) = \dfrac{P^*(a_1)}{\lambda_B(a_1)} = \dfrac{.55}{1} = .55.$

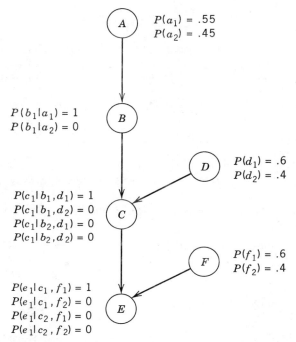

FIGURE 8.2 The causal network which represents the logic circuit described in Example 8.2.

Similarly, we obtain

$$\pi_B(A) = (.55, .45)$$
$$\pi_C(D) = (.6, .4)$$
$$\pi_E(F) = (.6, .4).$$

{Updating procedure C now takes over.}

When B receives a new π message from A,

C.1. $\pi(b_1) = \max_{1 \leq j \leq 2} (\pi_B(a_j)P(b_1 \mid a_j))$

$$= \max(.55(1), .45(0)) = .55$$

$\pi(b_2) = \max_{1 \leq j \leq 2} (\pi_B(a_j)P(b_2 \mid a_j))$

$$= \max(.55(0), .45(1)) = .45.$$

C.2. $P^*(b_1) = \lambda(b_1)\pi(b_1) = (1)(.55) = .55$
$P^*(b_2) = \lambda(b_2)\pi(b_2) = (1)(.45) = .45.$

C.3. $\pi_C(b_1) = \dfrac{P^*(b_1)}{\lambda_C(b_1)} = \dfrac{.55}{1} = .55$

$\pi_C(b_2) = \dfrac{P^*(b_2)}{\lambda_C(b_2)} = \dfrac{.45}{1} = .45.$

C.4. $b^* = b_1$, since $P^*(b_1) > P^*(b_2)$.
C.5. B has no other parents to which to send new λ messages.

When C receives new π messages from B and D,

C.1. $\pi(c_1) = \max\limits_{\substack{1 \le j \le 2 \\ 1 \le p \le 2}} (P(c_1 \mid b_j, d_p)\pi_C(b_j)\pi_C(d_p))$

$\qquad = \max((1)(.55)(.6), 0, 0, 0) = .33$

$\pi(c_2) = \max\limits_{\substack{1 \le j \le 2 \\ 1 \le p \le 2}} (P(c_2 \mid b_j, d_p)\pi_C(b_j)\pi_C(d_p))$

$\qquad = \max((0, (1)(.55)(.4), (1)(.45)(.6), (1)(.45)(.4)) = .27.$

C.2. $P^*(c_1) = \lambda(c_1)\pi(c_1) = (1)(.33) = .33$
$P^*(c_2) = \lambda(c_2)\pi(c_2) = (1)(.27) = .27.$

C.3. $\pi_E(c_1) = \dfrac{P^*(c_1)}{\lambda_E(c_1)} = \dfrac{.33}{1} = .33$

$\pi_E(c_2) = \dfrac{P^*(c_2)}{\lambda_E(c_2)} = \dfrac{.27}{1} = .27.$

C.4. $c^* = c_1$, since $P^*(c_1) > P^*(c_2)$.

C.5. $\lambda_C(b_1) = \max\limits_{\substack{1 \le i \le 2 \\ 1 \le p \le 2}} (\pi_C(d_p)P(c_i \mid b_1, d_p)\lambda(c_i))$

$\qquad = \max((.6)(1)(1), (.6)(0)(1), (.4)(0)(1), (.4)(1)(1))$

$\qquad = .6$

$\lambda_C(b_2) = \max\limits_{\substack{1 \le i \le 2 \\ 1 \le p \le 2}} (\pi_C(d_p)P(c_i \mid b_2, d_p)\lambda(c_i))$

$\qquad = \max((.6)(0)(1), (.6)(1)(1), (.4)(0)(1), (.4)(1)(1))$

$\qquad = .6.$

C.5. $\lambda_C(d_1) = \max\limits_{\substack{1 \le i \le 2 \\ 1 \le p \le 2}} (\pi_C(b_p)P(c_i \mid b_p, d_1)\lambda(c_i))$

$\qquad = \max((.55)(1)(1), (.55)(0)(1), (.45)(0)(1), (.45)(1)(1))$

$\qquad = .55$

$\lambda_C(d_2) = \max\limits_{\substack{1 \le i \le 2 \\ 1 \le p \le 2}} (\pi_C(b_p)P(c_i \mid b_p, d_2)\lambda(c_i))$

$\qquad = \max((.55)(0)(1), (.55)(1)(1), (.45)(0)(1), (.45)(1)(1))$

$\qquad = .55.$

We do not propagate these λ messages because they are both unchanged except for multiplicative constants.

When E receives new π messages from C and F,

C.1. $\pi(e_1) = \max\limits_{\substack{1 \le j \le 2 \\ 1 \le p \le 2}} (P(e_1 \mid c_j, f_p)\pi_E(c_j)\pi_E(f_p))$

$\qquad = \max((1)(.33)(.6), 0, 0, 0) = .198$

$\pi(e_2) = \max\limits_{\substack{1 \le j \le 2 \\ 1 \le p \le 2}} (P(e_2 \mid c_j, f_p)\pi_E(c_j)\pi_E(f_p))$

$\qquad = \max(0, (1)(.33)(.4), (1)(.27)(.6), (1)(.27)(.4)) = .162.$

C.2. $P^*(e_1) = \lambda(e_1)\pi(e_1) = (1)(.198) = .198$

$\qquad P^*(e_2) = \lambda(e_2)\pi(e_2) = (1)(.162) = .162.$

C.3. E has no children to which to send new π messages.

C.4. $e^* = e_1$, since $P^*(e_1) > P^*(e_2)$.

C.5. $\lambda_E(f_1) = \max\limits_{\substack{1 \le i \le 2 \\ 1 \le p \le 2}} (\pi_E(c_p)P(e_i \mid c_p, f_1)\lambda(e_i))$

$\qquad = \max((.33)(1)(1), (.33)(0)(1), (.27)(0)(1), (.27)(1)(1))$

$\qquad = .33$

$\lambda_E(f_2) = \max\limits_{\substack{1 \le i \le 2 \\ 1 \le p \le 2}} (\pi_E(c_p)P(e_i \mid c_p, f_2)\lambda(e_i))$

$\qquad = \max((.33)(0)(1), (.33)(1)(1), (.27)(0)(1), (.27)(1)(1))$

$\qquad = .33.$

C.5. $\lambda_E(c_1) = \max_{\substack{1 \le i \le 2 \\ 1 \le p \le 2}} (\pi_E(f_p)P(e_i \mid c_1, f_p)\lambda(e_i))$

   $= \max((.6)(1)(1), (.6)(0)(1), (.4)(0)(1), (.4)(1)(1))$

   $= .6$

$\lambda_E(c_2) = \max_{\substack{1 \le i \le 2 \\ 1 \le p \le 2}} (\pi_E(f_p)P(e_i \mid c_2, f_p)\lambda(e_i))$

   $= \max((.6)(0)(1), (.6)(1)(1), (.4)(0)(1), (.4)(1)(1))$

   $= .6.$

We do not propagate these λ messages because they are both unchanged except for multiplicative constants.
Propagation ends.

{Initialization now resumes.}

D. E posts λ messages to both C and F. These will be the same messages computed in C.5 above. It is left as an exercise to absorb these λ messages at F and at C and to propagate new λ messages from C up the network.

The state of the causal network after initialization is depicted in Figure 8.3. The most probable initial set of values is

$$\{a_1, b_1, c_1, d_1, e_1, f_1\}.$$

Next we instantiate D for d_1. When D is instantiated,

A.1. $P^*(d_1) = 1$ and $P^*(d_2) = 0$.

A.2. $\lambda(d_1) = 1$ and $\lambda(d_2) = 0$.

A.3. D has no parents to which to send new λ messages.

A.4. $\pi_C(D) = (1, 0)$.

A.5. $d^* = d_1$.

When C receives a new π message from D,

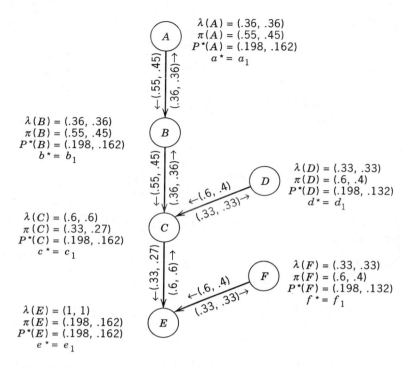

FIGURE 8.3 The causal network for Example 8.2 after initialization.

C.1. $\pi(c_1) = \max\limits_{\substack{1 \le j \le 2 \\ 1 \le p \le 2}} (P(c_1 \mid b_j, d_p)\pi_C(b_j)\pi_C(d_p))$

$= \max((1)(.55)(1), 0, 0, 0)$

$= .55$

$\pi(c_2) = \max\limits_{\substack{1 \le j \le 2 \\ 1 \le p \le 2}} (P(c_2 \mid b_j, d_p)\pi_C(b_j)\pi_C(d_p))$

$= \max(0, (1)(.55)(0), (1)(.45)(1), (1)(.45)(0))$

$= .45.$

C.2. $P^*(c_1) = \lambda(c_1)\pi(c_1) = (.6)(.55) = .33$

$P^*(c_2) = \lambda(c_2)\pi(c_2) = (.6)(.45) = .27.$

C.3. $\pi_E(c_1) = \dfrac{P^*(c_1)}{\lambda_E(c_1)} = \dfrac{.33}{.6} = .55$

$\pi_E(c_2) = \dfrac{P^*(c_2)}{\lambda_E(c_1)} = \dfrac{.27}{.6} = .45.$

C.4. $c^* = c_1$.

C.5. $\lambda_C(b_1) = \max_{\substack{1 \leq i \leq 2 \\ 1 \leq p \leq 2}} (\pi_C(d_p)P(c_i \mid b_1, d_p)\lambda(c_i))$

$\qquad = \max((1)(1)(.6), (1)(0)(.6), (0)(0)(.6), (0)(1)(.6))$

$\qquad = .6$

$\quad \lambda_C(b_2) = \max_{\substack{1 \leq i \leq 2 \\ 1 \leq p \leq 2}} (\pi_C(d_p)P(c_i \mid b_2, d_p)\lambda(c_i))$

$\qquad = \max((1)(0)(.6), (1)(1)(.6), (0)(0)(.6), (0)(1)(.6))$

$\qquad = .6.$

We do not propagate this λ message because it is unchanged except for a multiplicative constant.

When E receives a new π message from C,

C.1. $\pi(e_1) = \max_{\substack{1 \leq j \leq 2 \\ 1 \leq p \leq 2}} (P(e_1 \mid c_j, f_p)\pi_E(c_j)\pi_E(f_p))$

$\qquad = \max((1)(.55)(.6), 0, 0, 0) = .33$

$\quad \pi(e_2) = \max_{\substack{1 \leq j \leq 2 \\ 1 \leq p \leq 2}} (P(e_2 \mid c_j, f_p)\pi_E(c_j)\pi_E(f_p))$

$\qquad = \max(0, (1)(.55)(.4), (1)(.45)(.6), (1)(.45)(.4)) = .27.$

C.2. $P^*(e_1) = \lambda(e_1)\pi(e_1) = (1)(.33) = .33$

$\quad P^*(e_2) = \lambda(e_2)\pi(e_2) = (1)(.27) = .27.$

C.3. E has no children to which to send new π messages.

C.4. $e^* = e_1$.

C.5. $\lambda_E(f_1) = \max_{\substack{1 \leq i \leq 2 \\ 1 \leq p \leq 2}} (\pi_E(c_p)P(e_i \mid c_p, f_1)\lambda(e_i))$

$\qquad = \max((.55)(1)(1), (.55)(0)(1), (.45)(0)(1), (.45)(1)(1))$

$\qquad = .55$

$\quad \lambda_E(f_2) = \max_{\substack{1 \leq i \leq 2 \\ 1 \leq p \leq 2}} (\pi_E(c_p)P(e_i \mid c_p, f_2)\lambda(e_i))$

$\qquad = \max((.55)(0)(1), (.55)(1)(1), (.45)(0)(1), (.45)(1)(1))$

$\qquad = .55.$

We do not propagate this λ message, since it is unchanged except for a multiplicative constant.

Propagation ends.

The state of the causal network after D is instantiated for d_1 is depicted in Figure 8.4. Note that the most probable set of remaining values is still

$$\{a_1, b_1, c_1, e_1, f_1\}.$$

Next E is instantiated for e_2. When E is instantiated,

A.1. $P^*(e_1) = 0$ and $P^*(e_2) = 1$.

A.2. $\lambda(e_1) = 0$ and $\lambda(e_2) = 1$.

A.3. $\lambda_E(f_1) = \max_{\substack{1 \le i \le 2 \\ 1 \le p \le 2}} (\pi_E(c_p)P(e_i \mid c_p, f_1)\lambda(e_i))$

$\qquad\quad = \max((.55)(1)(0), (.55)(0)(1), (.45)(0)(0), (.45)(1)(1))$

$\qquad\quad = .45$

$\quad\ \lambda_E(f_2) = \max_{\substack{1 \le i \le 2 \\ 1 \le p \le 2}} (\pi_E(c_p)P(e_i \mid c_p, f_2)\lambda(e_i))$

$\qquad\quad = \max((.55)(0)(0), (.55)(1)(1), (.45)(0)(0), (.45)(1)(1))$

$\qquad\quad = .55.$

A.3. $\lambda_E(c_1) = \max_{\substack{1 \le i \le 2 \\ 1 \le p \le 2}} (\pi_E(f_p)P(e_i \mid c_1, f_p)\lambda(e_i))$

$\qquad\quad = \max((.6)(1)(0), (.6)(0)(1), (.4)(1)(0), (.4)(1)(1))$

$\qquad\quad = .4$

$\quad\ \lambda_E(c_2) = \max_{\substack{1 \le i \le 2 \\ 1 \le p \le 2}} (\pi_E(f_p)P(e_i \mid c_2, f_p)\lambda(e_i))$

$\qquad\quad = \max((.6)(0)(0), (.6)(1)(1), (.4)(0)(0), (.4)(1)(1))$

$\qquad\quad = .6.$

A.4. E has no children to which to send new π messages.

A.5. $e^* = e_2$.

When C receives a new λ message from E,

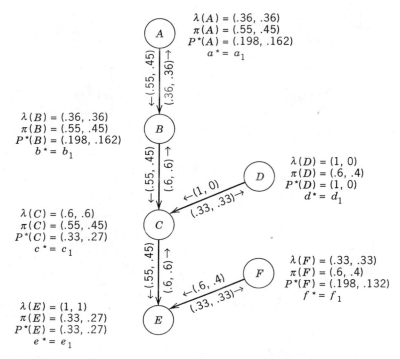

FIGURE 8.4 The causal network for Example 8.2 after D is instantiated.

B.1. $\lambda(c_1) = \lambda_E(c_1) = .4$

$\lambda(c_2) = \lambda_E(c_2) = .6.$

B.2. $P^*(c_1) = \lambda(c_1)\pi(c_1) = (.4)(.55) = .22$

$P^*(c_2) = \lambda(c_2)\pi(c_2) = (.6)(.45) = .27.$

B.3. $\lambda_C(b_1) = \max_{\substack{1 \le i \le 2 \\ 1 \le p \le 2}} (\pi_C(d_p)P(c_i \mid b_1, d_p)\lambda(c_i))$

$= \max((1)(1)(.4), (1)(0)(.6), (0)(0)(.4), (0)(1)(.6))$

$= .4$

$\lambda_C(b_2) = \max_{\substack{1 \le i \le 2 \\ 1 \le p \le 2}} (\pi_C(d_p)P(c_i \mid b_2, d_p)\lambda(c_i))$

$= \max((1)(0)(.4), (1)(1)(.6), (0)(0)(.4), (0)(1)(.6))$

$= .6.$

B.3. $\lambda_C(d_1) = \max_{\substack{1 \le i \le 2 \\ 1 \le p \le 2}} (\pi_C(b_p)P(c_i \mid b_p, d_1)\lambda(c_i))$

$= \max((.55)(1)(.4), (.55)(0)(.6), (.45)(0)(.4), (.45)(1)(.6))$

$= .27$

$\lambda_C(d_2) = \max_{\substack{1 \le i \le 2 \\ 1 \le p \le 2}} (\pi_C(b_p)P(c_i \mid b_p, d_2)\lambda(c_i))$

$= \max((.55)(0)(.4), (.55)(1)(.6), (.45)(0)(.4), (.45)(1)(.6))$

$= .33.$

B.4. C has no other children to which to send new π messages.

B.5. $c^* = c_2$.

When F receives a new λ message from E,

B.1. $\lambda(f_1) = \lambda_E(f_1) = .45$

$\lambda(f_2) = \lambda_E(f_2) = .55.$

B.2. $P^*(f_1) = \lambda(f_1)\pi(f_1) = (.45)(.6) = .27$

$P^*(f_2) = \lambda(f_2)\pi(f_2) = (.55)(.4) = .22.$

B.3. F has no parents to which to send new λ messages.

B.4. F has no other children to which to send new π messages.

B.5. $f^* = f_1$.

When B receives a new λ message from C,

B.1. $\lambda(b_1) = \lambda_C(b_1) = .4$

$\lambda(b_2) = \lambda_C(b_2) = .6.$

B.2. $P^*(b_1) = \lambda(b_1)\pi(b_1) = (.4)(.55) = .22$

$P^*(b_2) = \lambda(b_2)\pi(c_2) = (.6)(.45) = .27.$

B.3. $\lambda_B(a_1) = \max_{1 \le i \le 2} (P(b_i \mid a_1)\lambda(b_i))$

$= \max((1)(.4), (0)(.6))$

$= .4$

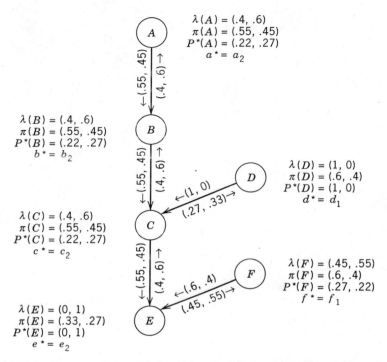

FIGURE 8.5 The causal network for Example 8.2 after both D and E are instantiated.

$$\lambda_B(a_2) = \max_{1 \le i \le 2}(P(b_i \mid a_2)\lambda(b_i))$$

$$= \max((0)(.4),(1)(.6))$$

$$= .6.$$

B.4. B has no other children to which to send new π messages.

B.5. $b^* = b_2$.

When A receives a new λ message from B,

B.1. $\lambda(a_1) = \lambda_B(a_1) = .4$

$\lambda(a_2) = \lambda_B(a_2) = .6$.

B.2. $P^*(a_1) = \lambda(a_1)\pi(a_1) = (.4)(.55) = .22$

$P^*(a_2) = \lambda(a_2)\pi(a_2) = (.6)(.45) = .27$.

B.3. A has no parents to which to send new λ messages.

B.4. A has no other children to which to send new π messages.

B.5. $a^* = a_2$.

When D receives a new λ message from C, nothing happens, since D is already instantiated.

Propagation ends.

The state of the causal network after E is instantiated for e_2 is depicted in Figure 8.5.

We see that the most probable set of values of the remaining variables given that D has taken value d_1 and E has taken value e_2 is

$$\{a_2, b_2, c_2, f_1\}.$$

As mentioned previously, in general, the most probable set of values of the remaining variables does not contain the most probable set of values of variables in the explanation set (the explanation set here is $\{A, F\}$). But, in this particular problem it does. Therefore it is most probable that A has failed but F has not. If we performed probability propagation as described in chapter 6, we would obtain

$$P'(a_2) = .672 \quad \text{and} \quad P'(f_1) = .403.$$

It is left as an exercise to actually do this. We see again that the most probable explanation is not obtained by taking the most probable values of each individual variable.

8.1.2. Restricting the Explanation Set

The algorithm in the previous subsection determines the most probable combination of all the uninstantiated variables in the network. However, as noted in the introduction to this chapter, in practice we actually want to compute the most probable combination of all variables in the explanation set, and the explanation set rarely includes all the uninstantiated variables in the network. For example, in Example 8.2, if A, D, and F are outputs of complex digital circuits and we are only interested in whether any of these circuits should be replaced, then we would not bother to propagate messages to those ancestors beyond A, D, and F.

We determine the most probable combination of variables in the explanation set as follows. Let X' be a set of instantiated values of instantiated variables, let Y include all the instantiated variables and all the variables in the explanation set, let Z be equal to $V - Y$, (i.e., the remaining variables in the network), and for a given variable, $B \in Y$, let $W = Y - \{B\}$. Then we would compute $P^*(b_i)$ as follows:

$$P^*(b_i) = \max_W P(W \cup \{b_i\} \mid X')$$

$$= \max_W \left(\sum_Z P(W \cup \{b_i\} \mid Z \cup X') P(Z \mid X') \right).$$

Thus we perform abductive inference according to the algorithm given in section 8.1.1 with the following modifications: We only use the maximization formulas, given in this chapter, on variables in Y. For variables in Z, we use the addition formulas for probability propagation given in chapter 6. It is left as an exercise to actually modify the abductive inference algorithm.

8.1.3. Handling Ties

In Example 8.2, for each variable there was always a unique value of that variable which maximized P^*. This need not always be the case, however. That is to say, it is possible that two values could yield the same maximal value. The example below illustrates that we cannot simply break ties arbitrarily:

Example 8.3. Suppose that we have the exact same causal network as in Figure 8.2 except that

$$P(f_1) = .55 \qquad \text{and} \qquad P(f_2) = .45.$$

Further suppose that E is instantiated for e_2 and D is instantiated for d_1 as before, and that the explanation set is equal to $\{A, F\}$. The problem is sufficiently simple that we can compute the conditional probabilities of the possible explanations directly (notice that we could have also done this in Example 8.2). We have

$$P(a_1, f_1 \mid d_1, e_2) = 0$$

$$P(a_1, f_2 \mid d_1, e_2) = \frac{P(a_1, f_2, d_1, e_2)}{P(d_1, e_2)}$$

$$= \frac{(.55)(.45)}{(1 - (.55)(.55))} = .3548$$

$$P(a_2, f_1 \mid d_1, e_2) = \frac{P(a_2, f_1, d_1, e_2)}{P(d_1, e_2)}$$

$$= \frac{(.55)(.45)}{(1 - (.55)(.55))} = .3548$$

$$P(a_2, f_2 \mid d_1, e_2) = \frac{P(a_2, f_2, d_1, e_2)}{P(d_1, e_2)}$$

$$= \frac{(.45)(.45)}{(1 - (.55)(.55))} = .2903.$$

We see that there are two most probable explanations in Example 8.3: $\{a_1, f_2\}$ and $\{a_2, f_1\}$. If, when applying the abductive inference algorithm, we arbitrarily broke ties, we could end up pairing a_2 with f_2 and concluding that $\{a_2, f_2\}$ was a most probable explanation. In fact, we could end up pairing a_1 with f_1 and concluding the impossible explanation $\{a_1, f_1\}$ as being the most probable one. We will now show what must be done to avoid these mistakes.

Consider the vertex C in the causal network in Figure 8.2. From operative formula 4, the π value of that vertex is given by

$$\pi(c_i) = \max_{\substack{1 \le j \le 2 \\ 1 \le p \le 2}} (P(c_i \mid b_j, d_p) \pi_C(b_j) \pi_C(d_p)).$$

That maximum is obtained from a specific pair of values of b_j and d_p. This pair must be the values of B and D in the most probable explanation which has C instantiated for c_i. Suppose now that it turns out that two values of C, say c_1 and c_2, both maximize $P^*(C)$. Then the most probable explanation which has C equal to c_1 is one containing a specific pair of values of b_j and d_p, while the most probable explanation which has C equal to c_2 is one containing another specific pair. Consequently, we cannot arbitrarily group c_1 with any maximizing values of B and D. To keep track of the proper values, whenever we compute a π value, we pass a pointer from each value of the variable back to the values of its parents which yield the π value (i.e., the values which maximize the above expression). If the maximizing values for c_1 were b_1 and d_2 whereas those for c_2 were b_2 and d_2, we would have

$$c_1 \to (b_1, d_2)$$

$$c_2 \to (b_2, d_2).$$

The same situation arises when a vertex receives a λ message. That λ message is the maximum value relative to a specific value of the child which sent the message and to specific values of the other parents of that child. For instance, consider a λ message which D receives from C:

$$\lambda_C(d_j) = \max_{\substack{1 \le p \le 2 \\ 1 \le i \le 2}} (\pi_C(b_p) P(c_i \mid d_j, b_p) \lambda(c_i)).$$

The maximizing value for a particular d_j will occur at some specific pair of values of c_i and b_p. Thus, when we pass this λ message, we must set up a pointer to those values. For example, if the maximizing value for d_1 occurs with $C = c_1$ and $B = b_1$ while the maximizing value for d_2 occurs with $C = c_2$ and $B = b_1$, we would have

$$d_1 \to (c_1, b_1)$$

$$d_2 \to (c_2, b_1).$$

By initially propagating λ messages from the leaves and π messages from the roots (and thereby setting up these pointers), we are guaranteed that we

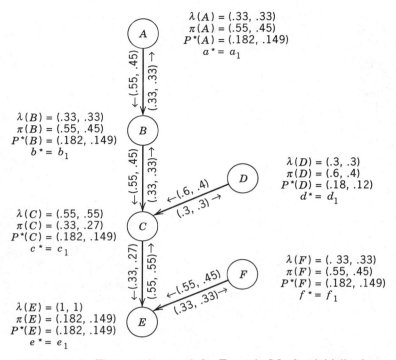

FIGURE 8.6   The causal network for Example 8.3 after initialization.

can take any variable, say B, and any value of that variable, say b_1, and re-trieve, by following the pointers, all the values of the other variables in the most probable initial explanation which contains b_1. As variables are instan-tiated, some of these pointers are changed as λ and π messages are passed. After instantiations, we can retrieve the most probable explanation containing b_1 based on those instantiations.

The instructions that follow must be added to the abductive inference algo-rithm in order to handle the creation of these pointers:

1. When computing a new π value for a variable B, for each alternative b_i of B, create a pointer from b_i to the values of the parents of b_i which determine $\pi(b_i)$ through maximization.

2. When a variable B receives a new λ message $\lambda_C(B)$ from a child C, for each alternative b_i of B, create a pointer from b_i to the value of C and the values of the other parents of C which determine $\lambda_C(b_i)$ through maximiza-tion.

We will now illustrate these concepts by continuing Example 8.3. Figure 8.6 contains the causal network after initialization. Many pointers have been set up as a result of this initialization. It is left as an exercise to perform this initialization and set up these pointers. We will follow the change in the pointers as variables are instantiated. Specifically, we will follow those pointers relevant to the most probable explanation containing b_1. After initialization, we have the following pointers relevant to that explanation:

$$b_1 \rightarrow a_1, \qquad b_1 \rightarrow (c_1, d_1), \qquad c_1 \rightarrow (e_1, f_1).$$

Therefore, by following the pointers emanating from b_1, we can retrieve

$$\{a_1, b_1, c_1, d_1, e_1, f_1\}$$

as the most probable explanation containing b_1. Since $b^* = b_1$, this explanation is the most probable overall initial explanation. Suppose we now instantiate E for e_2:

When E is instantiated,

A.1. $P^*(e_1) = 0$ and $P^*(e_2) = 1$.

A.2. $\lambda(e_1) = 0$ and $\lambda(e_2) = 1$.

A.3. $\lambda_E(f_1) = \max_{\substack{1 \leq i \leq 2 \\ 1 \leq p \leq 2}} (\pi_E(c_p)P(e_i \mid c_p, f_1)\lambda(e_i))$

$$= \max((.33)(1)(0), (.33)(0)(1), (.27)(0)(0), (.27)(1)(1))$$

$$= .27.$$

This λ message was obtained from $P(e_2 \mid c_2, f_1)$; that is, with C having a value of c_2 and E having the value of e_2. Therefore we change the pointer emanating from f_1 to

$$f_1 \rightarrow (c_2, e_2).$$

$$\lambda_E(f_2) = \max_{\substack{1 \leq i \leq 2 \\ 1 \leq p \leq 2}} (\pi_E(c_p)P(e_i \mid c_p, f_2)\lambda(e_i))$$

$$= \max((.33)(0)(0), (.33)(1)(1), (.27)(0)(0), (.27)(1)(1))$$

$$= .33.$$

This λ message is obtained from $P(e_2 \mid c_1, f_2)$. Thus we change the pointer emanating from f_2 to

$$f_2 \rightarrow (c_1, e_2).$$

A.3. $\lambda_E(c_1) = \max_{\substack{1 \leq i \leq 2 \\ 1 \leq p \leq 2}} (\pi_E(f_p)P(e_i \mid c_1, f_p)\lambda(e_i))$

$$= \max((.55)(1)(0), (.55)(0)(1), (.45)(0)(0), (.45)(1)(1))$$

$$= .45.$$

This λ message is obtained from $P(e_2 \mid c_1, f_2)$. Hence we change the pointer emanating from c_1 to

$$c_1 \rightarrow (e_2, f_2).$$

Notice that f_2 now points to c_1, and c_1 points to f_2. This need not always be the case. In other words, it is possible that the most probable explanation which has F equal to f_2 has C equal to c_2 even though the most probable explanation which has C equal to c_1 has F equal to f_2. (You will be asked to substantiate this fact in an exercise.)

$$\lambda_E(c_2) = \max_{\substack{1 \le i \le 2 \\ 1 \le p \le 2}} (\pi_E(f_p)P(e_i \mid c_2, f_p)\lambda(e_i))$$

$$= \max((.55)(0)(0), (.55)(1)(1), (.45)(0)(0), (.45)(1)(1))$$

$$= .55.$$

This λ message is obtained from $P(e_2 \mid c_2, f_1)$. Consequently, we change the pointer emanating from c_2 to

$$c_2 \rightarrow (e_2, f_1).$$

A.4. E has no children to which to send new π messages.

A.5. $e^* = e_2$.

When C receives a new λ message from E,

B.1. $\lambda(c_1) = \lambda_E(c_1) = .45$

$\lambda(c_2) = \lambda_E(c_2) = .55.$

B.2. $P^*(c_1) = \lambda(c_1)\pi(c_1) = (.45)(.33) = .149$

$P^*(c_2) = \lambda(c_2)\pi(c_2) = (.55)(.27) = .149.$

B.3. $\lambda_C(b_1) = \max_{\substack{1 \le i \le 2 \\ 1 \le p \le 2}} (\pi_C(d_p)P(c_i \mid b_1, d_p)\lambda(c_i))$

$$= \max((.6)(1)(.45), (.6)(0)(.55), (.4)(0)(.45), (.4)(1)(.55))$$

$$= .27.$$

This λ message was obtained from $P(c_1 \mid b_1, d_1)$. Therefore we change the pointer emanating from b_1 to

$$b_1 \to (c_1, d_1).$$

$$\lambda_C(b_2) = \max_{\substack{1 \le i \le 2 \\ 1 \le p \le 2}} (\pi_C(d_p) P(c_i \mid b_2, d_p) \lambda(c_i))$$

$$= \max((.6)(0)(.45), (.6)(1)(.55), (.4)(0)(.45), (.4)(1)(.55))$$

$$= .33.$$

This λ message was obtained from $P(c_2 \mid b_2, d_1)$. Hence we change the pointer emanating from b_2 to

$$b_2 \to (c_2, d_1).$$

B.3. $$\lambda_C(d_1) = \max_{\substack{1 \le i \le 2 \\ 1 \le p \le 2}} (\pi_C(b_p) P(c_i \mid b_p, d_1) \lambda(c_i))$$

$$= \max((.55)(1)(.45), (.55)(0)(.55), (.45)(0)(.45), (.45)(1)(.55))$$

$$= .248.$$

This λ message was obtained from $P(c_1 \mid b_1, d_1)$ or from $P(c_2 \mid b_2, d_1)$. Therefore we create two pointers emanating from d_1 as follows:

$$d_1 \to (c_1, b_1) \qquad \text{and} \qquad d_1 \to (c_2, b_2).$$

This fact means that the most probable explanation containing d_1 is not unique.

$$\lambda_C(d_2) = \max_{\substack{1 \le i \le 2 \\ 1 \le p \le 2}} (\pi_C(b_p) P(c_i \mid b_p, d_2) \lambda(c_i))$$

$$= \max((.55)(0)(.45), (.55)(1)(.55), (.45)(0)(.45), (.45)(1)(.55))$$

$$= .303.$$

This λ message was obtained from $P(c_2 \mid b_1, d_2)$. Thus we change the pointer emanating from d_2

$$d_2 \to (c_2, b_1).$$

B.4. C has no other children to which to send new π messages.
B.5. $c_1^* = c_2$ and $c_2^* = c_2$.

We leave it as an exercise to absorb these λ messages, to propagate the remaining λ messages, and to set up the remaining pointers. After this is done, the network will be as shown in Figure 8.7 and we will have the following additional pointer changes:

$$a_1 \to b_1 \qquad \text{and} \qquad a_2 \to b_2.$$

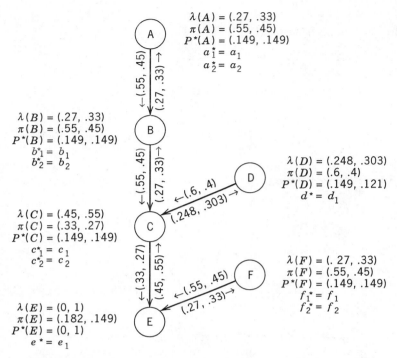

FIGURE 8.7 The causal network for Example 8.3 after E is instantiated.

Hence the pointer $b_1 \rightarrow a_1$ will be unchanged. The pointers relevant to retrieving the most probable explanation containing b_1 now are

$$b_1 \rightarrow a_1, \qquad b_1 \rightarrow (c_1, d_1), \qquad c_1 \rightarrow (e_2, f_2).$$

Following these pointers we now see that the most probable explanation containing b_1 is

$$\{a_1, b_1, c_1, d_1, f_2\}.$$

Since b_1 is one value of b^*, this explanation is a most probable one. Since b_2 is also a value of b^*, we can follow the pointers emanating from b_2 to retrieve the other most probable explanation. The relevant pointers are

$$b_2 \rightarrow a_2, \qquad b_2 \rightarrow (c_2, d_1), \qquad c_2 \rightarrow (e_2, f_1).$$

Therefore the other most probable explanation is

$$\{a_2, b_2, c_2, d_1, f_1\}.$$

We see then that it is always possible to retrieve all the most probable explanations.

Notice that these explanations are the most probable ones based on E being instantiated for e_2. If we next instantiate D for d_1, (possibly) new optimal values and new pointers will be generated. It is left as an exercise to follow the instantiation of D.

8.1.4. The Probability of an Explanation

Merely knowing that an explanation is the most probable one is often insufficient. For example, in medical diagnosis, we not only need to know that it is most probable that a patient has a given set of diseases, but also the actual probability that he has that set. If that probability is .1, the physician will need additional information, while if the probability is .95, he may be able to make a diagnosis. In this subsection we show how to determine the probability of an explanation.

Let $Y = \{v_1, v_2, \ldots, v_k\}$ be an arbitrary subset of k variables and let $Y' = \{v'_1, v'_2, \ldots, v'_k\}$ be a set of instantiated values of the variables in Y. If for that set of values of Y, $P(Y) \neq 0$, we can use the chain rule to obtain the probability of that set of values as follows:

$$P(Y') = P(v'_k \mid v'_1, v'_2, \ldots, v'_{k-1}) \ldots P(v'_2 \mid v'_1) P(v'_1).$$

We therefore need to compute k conditional probabilities in order to determine the $P(Y')$. We can do this by applying the methods for probability propagation, discussed in chapters 6 and 7, k times. For instance, we can instantiate $v_1, v_2, \ldots$, and v_{k-1} for their values and perform probability propagation to obtain $P(v'_k \mid v'_1, v'_2, \ldots, v'_{k-1})$.

Now suppose that X' is a set of instantiated values of the instantiated variables, and $\mathcal{E} = \{v'_1, v'_2, \ldots, v'_k\}$ is an explanation. We can compute the $P(\mathcal{E} \mid X')$ in the same way as follows:

$$P(\mathcal{E} \mid X') = P(v'_k \mid \{v'_1, v'_2, \ldots, v'_{k-1}\} \cup X') \ldots P(v'_2 \mid \{v'_1\} \cup X') P(v'_1 \mid X').$$

We illustrate this technique with an example:

Example 8.4. Consider the most probable explanation, $\{a_2, b_2, c_2, f_1\}$, determined in Example 8.2, for the case where D is instantiated for d_1 and E is instantiated for e_2. We have that

$$P(a_2, b_2, c_2, f_1 \mid d_1, e_2) = P(a_2 \mid b_2, c_2, f_1, d_1, e_2) \ldots P(f_1 \mid d_1, e_2).$$

The four conditional probabilities on the right can be obtained by performing probability propagation. For example, we can obtain $P(a_2 \mid b_2, c_2, f_1, d_1, e_2)$ by instantiating B, C, F, D, and E for b_2, c_2, f_1, d_1, and e_2, respectively, and propagating.

In a case where the explanation set includes all the uninstantiated variables and the set of instantiated variables is small, we can obtain the $P(\mathcal{E} \mid X')$ more easily as follows. We have that

$$P(\mathcal{E} \mid X') = \frac{P(\mathcal{E} \cup X')}{P(X')}.$$

Since $\mathcal{E}$ includes values of all the uninstantiated variables and X' includes values of all the instantiated variables, we can compute the numerator by appealing directly to the specified conditional distributions. In the current example, we have

$$P(a_2, b_2, c_2, f_1, d_1, e_2) = P(e_2 \mid c_2, f_1)P(c_2 \mid b_2, d_1)P(b_2 \mid a_2)P(f_1)P(d_1)p(a_2)$$

$$= (1)(1)(1)(.6)(.6)(.45) = .162.$$

To compute the denominator, we use the chain rule

$$P(e_2, d_1) = P(e_2 \mid d_1)P(d_1).$$

Using probability propagation, we then determine that

$$P(d_1) = .6 \qquad \text{and} \qquad P(e_2 \mid d_1) = .67.$$

Actually, since D is a root, we can obtain its a priori probabilities directly from the causal network. We now have that

$$P(e_2, d_1) = (.67)(.6)$$

and

$$P(a_2, b_2, c_2, f_1 \mid d_1, e_2) = \frac{.162}{(.6)(.67)}$$

$$= .403.$$

If the probability of the most probable explanation is sufficiently high, then we may not need any additional information beyond that explanation and its probability. For example, in medical diagnosis, if the probability of the most probable explanation is equal to .95, this information may be sufficient for a physician to make a decision. However, if the probability of the most probable explanation is equal to .63, he may not feel comfortable with only the most probable explanation. Therefore he would need to run additional tests. On the other hand, if he knew the second most probable explanation and knew that its probability was equal to .34, he might be able to reach a decision without running these additional tests. However, if the probability of the second most probable explanation were only .2, he may also need to know the third most probable and so on explanations. Hence, in some cases it would be beneficial to know more than just the most probable explanation.

One way to determine the second, third, and so on most probable explanations would be to compute the probabilities of all possible explanations using the method described in this subsection. This would be possible as long as the number of variables in the explanation set was small. However, in medicine, there are often hundreds of diseases in the explanation set of which somewhere between five and ten are possibly present. For example, in the case of INTERNIST [Pople, 1982] there are around 600 possible diseases. Thus there are approximately between $\binom{600}{5}$ and $\binom{600}{10}$ possible explanations; i.e., there are at least around 6×10^{11} possible explanations.

In section 8.3, we will discuss a method for obtaining the second, third, and so on most probable explanations for a special class of problems. In the next subsection, we will see that Pearl's method automatically gives us the second most probable explanation, but, in general, it cannot supply us with the third, fourth, and so on ones.

The method for determining the probability of an explanation, shown in this subsection, as discussed in Cooper [1987], is a special case of a method for computing the probability of an arbitrary proposition. For example, if A, B, C, D, and E are binary-valued propositional variables, Cooper's algorithm can yield

$$P(a_1 \wedge b_2 \mid ((c_2 \vee d_1) \wedge e_1)).$$

In general, the algorithm requires the determination of an exponentially large (relative to the number of variables in the probability expression) number of conditional probabilities. Thus, in general, the algorithm is only applicable to relatively small expressions. Fortunately, the special case presented here only requires the determination of a linear number of conditional probabilities. Furthermore, in this special case, we are not restricted to binary-valued propositional variables.

8.1.5. The Second Most Probable Explanation

Using the pointers, we can also retrieve the second most probable explanation in the following way. Let $\mathcal{E}_1$ and $\mathcal{E}_2$ be the first and second most probable explanations, and let B be a variable which has a different value in $\mathcal{E}_1$ than in $\mathcal{E}_2$. Clearly, there is at least one such B. Let $\hat{b}$ be the value of B which yields the second largest value of $P^*(b_i)$. Then the most probable explanation containing $\hat{b}$ has a greater probability than any explanation containing any other value of b_i (other than b^*), and therefore $\hat{b}$ must be the value of B in $\mathcal{E}_2$. Consequently, we can retrieve the second most probable explanation by following the pointers emanating from $\hat{b}$. Unfortunately, we do not know for which values $\mathcal{E}_1$ and $\mathcal{E}_2$ differ until we know $\mathcal{E}_2$, which is what we are trying to determine. We can overcome this circular problem by finding the most probable explanation containing the value which yields the second largest value of P^* for each variable in the network. That is, for every variable B in the network, we determine $\hat{b}$ and follow the pointers to retrieve an explanation. The most probable of these explanations is then the second most probable explanation. Since the multiplicative constant in $P^*(B)$ is not independent of B, we cannot appeal to P^* to rank these explanations. But, we can rank them by computing their probabilities using the method from the last subsection.

It may seem that the third most probable explanation is the one that ranks third, and so on down the line. Unfortunately, this is not the case. In the exercises, we will show that, in general, it is not possible to obtain any explanations other than the first two using this method.

Before ending this section, we note that, although not mentioned specifically, the abductive inference method presented here is applicable when the evidence is also in the form of virtual evidence.

PROBLEMS 8.1

1. Suppose that there are three variables, A, B, and C, in the network and that

$$P(a_1,b_1,c_1) = .2 \qquad P(a_1,b_1,c_2) = 0$$
$$P(a_1,b_2,c_1) = .15 \qquad P(a_1,b_2,c_2) = .15,$$

and that all other probabilities in the joint distribution are equal to .125. Show that the most probable initial set of all values does not contain the most probable initial set of values in the explanation set if the explanation set is $\{A,B\}$.

2. Perform the calculations which were left as exercises in Example 8.2.

3. Using probability propagation, show, in Example 8.2., that $P'(a_2) = .672$ and $P'(f_1) = .403$ after D is instantiated for d_1 and E is instantiated for e_2.

4. Perform the initialization in Example 8.3 and create all the initial pointers. Use the pointers to retrieve the second most probable initial explanation. Compute the probabilities of the two most probable initial explanations.

5. Suppose we have only two variables, C and F, in the network and we have the following values for the joint distribution:

$$P(c_1,f_1) = .1 \qquad P(c_1,f_2) = .3$$
$$P(c_2,f_1) = .2 \qquad P(c_2,f_2) = .4.$$

Show that initially $f_2 \rightarrow c_2$ and $c_1 \rightarrow f_2$, thereby proving that the pointer relationship is not commutative.

6. In Example 8.3, perform the propagation, including the changing of the pointers, when D is instantiated for d_1. Use the pointers to retrieve the second most probable explanation (in this case it is the only remaining explanation). Compute the probabilities of the two most probable explanations.

7. Suppose we have only two variables, C and F, in the network and we have the following values for the joint distribution:

$$P(c_1,f_1) = .1 \qquad P(c_1,f_2) = .4$$
$$P(c_2,f_1) = .3 \qquad P(c_2,f_2) = .2.$$

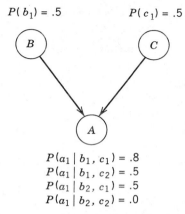

$P(b_1) = .5 \qquad\qquad P(c_1) = .5$

$P(a_1 \mid b_1, c_1) = .8$
$P(a_1 \mid b_1, c_2) = .5$
$P(a_1 \mid b_2, c_1) = .5$
$P(a_1 \mid b_2, c_2) = .0$

FIGURE 8.8 The causal network for Problem 8.1.8.

Show that it is not possible to obtain the third most probable initial set of values from the pointers.

8. Consider the causal network in Figure 8.8. If the explanation set is $\{A, B\}$, show that the initial most probable explanation is $\{a_2, b_2\}$; however, given that C is instantiated for c_1, the most probable explanation is $\{a_1, b_1\}$. If these variables represent the situation concerning Mr. Holmes' alarm, a burglary, and earthquakes, as described in Example 5.3, this would mean that initially it is most probable that his residence has not been burglarized and his alarm has not sounded. However, when it is learned that there has been an earthquake, it becomes most probable that he has been burglarized and that his alarm has sounded. One might feel that $\{a_1, b_2\}$ should become the most probable explanation. Show that indeed the probability of this explanation has increased the most. However, due to its low a priori probability, it has not overtaken $\{a_1, b_1\}$, while the probabilities of the other explanations have decreased.

9. In Problem 8,

$$P(a_1, b_2) < P(a_1, b_1) \qquad \text{and} \qquad P(a_1, b_2 \mid c_1) < P(a_1, b_1 \mid c_1).$$

One would expect this result. Show, however, that if we have the probabilities below,

$$P(b_1) = .3 \qquad\qquad P(c_1) = .1$$
$$P(a_1 \mid b_1, c_1) = .9 \qquad P(a_1 \mid b_1, c_2) = .25$$
$$P(a_1 \mid b_2, c_1) = .25 \qquad P(a_1 \mid b_2, c_2) = .21,$$

then

$$P(a_1, b_2) > P(a_1, b_1) \qquad \text{and} \qquad P(a_1, b_2 \mid c_1) < P(a_1, b_1 \mid c_1).$$

Perhaps this unintuitive result is due to the fact that we have allowed $P(a_1 \mid b_2, c_2)$ to be > 0. Show that if $P(a_1, b_2, c_2) = 0$, we cannot have this result.

10. In reality, in Problem 9, $P(a_1 \mid b_2, c_2)$ can be > 0, since other causes (e.g., dampness or radiation) could trigger the alarm. If we group all these possibilities as a third cause, then we could have the unintuitive result in Problem 9. It seems, however, that the assumptions in the probabilistic causal method are in order in this problem. Show that, with these assumptions, if

$$P(a_1, b_2) > P(a_1, b_1) \qquad \text{then} \qquad P(a_1, b_2 \mid c_1) > P(a_1, b_1 \mid c_1),$$

even if we have additional causes.

11. Create an algorithm which performs abductive inference in a singly connected causal network in the case where the explanation set does not include all the uninstantiated variables (see subsection 8.1.2).

COMPUTER PROBLEM 8.1

Write a program which performs abductive inference in singly connected networks. The specifications are essentially the same as those for Computer Problem 6.2. Remember to set up the pointers as messages are passed.

8.2. ABDUCTIVE INFERENCE IN AN ARBITRARY CAUSAL NETWORK

We can perform abductive inference in causal networks which are not singly connected by using the "conditioning" technique given at the end of chapter 6 along with the algorithm in section 8.1. In this text, however, we will concentrate on a method which combines some of the techniques in the propagation method described in chapter 7 with the algorithm in section 8.1. Pearl [1988] has called this method "clustering." We concentrate on this method because it gives us a closed-form solution; that is, we need not be concerned with determining variables on which to condition. As in the case of probability propagation, there are times when conditioning would be more efficient. In a particular instance, we should apply the method which is most appropriate. We will focus on Example 7.5 to describe the 'clustering' method.

Initially, we marry parents, triangulate the graph, and identify the cliques and their parents just as we do in probability propagation in trees of cliques. In this way we create a permanent tree of cliques as depicted in Figure 7.5. This tree becomes a permanent part of the expert system. We will show that this tree is contained in a causal network which contains the same probability distribution as the original network. We can then perform abductive inference on this new causal network using the method described in subsection 8.1.1.

First, the variables in this new network are $Clq_1, Clq_2, \ldots,$ and Clq_6. The possible values of the variable Clq_i are the conjunctions of the variables, from the original network, which are contained in Clq_i. For instance, the possible values of Clq_2 are

$$(b_1, e_1, c_1) \quad (b_1, e_1, c_2) \quad (b_1, e_2, c_1) \quad (b_1, e_2, c_2)$$
$$(b_2, e_1, c_1) \quad (b_2, e_1, c_2) \quad (b_2, e_2, c_1) \quad (b_2, e_2, c_2).$$

In order to create a causal network we need to specify, for each i,

$$\hat{P}(Clq_i \mid c(Clq_i)),$$

where $c(Clq_i)$ is the single parent of Clq_i. For example, we need to specify

$$\hat{P}(Clq_5 \mid Clq_3).$$

We specify these distributions by *defining* for each i

$$\hat{P}(Clq_i \mid c(Clq_i)) = P(R_i \mid S_i)$$

for all cases where Clq_i and $c(Clq_i)$ do not contain contradictory values of a variable. In this latter case, the conditional probability is defined to be 0. R_i and S_i are as defined in Definition 7.2.

To obtain the values of $P(R_i \mid S_i)$, we perform the propagation up the tree, described in chapter 7, which determines the a priori probability, $P(Clq_i)$, and the value of the function, $\psi'(Clq_i)$, for each i. Recall that when we perform this propagation, we end up with

$$\psi'(Clq_i) = P(R_i \mid S_i).$$

These values, along with the values of $P(Clq_i)$, are listed in Table 7.6 (in that table we refer to Clq_i' and $P'(Clq_i')$; however, when the a priori probabilities are computed, they equal Clq_i and $P(Clq_i)$, respectively, and therefore we drop the primes here). Below are examples of the values in the specified distributions:

$$\hat{P}(b_1, e_1, c_1 \mid a_1, b_1) = P(c_1, e_1 \mid b_1) = \psi'(b_1, e_1, c_1) = .055$$
$$\hat{P}(b_1, e_1, c_1 \mid a_1, b_2) = 0,$$

since Clq_2 contains b_1 and Clq_1 contains the contradictory value b_2.

For the root Clq_1,

$$\hat{P}(Clq_1) = P(R_1 \mid \varnothing) = P(Clq_1 \mid \varnothing) = \psi'(Clq_1) = P(Clq_1).$$

We now have a joint probability distribution $\hat{P}$ defined on $\{Clq_1, Clq_2, \ldots, Clq_6\}$, and, owing to Theorem 5.2, the tree in Figure 7.5, together with this joint distribution, constitutes a causal network.

Since

$$\hat{P}(Clq_1, Clq_2, \ldots, Clq_6) = \hat{P}(Clq_1) \prod_{i \neq 1} \hat{P}(Clq_i \mid c(Clq_i)) = P(Clq_1) \prod_{i \neq 1} P(R_i \mid S_i),$$

owing to Theorem 7.4, the joint probability distribution determined by the specified distributions, $\hat{P}(\text{Clq}_i \mid c(\text{Clq}_i))$, is equal to $P(V)$, where V is the set of variables in our original causal network (in this case, $V = \{A, B, C, D, E, F, G, H\}$). Therefore we can find the most probable explanation containing values of the variables in V by determining the most probable explanation containing values of the cliques.

We find this most probable explanation by simply passing messages and setting up pointers as we do in any singly connected causal network. In fact, since the network in this case is always a tree, we can use a simplified version of the algorithm in the previous section which works only for trees. This algorithm would be similar to the one developed in section 6.2. It is left as an exercise to develop this special case.

Notice that when a variable from V is instantiated, a variable in the tree of cliques is not instantiated. For instance, if B is instantiated for b_2, no variable in the tree of cliques is instantiated. We solve this problem by treating the instantiation of the variable as virtual evidence for a clique which contains that variable. For example,

$$P(b_2 \mid a_1, b_1) = 0 \qquad P(b_2 \mid a_1, b_2) = 1$$
$$P(b_2 \mid a_2, b_1) = 0 \qquad P(b_2 \mid a_2, b_2) = 1.$$

Thus, when b_2 is instantiated, we pass Clq_1 a λ message equal to $(0,1,0,1)$, as is done in the case of ordinary virtual evidence. Since b_2 is also in Clq_2, instead we could pass Clq_2 a λ message equal to $(0,0,0,0,1,1,1,1)$. The choice is arbitrary.

A problem occurs with clustering if the explanation set does not include all the uninstantiated variables. That is, variables from the explanation set and other uninstantiated variables could end up in the same clique, making it impossible to use the revised algorithm described in section 8.1.2. Generally, if this happens, we could try to restructure the cliques or use conditioning. However, as we shall see in Problem 8.2.2, there are some cases, in which the explanation set does not include all the uninstantiated variables, in which we do not need the revised algorithm.

PROBLEMS 8.2

1. Create an algorithm which performs abductive inference in the special case where the network is a tree.

2. Example 7.5 is a bit complex to solve by hand. Consider the causal network concerning metastatic cancer in Example 7.4. The tables pertaining to that example are Tables 7.2 and 7.3. Suppose it is learned that the patient has increased total serum calcium and is not in a coma. Determine the most probable and second most probable explanations, and compute the probabilities of these explanations. Assume the explanation set includes a brain tumor and metastatic cancer. Note that, since papilledema is a leaf, we can

just remove it from the network when determining the most probable explanation. That is, we need not use the revised algorithm described in section 8.1.2. In the case of the classical diagnostic problem, the variables which are not in the explanation set are always leaves. Therefore, in that case, we would never need the revised algorithm described in section 8.1.2.

COMPUTER PROBLEM 8.2

Write a program which performs abductive inference in an arbitrary causal network. The specifications are the same as those in Computer Problem 8.1.

8.3. DETERMINING MORE EXPLANATIONS

As mentioned in subsection 8.1.4, there are many cases in which we want more than just the two most probable explanations. Cooper [1984] has obtained a method for obtaining the third, fourth, and so on most probable explanations for a special class of problems. As far as this author knows, there is no current method for obtaining these other explanations for an arbitrary causal network. This is certainly a fertile area for research. In this section we discuss Cooper's method as it now exists. He is currently investigating ways of making the method more general.

In order to define the class of problems addressed by Cooper's method, we first need to introduce some new notation. If $\mathcal{E}$ is an explanation set which contains all binary-valued variables, and H is a subset of $\mathcal{E}$, by

$$H'$$

we shall mean the event that the variables in H are instantiated for their first value and the variables in $\mathcal{E} - H$ are instantiated for their second value. Similarly, if H_1 and H_2 are two subsets of $\mathcal{E}$, by

$$H_1' \cup H_2'$$

we shall mean the event that all the variables in $H_1 \cup H_2$ are instantiated for their first value and all those in $\mathcal{E} - (H_1 \cup H_2)$ are instantiated for their second value.

Since the variables in $\mathcal{E}$ are binary valued, the first value of each variable stands for the fact that some attribute (like a disease) is present, whereas the second stands for the fact that it is not. Thus we see that H' is the hypothesis that all the attributes represented by variables in H are present, while all those represented by variables in $\mathcal{E} - H$ are not present.

If X is a set of instantiated variables, by X' we will continue to mean the set of instantiated values. That is, when representing an event, X' stands for the event that the variables in X are instantiated for the values in X', and no other variables are instantiated for any values.

Since Cooper's method was originally developed for a medical diagnostic aid called NESTOR, we shall call the class of problems addressed by that method NESTOR diagnostic problems.

Definition 8.3. (*NESTOR Diagnostic Problem*). Let $C = (V, E, P)$ be a causal network, let $\mathcal{E}$ be an explanation set in V, and suppose $\mathcal{E}$ satisfies the following conditions:

1. All variables in $\mathcal{E}$ are binary valued.
2. If H_1 and H_2 are two subsets of $\mathcal{E}$, then for $i = 1, 2$,

$$P(H_1' \cup H_2') \le P(H_i').$$

Then the problem of performing abductive inference in this network (i.e., the act of determining the most probable, second most probable, third most probable, and so on sets of values in $\mathcal{E}$, given a set of instantiated values X' of instantiated variables X) is called the NESTOR diagnostic problem.

Cooper originally conceived this problem for the case where $\mathcal{E}$ is a set of diseases. We will thus adopt some of the notation from the classical diagnostic problem in the NESTOR diagnostic problem. That is, we will denote a variable in $\mathcal{E}$ by d_i, and the values of the variable by $+d_i$ and $\neg d_i$. Note that it is only required that the variables in $\mathcal{E}$ be binary valued. Therefore we will continue to denote other variables in their usual way.

We immediately have the following theorem and corollary which show that the NESTOR diagnostic problem includes a large class of problems:

Theorem 8.1. Suppose $C = (V, E, P)$ is a causal network and that the explanation set $\mathcal{E}$, satisfies the conditions below:

1. All variables in $\mathcal{E}$ are binary valued.
2. All variables in $\mathcal{E}$ are mutually independent.
3. If $d_i \in \mathcal{E}$, then $P(+d_i) \le P(\neg d_i)$.

Then V and $\mathcal{E}$ satisfy the conditions in the NESTOR diagnostic problem.

Proof. We need only show that if H_1 and H_2 are two subsets of $\mathcal{E}$, then for $i = 1, 2$,
$$P(H_1' \cup H_2') \le P(H_i').$$
To that end, let
$$H_1 = \{d_1, d_2, ..., d_k\}$$
$$H_1 \cup H_2 = \{d_1, d_2, ..., d_m\},$$

where $k \le m$. For the sake of notational simplicity, we have simply denoted the members of H_1 as the first k members of $\mathcal{E}$ and the members of $H_2 - H_1$

as the next $(m - k)$ members of $\mathcal{E}$. Suppose that $\mathcal{E}$ has n members. Since the variables in $\mathcal{E}$ are independent, we then have that

$$P(H_1' \cup H_2') = P(+d_1, +d_2, \ldots, +d_k, +d_{k+1}, \ldots, +d_m, \neg d_{m+1}, \ldots, \neg d_n)$$

$$= P(+d_1)P(+d_2)\ldots P(+d_k)P(+d_{k+1})\ldots P(+d_m)P(\neg d_{m+1})\ldots P(\neg d_n)$$

$$\leq P(+d_1)P(+d_2)\ldots P(+d_k)P(\neg d_{k+1})\ldots P(\neg d_m)P(\neg d_{m+1})\ldots P(\neg d_n).$$

The inequality is due to the assumption that $P(+d_i) \leq P(\neg d_i)$ for all members of $\mathcal{E}$. Since

$$P(H_1') = P(+d_1)P(+d_2)\ldots P(+d_k)P(\neg d_{k+1})\ldots P(\neg d_m)P(\neg d_{m+1})\ldots P(\neg d_n),$$

the theorem is now proven. $\square$

The assumption that $P(+d_i) \leq P(\neg d_i)$ is valid in most medical applications, since most diseases have very small a priori probabilities. But even if this were not the case, we could satisfy the assumption by letting $+d_i$ represent the event that the disease is not present. Hence the only real restrictions in Theorem 8.1 are that the variables be binary valued and mutually independent. Because the roots of a causal network are always mutually independent, we have the following corollary:

Corollary to Theorem 8.1. If all the variables in the explanation set are binary valued and are roots in the causal network, then the conditions of the NESTOR diagnostic problem are satisfied.

As mentioned in the introduction to this chapter, often in medical diagnosis the variables in the explanation set are all roots. Even if they are not roots, by Theorem 8.1, the assumptions in the NESTOR diagnostic problem will be satisfied as long as no two members in the explanation set share a common cause (either direct or indirect). However, Example 7.5 shows that a medical diagnostic problem need not satisfy the conditions of Theorem 8.1. In that example the diseases of bronchitis and lung cancer can both be caused by smoking, and consequently they are not necessarily independent. Cooper is currently investigating the existence of weaker conditions (than those in Theorem 8.1) which imply the assumptions in the NESTOR diagnostic problem. The following example illustrates a medical diagnostic problem which does satisfy the conditions of Theorem 8.1.

Example 8.5. This example is an enhancement of Example 5.13 and contains a very simplified subset of the nodes and arcs in a causal network which represents hypercalcemic disorders. That network actually contains 100 vertices and

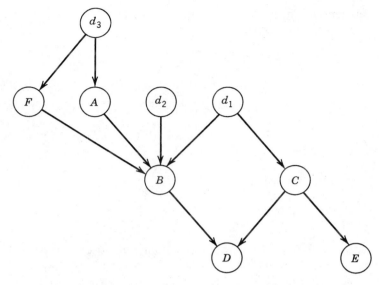

FIGURE 8.9   The causal network for Example 8.5.

200 arcs, and is the network which was used to test NESTOR. Let

$+d_1$ = metastatic cancer present

$+d_2$ = myeloma present

$+d_3$ = primary hyperparathyroidism

a_1 = increased bone reabsorption of calcium

b_1 = increased total serum calcium

c_1 = brain tumor present

d_1 = coma present

e_1 = papilledema present

f_1 = increased gastrointestinal absorption of calcium

and suppose that we have the causal relationships among these variables embodied in the causal network in Figure 8.9. Since

$$\mathcal{E} = \{d_1, d_2, d_3\},$$

and d_1, d_2, and d_3 are all roots and are binary valued, the conditions of Theorem 8.1 hold. Although, in this simplified example, the variables in $V - \mathcal{E}$ are all binary, they need not be. Indeed, in the actual causal network, variable B has six possible values, representing six levels of serum calcium.

The problem then is to determine the most probable, second most probable, and so on sets of values of the variables in $\mathcal{E}$ given that a subset of the other variables is instantiated for specific values. For instance, in Example 8.5 if a patient has a severe headache (e_1) and increased total serum calcium (b_1), we must determine the most probable combination of diseases. In this simplified example, since there are only $2^3 = 8$ possibilities, we can simply exhaustively compute the probabilities of all of them using the method described in section 8.1.4. However, as noted in subsection 8.1.4, $\mathcal{E}$ often contains hundreds of diseases, and therefore an exhaustive search is not usually feasible. Cooper's method is a best-first tree search with branch and bound pruning. In Cooper's words [Cooper, 1984],

> The central idea of best-first search is that hypotheses that are believed to be best are considered first. Thus, best-first search might be more clearly phrased as best-guesses-first search, since in general there is no guarantee that the hypotheses initially judged most likely are necessarily the best. Best-first search has at least two advantages. First, if there is insufficient time to locate the provably most probable hypothesis, the current best hypothesis can be reported. Second, locating highly probable hypotheses early in the search process allows the branch and bound pruning method to ignore entire areas of the search space.
>
> The branch and bound pruning method is a means of disregarding provably nonoptimal areas of the search space.

We will explain the branch and bound pruning method used in Cooper's algorithm after giving the following theorem, which is central to that method:

Theorem 8.2. Suppose $C = (V, E, P)$ is a causal network and $\mathcal{E}$ is an explanation set which satisfies the conditions of the NESTOR diagnostic problem. Let X' be a set of instantiated values of instantiated variables X, and let H_1 and H_2 be subsets of $\mathcal{E}$ such that

$$P(H_1') < P(H_2' \mid X')P(X').$$

Then for every subset H of $\mathcal{E}$, we have that

$$P(H_1' \cup H' \mid X') < P(H_2' \mid X').$$

Proof. For any subset K of $\mathcal{E}$, we have that

$$P(K' \mid X') = \frac{P(X' \mid K')P(K')}{P(X')}.$$

Because the denominator does not depend on K', to prove the theorem we need only show that for any subset H of $\mathcal{E}$,

$$P(X' \mid H_1' \cup H')P(H_1' \cup H') \leq P(X' \mid H_2')P(H_2').$$

To that end, since any probability value is less than or equal to 1, we have that

$$P(X' \mid H_1' \cup H')P(H_1' \cup H') \le P(H_1' \cup H')$$
$$\le P(H_1')$$
$$< P(H_2' \mid X')P(X').$$

The second inequality is one of the assumptions in the NESTOR diagnostic problem, while the third is the assumption in this theorem. Now by the definition of conditional probability,

$$P(H_2' \mid X')P(X') = P(X' \mid H_2')P(H_2'),$$

which proves the theorem.   □

Example 8.6. Consider the causal network in Example 8.5. Suppose that it is learned that the patient has a papilledema and increased total serum calcium. Then

$$X' = (b_1, e_1).$$

(X' stands for both the set $\{b_1, e_1\}$ and the event (b_1, e_1), which, as discussed in section 5.2, is the event $b_1 \wedge e_1$. We will denote X' by the event notation here, since that it is the context in which we are using X'.)

If we let

$$H_1 = \{d_1\} \qquad \text{and} \qquad H_2 = \{d_2\},$$

then

$$H_1' = (+d_1, \neg d_2, \neg d_3) \qquad \text{and} \qquad H_2' = (\neg d_1, +d_2, \neg d_3).$$

Suppose we compute that

$$P(H_1') = .001, \qquad P(H_2' \mid X') = .1, \qquad P(X') = .02.$$

Then

$$P(H_1') < P(H_2' \mid X')P(X'),$$

and the conditions of Theorem 8.2 are satisfied. If we let H be any subset of $\mathcal{E}$ (including the empty set), this implies that

$$P(H_1' \cup H' \mid X') < P(H_2' \mid X').$$

Therefore we not only can eliminate H_1' as a candidate for the most probable explanation, but we can also eliminate any explanation which has the variables in H_1 instantiated for their + values. Thus we have learned that

$$(+d_1, \neg d_2, \neg d_3), \qquad (+d_1, +d_2, +d_3),$$
$$(+d_1, +d_2, \neg d_3), \qquad (+d_1, \neg d_2, +d_3)$$

each cannot be the most probable explanation.

8.3.1. An Algorithm for Determining the Most Probable Explanation

We will now give a version of Cooper's algorithm [Cooper, 1984] for determining the most probable explanation. In section 8.3.2, we will show how the algorithm can be modified to find the second, third, and so on most probable explanations. The algorithm initially involves making a guess at the most probable explanation. Regardless of the guess, the algorithm will eventually find the most probable explanation. However, the better the guess, the faster the most probable explanation will be found. We will describe the algorithm with the initial guess being that the patient has no diseases; that is, $H = \emptyset$. Later we will discuss a heuristic method for making a possibly better guess.

The algorithm proceeds as follows: Although it is a tree search, we do not actually construct a tree. We need only maintain two variables, HBEST and EXPLANATION.LIST. HBEST is equal to the set of variables, such that, of all those explanations thus far investigated, $P(\text{HBEST}' \mid X')$ is maximal. Initially, HBEST is set equal to the guess, which we are currently assuming to be $\emptyset$. We then compute $P(X')$ and $P(\text{HBEST}' \mid X')$ (at this point HBEST' is the event that all the variables in $\mathcal{E}$ are instantiated for their $\neg$ values; i.e., no diseases are present). Next we place $\emptyset$ in EXPLANATION.LIST. EXPLANATION.LIST stores all those subsets H of $\mathcal{E}$ which we have investigated, but have not yet extended. A subset H is extended when we add another variable to H. A subset H is investigated when (1) we have determined whether

$$P(H' \mid X') > P(\text{HBEST}' \mid X')$$

and therefore whether H is better than HBEST and should replace HBEST; and (2) we have determined whether an extension of H could possibly be better than the current value of HBEST. In the next step of the algorithm we will see how to investigate a subset H.

The next step is to extend the root $\emptyset$ and investigate all subsets of $\mathcal{E}$ which contain exactly one disease. That is, if n is the number of variables in $\mathcal{E}$, for $1 \leq i \leq n$ we determine whether

$$P(H_i') < P(\text{HBEST}' \mid X')P(X'),$$

where $H_i = \{d_i\}$. If this inequality holds for a given i, we know, by Theorem 8.2, that any explanation which has d_i instantiated for its + value cannot be the most probable explanation. This fact leads to our pruning technique. After considering all single disease sets, we will then extend each single disease set to two disease sets and investigate these two disease sets; two disease sets will then be extended to three disease sets, and so on. However, by Theorem 8.2, any set $\{d_i\}$ for which the above inequality holds need not be extended. None of those extensions can possibly be better than the current value of HBEST. Thus we can prune the entire branch emanating from $\{d_i\}$. In general, each time we generate a new set of disease variables, H, we determine whether

$$P(H') < P(\text{HBEST}' \mid X')P(X'),$$

and, if the inequality holds, we know that we can prune the entire branch emanating from H owing to Theorem 8.2. In reality, we are not actually pruning the branches; we are simply never adding them.

The sets $\{d_i\}$ which are not pruned are placed in EXPLANATION.LIST ordered according to their a priori probabilities, and $\emptyset$ is removed from EXPLANATION.LIST, since it has been extended. For each of the new sets $\{d_i\}$ in the EXPLANATION.LIST, we then compute

$$P(H_i' \mid X'),$$

where $H_i = \{d_i\}$. If this value is greater than $P(\text{HBEST}' \mid X')$, we set HBEST equal to H_i. Whenever the value of HBEST is changed, we check all current members of EXPLANATION.LIST to see which additional ones can now be pruned.

Next we extend the top set in the EXPLANATION.LIST, say $\{d_k\}$, to all possible two disease sets containing d_k, and remove $\{d_k\}$ from EXPLANATION.LIST. We then compute the a priori probabilities of the extensions, prune those which should be pruned, add those remaining to EXPLANATION.LIST (we place them in the correct location ordered in descending order according to their a priori probabilities), determine whether any of these sets should replace HBEST, and prune whenever HBEST is replaced.

We then again take the top set in EXPLANATION.LIST (which now may contain two diseases), extend that set, and repeat the above procedure. We do this until the EXPLANATION.LIST is empty; in other words, until there are no more candidates for the most probable explanation. At this point, HBEST will be equal to the subset of $\mathcal{E}$ such that $P(\text{HBEST}' \mid X')$ is maximal. If there are two such sets, HBEST will be equal to one of them.

After giving pseudocode for this algorithm, we will work through an example.

ABDUCTIVE INFERENCE IN THE NESTOR DIAGNOSTIC PROBLEM

constant $\mathcal{E}$: the explanation set;
      n: the number of variables in $\mathcal{E}$;
      X': set of instantiated values of instantiated variables;

var EXPLANATION.LIST: ordered list of subsets of $\mathcal{E}$;
   HBEST, H, K: subset of $\mathcal{E}$;

begin
   HBEST := $\emptyset$;
   EXPLANATION.LIST := $\{\emptyset\}$;
   compute $P(X')$ and $P(\text{HBEST}' \mid X')$;
   while EXPLANATION.LIST $\neq \emptyset$, do
      begin
         H := top set in EXPLANATION.LIST;

          Extend (H)
      end;
   write (HBEST)
end.

Procedure Extend $(H$: subset of $\mathcal{E})$;
var K: subset of $\mathcal{E}$;
begin
   for all $d_i \in \mathcal{E} - H$, do
      begin
         $K := H \cup \{d_i\}$;
         Add(K)
      end;
   remove H from EXPLANATION.LIST
end;

Procedure Add $(H$: subset of $\mathcal{E})$;
 begin
   compute $P(H')$;
   if $(P(H') \geq P(\text{HBEST}' \mid X')P(X'))$ and $(H \notin \text{EXPLANATION.LIST})$,
   then
      begin
         add H to the EXPLANATION.LIST ordered according to $P(H')$;
         compute $P(H' \mid X')$;
         if $P(H' \mid X') > P(\text{HBEST}' \mid X')$, then
        begin
           HBEST := H;
           for all $K \in \text{EXPLANATION.LIST}$, do
             if $P(K') < P(\text{HBEST}' \mid X')P(X')$, then
               remove K from EXPLANATION.LIST
        end
      end
end;

Example 8.7. In this example we will simply assign a priori and conditional probabilities while illustrating the algorithm. In practice these probabilities are obtained from the causal network using the method described in section 8.1.4. The tree in Figure 8.10 is the search tree that is generated in this example. In that tree, if H is the set stored at a node, $P(H')$ appears above the node and $P(H' \mid X')$ appears below a node. Suppose now that

$$E = \{d_1, d_2, d_3, d_4\} \qquad \text{and} \qquad P(X') = .01.$$

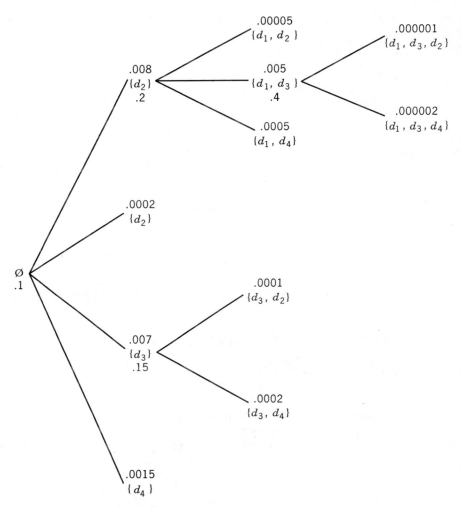

FIGURE 8.10   The search tree created in Example 8.7.

First we set

$$\text{HBEST} = \emptyset \qquad \text{and} \qquad \text{EXPLANATION.LIST} = \{\emptyset\}$$

and determine that

$$P(\text{HBEST}' \mid X') = .1.$$

Thus

$$P(\text{HBEST}' \mid X')P(X') = (.1)(.01) = .001.$$

Next we extend $\emptyset$. In the algorithm we add the extensions to EXPLANA-
TION.LIST in sequence and determine whether a given extension should re-

place HBEST before possibly adding the next extension. For the sake of simplicity, here we will add all the extensions to EXPLANATION.LIST simultaneously. The only resulting difference is that some additional extensions may be placed temporarily into EXPLANATION.LIST; however, they will be eliminated as soon as HBEST is changed. We obtain the following extensions of $\emptyset$:

$$P(\{d_1\}') = .008, \qquad P(\{d_2\}') = .0002, \qquad P(\{d_3\}') = .007, \qquad P(\{d_4\}') = .0015.$$

Since

$$P(\{d_2\}') < P(\text{HBEST}' \mid X')P(X'),$$

we can prune $\{d_2\}$. The other three sets we add to EXPLANATION.LIST. At this point we have that

$$\text{EXPLANATION.LIST} = \{\{d_1\}, \{d_3\}, \{d_4\}\}.$$

Since $\{d_1\}$ is the top set in EXPLANATION.LIST, we next determine that

$$P(\{d_1\}' \mid X') = .2.$$

Since

$$P(\{d_1\}' \mid X') > P(\text{HBEST}' \mid X'),$$

we set

$$\text{HBEST} = \{d_1\}$$

and determine that

$$P(\text{HBEST}' \mid X')P(X') = (.2)(.01) = .002.$$

Therefore we can now prune $\{d_4\}$ from the tree. That is, we remove it from EXPLANATION.LIST. Next we determine that

$$P(\{d_3\}' \mid X') = .15.$$

Since .15 is less than the current value of $P(\text{HBEST}' \mid X')$, we do not replace HBEST by $\{d_3\}$. Because $\emptyset$ has now been extended, we remove it from EXPLANATION.LIST. At this point we have that

$$\text{HBEST} = \{d_1\} \qquad \text{and} \qquad \text{EXPLANATION.LIST} = \{\{d_1\}, \{d_3\}\}$$

$$P(\text{HBEST}' \mid X') = .2 \qquad \text{and} \qquad P(\text{HBEST}' \mid X')P(X') = .002.$$

Next we extend the top set in EXPLANATION.LIST, $\{d_1\}$, and determine that

$$P(\{d_1, d_2\}') = .00005, \quad P(\{d_1, d_3\}') = .005, \quad P(\{d_1, d_4\}') = .0005.$$

Owing to pruning, we only add $\{d_1, d_3\}$ to EXPLANATION.LIST. We then determine that

$$P(\{d_1, d_3\} \mid X') = .4$$

and thus HBEST is set equal to $\{d_1, d_3\}$. We then have that

$$P(\text{HBEST}' \mid X')P(X') = (.4)(.01) = .004.$$

Since $\{d_1\}$ has now been extended, we next remove $\{d_1\}$ from EXPLANA-TION.LIST. At this point we have that

$$\text{HBEST} = \{d_1, d_3\} \qquad \text{and} \qquad \text{EXPLANATION.LIST} = \{\{d_3\}, \{d_1, d_3\}\}$$

$$P(\text{HBEST}' \mid X') = .4 \qquad \text{and} \qquad P(\text{HBEST}' \mid X')P(X') = .004.$$

Next we extend the top set in EXPLANATION.LIST, $\{d_3\}$, and determine that

$$P(\{d_3, d_1\}') = .005, \qquad P(\{d_3, d_2\}') = .0001, \qquad P(\{d_3, d_4\}') = .0002.$$

Owing to pruning, we would only add $\{d_3, d_1\}$ to EXPLANATION.LIST, but since that set is already in EXPLANATION.LIST, we do not add it. Because we have now extended $\{d_3\}$, we remove $\{d_3\}$ from EXPLANATION.LIST. Next we extend the top set in EXPLANATION.LIST, $\{d_1, d_3\}$, and determine that

$$P(\{d_1 d_3, d_2\}') = .000001 \qquad \text{and} \qquad P(\{d_1, d_3, d_4\}') = .000002.$$

Owing to pruning, we do not add any sets to EXPLANATION.LIST. Since $\{d_1, d_3\}$ has now been extended, we remove it from EXPLANATION.LIST. Because EXPLANATION.LIST is now empty, we are done. The best explanation is $\{d_1, d_3\}'$, which is the event

$$(+d_1, \neg d_2, +d_3, \neg d_4),$$

and its probability is equal to .4. That is, it is most probable that the patient has diseases d_1 and d_3 and does not have diseases d_2 and d_4.

Notice that we have created a number of unnecessary nodes. For example, we learned early that any set which contains d_2 cannot be a candidate to re-place HBEST. Yet later we bother to extend $\{d_1\}$ to $\{d_1, d_2\}$. The algorithm could be modified in the following way: Maintain a set, REJECTS, which con-tains all those sets which have already been investigated and whose extensions, we have learned, cannot be better than the current value of HBEST. When we are about to extend $\{d_1\}$, we would have that

$$\text{REJECTS} = \{\{d_2\}, \{d_4\}\}.$$

Before investigating a possible extension, we then check to see if any member of REJECTS is a subset of the extension. The problem with this modification is that, if REJECTS became large, it may take more time to check REJECTS than it would to just investigate the hopeless extensions. Recall that we need only compute the a priori probability of an extension to determine whether it should be pruned. Since all members of $\mathcal{E}$ are roots, these a priori probabilities are easily computed.

There are two aspects to the best-first search part of the algorithm. The first is the guess at the best set. We simply took that set to be Ø in the above algorithm. In practice this would not be a very good guess, since it means that the patient has no diseases; that is, that he has normal physiology. Cooper [1984] suggests using a heuristic guessing technique similar in spirit to the one used in INTERNIST [Pople, 1982]. The technique is to cycle through the diseases in the order of their a priori probabilities and to add a disease to the initial guess if there is a path from that disease to a variable in X. Any members of X for which there is a path from that disease are disregarded by later diseases in the sequence. Such a set of diseases is a "set cover," as discussed by Reggia, Nau, and Wang [1983]. If Ø is not used as the initial guess, it would be necessary to check, at the beginning of the algorithm, if Ø is better than the initial guess.

The second aspect of best-first search is the method of ordering the sets in EXPLANATION.LIST. We ordered them according to their a priori probabilities, thereby assuming that an explanation with a higher a priori probability is more likely to lead to better extensions. Other strategies can be used to order these sets. One possibility would be to always move HBEST to the top of the EXPLANATION.LIST. That way we always extend the current best set first.

In section 8.3.3, we will give some results of an evaluation of Cooper's best-first search with branch and bound pruning algorithm [Cooper, 1984]. First, in section 8.3.2, we show how to modify the algorithm to provide more than just the best explanation.

8.3.2. Modifying the Algorithm to Provide Additional Explanations

Cooper [1984] shows that it is straightforward to modify the algorithm to provide the first N explanations, where N is a value supplied by the user. The variable HBEST is replaced by an ordered list HBEST.LIST of length N. The sets in this list are ordered according to the probabilities of the explanations. The set on the bottom of the list is then used for pruning and for determining whether a newly investigated set should replace a set already in HBEST.LIST. In other words, the set on the bottom of HBEST.LIST replaces HBEST in the algorithm.

At the time the user specifies N, he has no way of knowing the probability that at least one of the N best explanations is correct. For example, suppose he feels that five explanations are sufficient and he accordingly specifies a value of N equal to 5. After determining the five best explanations and their probabilities, the system might inform him that the total of the five probabilities is equal to .7. A physician would probably not be comfortable with only those five explanations given this relatively low probability. He might thus repeat the procedure with $N = 6$. Rather than making the user guess at a value of N which might yield a comfortable probability value, Peng and Reggia [1987b] suggest allowing the user to specify a minimal probability value with which he would be comfortable. They call such a value a "comfort measure." For instance, if the user enters a comfort measure equal to .95, the system should produce the minimum number of explanations such that the probability that at least one of them is correct is greater than .95.

It is also easy to modify the algorithm to allow the user to enter such a comfort measure. We simply let HBEST.LIST be a variable-length list. We always maintain in HBEST.LIST the minimum number of explanations such that sums of their probabilities are greater than the comfort measure. HBEST.LIST will initially simply grow until that sum is greater than the comfort measure. After that point is reached, when a new member is added to HBEST.LIST, more than one current member might be removed.

8.3.3. An Evaluation of the Algorithm

Cooper [1984] evaluated the algorithm by testing it on a causal network which represents hypercalcemic disorders. In that particular network, there are seven possible diseases and hence $2^7 = 128$ possible explanations. He used five different methods to generate the hypotheses, and he obtained the results of several efficiency measures for each method. The methods included combinations of the following parameters:

1. The initial guess:
    $H0$: Assume patient has no diseases as the initial guess;
    $H1$: Use the method from INTERNIST to make the initial guess (this method is discussed in section 8.3.1).
2. The best-first search strategy:
    $B0$: Place any new set at the head of the EXPLANATION.LIST;
    $B1$: Use the a priori probabilities to order the sets.
3. The pruning method:
    $P0$: Never prune any sets;
    $P1$: Use the probability of the current best set to prune.

The following combinations of these parameters were evaluated: $(H0, B0, P0)$, $(H0, B0, P1)$, $(H1, B0, P1)$, $(H0, B1, P1)$, and $(H1, B1, P1)$. Clearly, combination $(H0, B0, P0)$ produces an exhaustive search, while combination $(H1, B1, P1)$ is the algorithm given in this text, with the modification that the method from INTERNIST is used to make the initial guess. The methods for propagating probabilities, described in chapters 6 and 7, were not available when Cooper developed NESTOR. At that time, he developed a method which determines upper and lower bounds for the probabilities of the explanations, and it is this method which was used in the evaluation. The method, described in Cooper [1984], also permits the expert to specify ranges for the probability values which are stored in the causal network, instead of point values. The idea is that the interval represents the expert's uncertainty in the probability values themselves. The representation of uncertainty in probability values will be discussed in chapter 10. The point here is that the techniques from chapters 6 and 7 were not used to determine probabilities when the evaluation was performed; however, there is no reason to expect the results to be substantially different if they were used instead of the approximation method.

Cooper's evaluation yielded the following averages for the 41 cases that were studied:

Combination $(H0, B0, P0)$: 128.0 sets generated

: 80.2 sets have their probabilities (i.e., $P(H' \mid X')$) computed

: 204.2 CPU seconds consumed.

Combination $(H1, B1, P1)$: 36.0 sets generated

: 5.2 sets have their probabilities (i.e., $P(H' \mid X')$) computed

: 14.3 CPU seconds consumed.

The results of using the other combinations were all significantly better than those for combination $(H0, B0, P0)$ and were slightly worse than those for combination $(H1, B1, C1)$. To summarize these results, the use of the pruning technique significantly reduced the search time required to obtain the best explanation, while the use of a heuristic method to obtain an initial guess and the use of a best-first search strategy did little to decrease the search time. The fact that these latter techniques provided little additional savings may be due to the small domain used in the test.

The question remains as to how increases in the domain size, the number of instantiated variables, and the number of diseases in the explanation set will affect the search time.

In order to analyze these affects, we introduce the following quantities:

$$c_0 = \max_{\substack{c(v) \\ v \in X}} [P(v' \mid c(v))]$$

$$c_1 = \min_{\substack{c(v) \\ v \in X}} [P(v' \mid c(v))]$$

$$c_2 = \min_{d_i \in \mathcal{E}} [P(+d_i)]$$

$$c_3 = \max_{d_i \in \mathcal{E}} [P(+d_i)]$$

n = number of members of $\mathcal{E}$

$|X|$ = number of members of X

|HBEST| = number of members of the most probable explanation.

In the expressions for c_0 and c_1, v' stands for the instantiated value of variable v. The maximum and minimum are over all combinations of values of variables in $c(v)$ for all $v \in X$. If we let

$$L = \frac{|X| \ln c_0 + |\text{HBEST}|(\ln c_3 - \ln(1 - c_2)) + n(\ln(1 - c_2) - \ln(1 - c_3))}{\ln c_2 - \ln(1 - c_3)},$$

then Cooper [1989] has shown that a lower bound on the complexity of hypothesis search using any search strategy is

$$\Omega \left(\sum_{i=1}^{L} \binom{n}{i} \right).$$

This bound is on the number of explanations considered by the algorithm in this section. The cost of computing the conditional probabilities using the method in section 8.1.4 and one of the algorithms in chapter 6 or 7 is treated as a fixed quantity in this analysis. Recall that the problem of computing these probabilities is NP-hard.

If c_0, c_2, and c_3 are all nonzero, then L is a linear function of $|X|$, $|\text{HBEST}|$, and n. That is, we can represent L by

$$K_1|X| + K_2|\text{HBEST}| + K_3 n,$$

where K_1, K_2, and K_3 are constants that can be readily calculated from c_0, c_2, and c_3.

It is easy to see that, if c_2 and c_3 are both about equal and close to zero (as is the case in some medical applications), then

$$\lim_{c_0 \to 1} L = |\text{HBEST}|.$$

Thus in such cases, if n is much larger than $|\text{HBEST}|$, the lower bound is exponential relative to HBEST.

It is also easy to see that, if c_0, c_2, and c_3 are all about equal and close to zero, then

$$L = |X| + |\text{HBEST}|.$$

Using out best-first search strategy, in the worst case, every explanation except the last one may have a conditional probability (relative to X') equal to zero. The last one is therefore HBEST. Thus the complexity in the worst case is

$$O(2^n).$$

Notice that it is not necessary that $|\text{HBEST}| = n$ for this to happen. For example, if explanations a priori probabilities are used in the best-first search strategy, it is possible that (1) $|\text{HBEST}| = 1$, (2) HBEST has the lowest a priori probability of all explanations; and (3) HBEST is the only explanation with a conditional probability greater than zero.

However, we can investigate a breadth-first search strategy to see how we can expect increases in n, $|\text{HBEST}|$, and $|X|$ to often affect best-first search. Although breadth-first search may perform poorly in the average case, Cooper

[1989] has shown that an upper bound on the time complexity of hypothesis search when a breadth-first search strategy is used is

$$O\left(\sum_{i=1}^{|\text{HBEST}|} \binom{n}{i} + \sum_{i=1}^{M} \binom{n}{i} \right),$$

where

$$M = \frac{|X|\ln c_1 + |\text{HBEST}|(\ln c_2 - \ln(1-c_3)) + n(\ln(1-c_3) - \ln(1-c_2))}{\ln c_3 - \ln(1-c_2)}.$$

If c_1, c_2, and c_3 are all nonzero, then M is a linear function of $|X|$, $|\text{HBEST}|$, and n. If c_2 and c_3 are both about equal and close to zero, then

$$\lim_{c_1 \to 1} M = |\text{HBEST}|.$$

If c_1, c_2, and c_3 are all about equal and close to zero, then

$$M = |X| + |\text{HBEST}|.$$

Recall that the problem of abductive inference is NP-hard, and thus, in general, it may not be possible to improve significantly on the algorithm. However, the exponential increase in this particular algorithm is due to the depth of the tree created and is hence directly affected by the pruning method used. In some cases we can possibly improve on the pruning method. The pruning method we used is obtained from the results of Theorem 8.2. Recall, in the proof of that theorem, we had the following step:

$$P(X' \mid H_1' \cup H')P(H_1' \cup H') \leq P(H_1' \cup H').$$

That is, we used the trivial upper bound of 1 as a bound for $P(X' \mid H_1' \cup H')$. In a particular system, a more meaningful bound might be found resulting in a decrease in searching. For instance, if we could find a subset Y of X, where X is the set of instantiated variables, such that no variable in $\mathcal{E} - H_1$ has a path leading to any variable in Y, then we would have for any $H \in \mathcal{E} - H_1$ that

$$P(X' \mid H_1' \cup H') \leq P(Y' \mid H_1' \cup H') = P(Y' \mid H_1').$$

The last equality is owing to d-separation. We could then use $P(Y' \mid H_1')$ as a nontrivial upper bound of $P(X' \mid H_1' \cup H')$, and therefore prune H_1 whenever

$$P(H_1')P(Y' \mid H_1') \leq P(\text{HBEST}' \mid X')P(X').$$

Cooper [1984] suggests other methods to possibly decrease the search time in particular cases. Independently, Peng and Reggia [1987b] have obtained a method which substantially decreases the search time in the case of performing abductive inference in the classical diagnostic problem, while making the assumptions in the probabilistic causal method (discussed in section 7.7.2). As originally developed, their method does not use the techniques described in chapters 6 and 7, but rather determines bounds for the probabilities of the explanations.

PROBLEMS 8.3

1. Revise the algorithm given in this section to determine the N best explanations, where N is a value entered by the user.

2. Revise the algorithm in this section to determine the best set of explanations which satisfies a comfort measure, where the comfort measure is a value entered by the user.

COMPUTER PROBLEM 8.3

Write a program which performs abductive inference in the case of the NESTOR diagnostic problem. The user should be allowed to enter a value N, where N is the number of best explanations desired, or a comfort measure, where the total probability of the computed best explanations exceeds the comfort measure.

8.4. AN APPLICATION TO MEDICINE

The physician wants to determine the overall most probable set of diseases in order to arrive at his diagnosis. However, in some cases, he is more interested in determining the best treatments. This would especially be the case when the patient is first examined and some treatments are in order while additional tests are being conducted. We can determine these treatments directly by including a vertex for each treatment in the causal network. If a treatment is given for a particular disease, we place an arc from that disease to that treatment. If the information is obtained from an expert, in many cases the probability of the treatment given the presence of a disease, which point to that treatment, will be 1. If we wished to determine the most probable set of treatments, our explanation set would include the set of all treatments. Alternatively, we could perform simple probability propagation to ascertain the individual probability of each treatment. If a particular treatment were highly probable, then the physician might recommend that treatment regardless of the overall most probable set of treatments. Of course, the side effects and the cost of the treatment are also important. Such considerations are discussed in the next chapter.

CHAPTER 9

APPLICATIONS TO DECISION THEORY

The previous three chapters were concerned only with computing the conditional probabilities of the values of variables given information. They did not explicitly address a method for using these probabilities to make a decision. Of course, the user does use the probabilities to arrive at decisions. For example, if a physician learned that the probability of a certain explanation were .99, he would most likely decide to treat the diseases in that explanation. However, it is left to the physician to analyze the probabilities and make a decision; that is, there is no recommended decision.

In this chapter we will focus on a method which recommends an explicit decision to the user. To do this, we must have some entity which we value and which we wish to maximize. That entity is called utility, and in many cases it is money. In such cases the decision is to take the course of action for which we can expect to make the most money. For example, suppose action a_1 can have three consequences, c_1, c_2, c_3, we will receive \$3, \$4, and $-\$5$, respectively, if each of these consequences occur, and, given that we take action a_1, the probabilities of c_1, c_2, and c_3 are .4, .5, and .1, respectively. Then we say that the utilities U of c_1, c_2, and c_3 are given by

$$U(c_1) = 3, \qquad U(c_2) = 4, \qquad U(c_3) = -5,$$

and the utility of action a_1 is the expected value of the utility if that decision is made. That is,

$$U(a_1) = (.4)(3) + (.5)(4) + (.1)(-5) = 2.7.$$

Note that the utility is a random variable on the probability space consisting of the consequences of action a_1. If the only other option is to take action a_2, and a similar analysis shows that

$$U(a_2) = 2.2,$$

then the recommended decision would be to take action a_1, since that action has a higher expected utility.

Of course, most decisions are far more complex than the above simple example. Even in that example, if the decision maker would not have enough money to pay his mortgage if he lost \$5, this result would have to be incorporated into the utility of consequence c_3. In more involved problems, one decision can lead to a second decision, which in turn can lead to a third decision, and so on. The end results of all these decisions must be analyzed. Another complication is that, although the pure use of money as the utility is convenient in textbook examples, in practice the utility often includes other entities, such as the welfare of a company's employees or that of society.

In section 9.1 we show how to solve problems in decision analysis using the classical decision tree approach, while in section 9.2 we show how to solve such problems by representing them in the more natural framework of causal networks. When used in such problems, causal networks are augmented with additional types of vertices (i.e., vertices which do not represent propositional variables), and they are traditionally called influence diagrams. In section 9.3 we give an example in which the utility is not money.

9.1. DECISION TREES

We will demonstrate the decision tree approach to solving problems in decision analysis by focusing on some examples. The examples given here are variations of an example given in Howard [1984].

Example 9.1. Suppose Joe wishes to buy a used car, and he finds a 1974 Starfire selling for \$1000. Joe knows that if this particular type of vehicle is in good shape, it is worth \$1100. Being an excellent mechanic, Joe is able to examine the vehicle, and he determines that it is in good shape. He must then make the decision of whether to buy the car. Assuming that the utility which Joe wants to maximize is money, if he buys the vehicle he will make

$$\$1100 - \$1000 = \$100,$$

whereas if he does not buy it he will make \$0. These facts are represented in the decision tree in Figure 9.1, where a decision node is represented by an ×, the arc marked with a B represents the decision to buy the vehicle, and the arc marked with an R represents the decision to reject buying the vehicle. Each arc is weighted with the amount of money which will be realized if that decision is made. At a leaf of the tree is the total amount of money which will

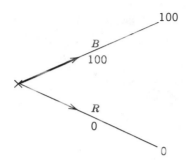

FIGURE 9.1 The decision tree in the used car example when there is no uncertainty.

be made if that leaf is reached. In this simple example, there is no uncertainty and there is only one decision. Thus the expected value of the utility if Joe buys the vehicle is 100, while the expected value is 0 if he does not. He therefore should choose to buy the vehicle.

Example 9.2. Suppose now that Joe is not an excellent mechanic. However, he learns that exactly 80% of the 1974 Starfires were manufactured in a plant which made good vehicles, while the other 20% were manufactured in a plant which made lemons. Furthermore, he learns that, of ten possible mechanical parts, each good vehicle has exactly one part which will fail, whereas each lemon has exactly six parts which will fail. Joe also finds out that this vehicle has never been repaired; hence it either has one part which will fail or six parts which will fail. A mechanic has agreed to charge $40 to repair one part and $200 to repair six parts. These facts are represented by the decision tree in Figure 9.2. The top of the tree in Figure 9.2 is the same as that in Figure 9.1. However, if Joe decides to purchase the vehicle, he reaches a chance node, which is represented by a small blackened circle. Given that he buys the vehicle, it could either turn out to be good (*G*) or a lemon (*L*). If it is good, we take the upper arc, which is weighted with the $40 Joe will need to spend in order to repair one part, while if it is a lemon we take the lower arc, which is weighted with the $200 he will need to spend to repair six parts. Again, at a leaf is stored the amount of money which will be realized if that leaf is reached.

The situation here is quite different from that in Example 9.1. Joe has no control over which arc is selected at the chance node. However, due to the available probabilities, he can determine how likely it is that each arc will be selected. That is, he associates a probability of .8 with the upper arc and a probability of .2 with the lower arc. These probabilities are depicted in Figure 9.3. Although it is not possible to determine a definite utility at the chance node, the expected value of the utility at that node is given by

$$(.8)(60) + (.2)(-100) = 28.$$

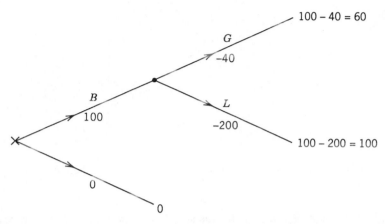

FIGURE 9.2 The original decision tree in the used car example when there is uncertainty.

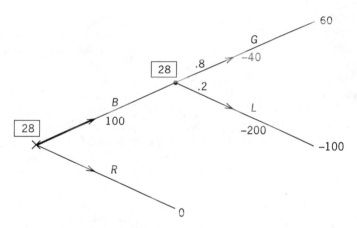

FIGURE 9.3 The original decision tree in the used car example with probabilities included.

That value is listed in a box above the chance node in Figure 9.3. The best decision at the decision node is then the arc which leads to the highest expected utility, and the expected utility at the decision node is the value of that highest expected utility. Again we see that the decision is to buy the vehicle.

Example 9.3. Suppose Joe learns that he has the option of performing some tests on the vehicle before purchasing it. Specifically, suppose he can have one part tested, and the cost of that test is $9, or he can have two parts tested and the cost of that test is $13. Joe now has two decisions: first he must decide whether to perform a test; second, he must decide whether to purchase the vehicle. If we let

T = the result of the first test is positive

S = the result of the second test is positive

D_0 = choose to perform no test

D_1 = choose to perform only one test

D_2 = choose to perform two tests,

then Figure 9.4 represents the facts now known by Joe. First he must decide whether to perform a test. If he chooses to perform only T, it will cost him $9. Thus that arc is weighted with a value of -9. Having made that choice, he reaches a chance node. That is, the test could turn out either positive (T) or negative ($\neg T$). We shall assume that a positive test result means that nothing was found to be wrong. After following these chance branches, the situation is then the same as that depicted in Figure 9.2. If he chooses to perform two tests, the situation is similar except for the fact that at the first chance node there are four possible branches.

To determine the best choice, we again need the probabilities at all the chance nodes. That is, we need the following a priori probabilities:

$$P(T), \quad P(\neg T), \quad P(T,S), \quad P(T,\neg S), \quad P(\neg T,S), \quad P(\neg T,\neg S),$$

and the following conditional probabilities:

$$P(G\,|\,T), \; P(G\,|\,\neg T), \; P(G\,|\,T,S), \; P(G\,|\,T,\neg S), \; P(G\,|\,\neg T,S), \; P(G\,|\,\neg T,\neg S).$$

Of course, we also need $P(L\,|\,T)$, etc. However, those values are uniquely determined by $P(G\,|\,T)$, etc. Unfortunately, none of these a priori or conditional probabilities are immediately available to Joe. Joe does know the a priori probability that the vehicle is good (i.e., $P(G) = .8$), and he does know the conditional probability that a test will be positive given that the vehicle is either good or is a lemon. That is to say, since a good vehicle has exactly one of ten parts bad, $P(T\,|\,G) = .9$, and since a lemon has exactly six of ten parts bad, $P(T\,|\,L) = .4$. These probabilities, which are known to Joe, are associated with chance nodes in another tree. In decision analysis this other tree is called Nature's tree. It is termed this because we can conceive of Nature making a selection at each node of the tree according to a chance mechanism. Nature's tree, for this particular example, is depicted in Figure 9.5. Notice that all the nodes in Nature's tree are chance nodes. At the root, Nature can select either to make a good vehicle or a lemon. She makes the first selection with probability .8 and the second selection with probability .2. Those values are weighted on the appropriate arcs. Given that she makes a good vehicle, she can then select either to make the first test positive or negative. She makes the first selection with probability .9 and the second selection with probability .1. Given that the first test is positive, there are then nine remaining parts, of

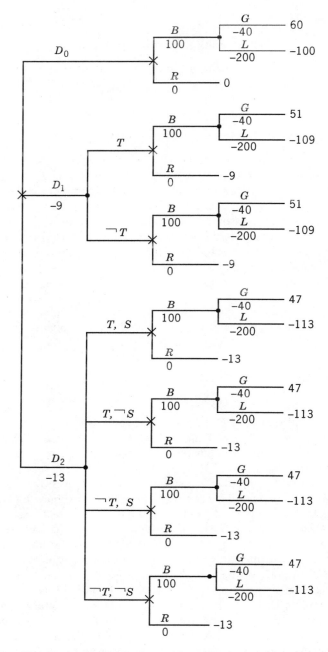

FIGURE 9.4 The decision tree in the used car example when there are two test options.

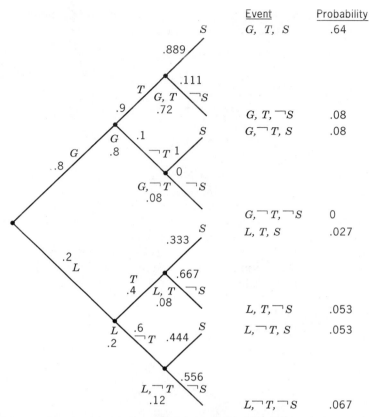

FIGURE 9.5   Nature's tree for the used car example when there are two test options.

which one is definitely bad. Therefore, given that the vehicle is good and the first test is positive, Nature then selects to make the second test positive with probability 8/9 or .889. The remaining probabilities, which are used by Nature to make selections, are weighted on the arcs in Figure 9.5. We need only multiply these probabilities to obtain the joint occurrence of any possible event. For instance,

$$P(G,T,S) = P(S \mid G,T)P(T \mid G)P(G)$$

$$= (.889)(.9)(.8) = .64,$$

and

$$P(G,T) = P(T \mid G)P(G)$$

$$= (.9)(.8) = .72.$$

The other joint probabilities have been computed and are stored at the leaves and the chance nodes in the decision tree in Figure 9.5.

The importance of Nature's tree is that we can compute all the probabilities needed in the decision tree from Nature's tree. For example,

$$P(T) = P(G,T) + P(L,T) = .72 + .08 = .8$$

$$P(T,S) = P(G,T,S) + P(L,T,S) = .64 + .027 = .667$$

$$P(G \mid T) = \frac{P(G,T)}{P(T)} = \frac{.72}{.8} = .9$$

$$P(G \mid T,S) = \frac{P(G,T,S)}{P(T,S)} = \frac{.64}{.667} = .96.$$

The other probability values have been computed and are stored on the appropriate arcs in Figure 9.6. Furthermore, the expected utility of the chance nodes have been computed and passed up the tree in that figure. The method by which this is done is similar to the mini-max method used in game theory except that expected utilities are passed to chance nodes and maximums are always passed to decision nodes. For instance, suppose decision D_1 is made, test T is positive, and decision B is made. We arrive at the chance node which has a box containing the value 35 over it. That value has been obtained as follows:

$$(.9)(51) + (.1)(-109) = 35.$$

The value above the decision node, which contains the decision whether to purchase the vehicle, is the maximum of 35 and −9. The value above the chance node, which is the chance of test T being positive or negative, is then obtained:

$$(.8)(35) + (.2)(-9) = 26.2.$$

Finally, the value above the decision node, which contains the decision of which, if any, tests to perform, is the maximum of 28, 26.2, and 22.8, which is 28. We see that the recommended decision is to perform no tests.

The above technique can be used regardless of the number of decisions or chance nodes. If there are other test options, we simply add another arc from the root. If there are other possible decisions before the test options, we place that decision node before the tree in Figure 9.6.

The decision tree representation is appealing when we have a naturally ordered sequence of decisions to be made. Notice, however, that we have a great deal of redundancy. In Figure 9.6, a subtree starting with the decision of whether to purchase the vehicle is repeated seven times. If we add another test option, it will be repeated at least two more times. Moreover, if we add a decision before the test options, the entire tree in Figure 9.6 may be repeated for each possible choice at that decision.

Another problem in the decision tree approach is that the probabilities which are available to the decision-maker (i.e., the ones in Nature's tree) are not the ones needed in the decision tree. Thus we have the task of converting the naturally occurring probabilities to the needed ones. Note, however, that these naturally occurring probabilities are the same ones which we assess in

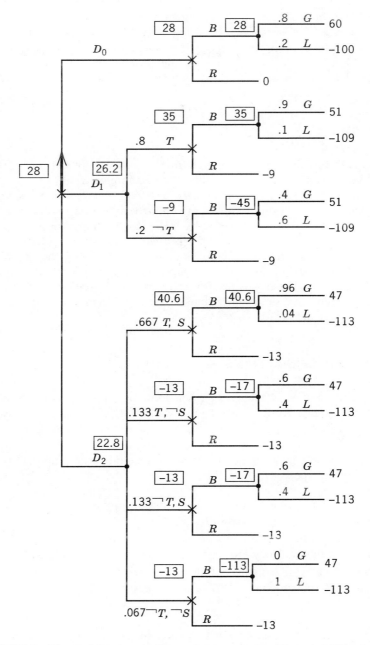

FIGURE 9.6 The decision tree in the used car example with probabilities included.

a causal network. Therefore, if we can represent the problem in the framework of a causal network, we can simply assess Nature's probabilities and let the propagation techniques in chapters 6 and 7 determine the needed probabilities. Additionally, by representing the problem in a causal network, we eliminate the redundancy referred to above. We shall see how to accomplish this in the next section.

PROBLEM 9.1

Suppose Joe has the option of purchasing exactly one of two cars. The first car is the Starfire referred to in the above examples, while the second is a 1976 Roadrunner. The Roadrunner sells for $1400, whereas its market value is $1600. Joe has the option of either performing one test on the Starfire for $20 or one test on the Roadrunner for $30. If the Roadrunner is of good quality, it will cost nothing to repair it, while if it is a lemon, it will cost $250 to repair it. The probability that the Roadrunner is of good quality is .7, the probability of the Roadrunner passing its test if it is of good quality is 1, while the probability of the Roadrunner passing its test if it is a lemon is .3. Create the decision tree for this problem, and determine whether Joe should perform no test, test the Starfire, or test the Roadrunner.

9.2. INFLUENCE DIAGRAMS

The description of an influence diagram, contained in this section, is obtained, for the most part, from Howard and Matheson [1984] and from Shachter [1986].

Definition 9.1. An influence diagram is a directed acyclic graph (DAG) consisting of three types of nodes:

1. Zero or more chance nodes, which contain propositional variables. They are represented by circles in the DAG.
2. Zero or more decision nodes, which contain choices available to the decision maker. They are represented by squares in the DAG.
3. One value node, which contains a random variable, whose value is the utility of the outcome. This node is represented by a diamond in the DAG.

As in a standard causal network, the arcs into chance nodes show the variables upon which that node is conditionally dependent. The arcs into decision nodes show exactly which variables will be known to the decision maker at the time the decision is made. The arcs into the value node show which variables enter into the calculation of the utility.

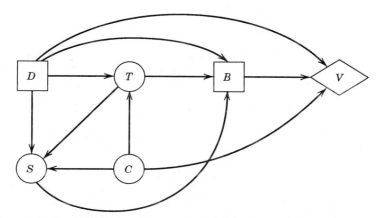

FIGURE 9.7   The influence diagram for the problem in Example 9.3.

Figure 9.7 contains an influence diagram which represents the problem in Example 9.3. We will show how that diagram has been created and explain the meaning of the nodes and the arcs.

If we let D contain the decision of which test option to choose, then the possible values of D are

d_0 = choose no test

d_1 = choose to run the first test

d_2 = choose to run both the first test and the second test.

If we let T contain the results of the first test, then the possible values of T are

t_0 = test is not run

t_1 = test is positive

t_2 = test is negative.

If we let S contain the results of the second test, then the possible values of S are

s_0 = test is not run

s_1 = test is positive

s_2 = test is negative.

If we let C contain the conditions of the vehicle, then the possible values of C are

c_1 = vehicle is in good condition

c_2 = vehicle is a lemon.

If we let B contain the decision of whether to buy the vehicle, then the possible values of B are

$$b_1 = \text{buy the vehicle}$$

$$b_2 = \text{reject buying the vehicle.}$$

The node V represents a function of the variables which have arcs leading into V. For example, if D has value d_1, B has value b_1, and C has value c_1, then we will spend \$9 to run test 1, make \$100 when we purchase the vehicle, and spend \$40 when we repair the vehicle. Accordingly,

$$V(d_1, b_1, c_1) = -9 + 100 - 40 = 51.$$

Actually, this node can be considered a propositional variable which can take exactly one value for each combination of values of its parents. The possible values which a variable contributes to V can be stored on the arc from that variable. For example, $(0, -9, -13)$ can be stored on the arc from D to V.

Each chance node is conditionally dependent on its parents as in a standard causal network. Since C has no parents, the values stored at C are

$$P(c_1) = .8 \qquad \text{and} \qquad P(c_2) = .2.$$

The result of test 1 depends not only on the condition of the vehicle, C, but also on the decision as to whether we run the test (i.e., there is no result if we do not run the test). Consequently, the values stored at node T are

$$P(t_0 \mid d_0, c_i) = 1 \qquad P(t_1 \mid d_0, c_i) = 0 \qquad P(t_2 \mid d_0, c_i) = 0$$
$$P(t_0 \mid d_1, c_1) = 0 \qquad P(t_1 \mid d_1, c_1) = .9 \qquad P(t_2 \mid d_1, c_1) = .1$$
$$P(t_0 \mid d_1, c_2) = 0 \qquad P(t_1 \mid d_1, c_2) = .4 \qquad P(t_2 \mid d_1, c_2) = .6$$
$$P(t_0 \mid d_2, c_1) = 0 \qquad P(t_1 \mid d_2, c_1) = .9 \qquad P(t_2 \mid d_2, c_1) = .1$$
$$P(t_0 \mid d_2, c_2) = 0 \qquad P(t_1 \mid d_2, c_2) = .4 \qquad P(t_2 \mid d_2, c_2) = .6.$$

The results of test 2 depend on the condition of the car, C, the results of test 1, T, and the decision of whether we run the test, D. Therefore the values stored at node S are

$$P(s_0 \mid d_0, c_i, t_i) = 1 \qquad P(s_1 \mid d_0, c_i, t_i) = 0 \qquad P(s_2 \mid d_0, c_i, t_i) = 0$$
$$P(s_0 \mid d_1, c_i, t_i) = 1 \qquad P(s_1 \mid d_1, c_i, t_i) = 0 \qquad P(s_2 \mid d_1, c_i, t_i) = 0$$
$$P(s_0 \mid d_2, c_1, t_1) = 0 \qquad P(s_1 \mid d_2, c_1, t_1) = .889 \qquad P(s_2 \mid d_2, c_1, t_1) = .111$$
$$P(s_0 \mid d_2, c_1, t_2) = 0 \qquad P(s_1 \mid d_2, c_1, t_2) = 1 \qquad P(s_2 \mid d_2, c_1, t_2) = 0$$
$$P(s_0 \mid d_2, c_2, t_1) = 0 \qquad P(s_1 \mid d_2, c_2, t_1) = .333 \qquad P(s_2 \mid d_2, c_2, t_1) = .667$$
$$P(s_0 \mid d_2, c_2, t_2) = 0 \qquad P(s_1 \mid d_2, c_2, t_2) = .444 \qquad P(s_2 \mid d_2, c_2, t_2) = .556.$$

Note that all these values are the same values which are stored in Nature's tree in Figure 9.5. In other words, the probabilities which we must access for an influence diagram are the ones in Nature's tree.

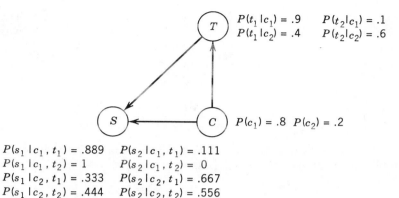

$P(t_1|c_1) = .9 \quad P(t_2|c_1) = .1$
$P(t_1|c_2) = .4 \quad P(t_2|c_2) = .6$

$P(c_1) = .8 \quad P(c_2) = .2$

$P(s_1|c_1,t_1) = .889 \quad P(s_2|c_1,t_1) = .111$
$P(s_1|c_1,t_2) = 1 \quad P(s_2|c_1,t_2) = 0$
$P(s_1|c_2,t_1) = .333 \quad P(s_2|c_2,t_1) = .667$
$P(s_1|c_2,t_2) = .444 \quad P(s_2|c_2,t_2) = .556$

FIGURE 9.8 The resultant causal network when D is instantiated for d_2 and B is instantiated for b_i in the influence diagram in Figure 9.7.

For each possible combination of values of the decision nodes, the chance nodes constitute a causal network. For example, if

$$D = d_2 \quad \text{and} \quad B = b_i,$$

the chance nodes constitute the causal network depicted in Figure 9.8. Therefore, given that the decision nodes are all instantiated, we can use the techniques in chapters 6 and 7 to propagate probabilities in the resultant causal network.

Before showing how to use the techniques in chapters 6 and 7 to solve the problem in Example 9.3, we state two assumptions which must be made in decision analysis:

1. There is a total ordering among the decision nodes.
2. If decision node A precedes decision node B in the total ordering, then A and all of the informational predecessors of A are informational predecessors of B. That is, there is an arc from A to B and, if there is an arc from a node C to A, then there is an arc from C to B.

The first restriction implies that the decisions must be made in order according to time, while the second implies that any information available at a given decision is still available at a later decision. This restriction is termed "no forgetting," as it implies that the decision maker recalls all past information and decisions.

We can now solve the problem in Example 9.3 using the influence diagram in Figure 9.7. First we order the decision nodes according to time to obtain the ordering

$$[D,B].$$

Next we let D equal its first value d_0 and proceed to the next decison node B. When we make that decision, the information at T and S is available. Recall that we need to compute an expected utility at chance occurrences. Thus we begin computing that expected utility by marking T and S together with a 0 and letting T equal its first value t_0 and S equal its first value s_0. Then we let B equal its first value b_1, and we compute the expected value of V given these instantiations. That is, we compute

$$E(V \mid d_0, t_0, s_0, b_1) = V(d_0, b_1, c_1)P(c_1 \mid d_0, t_0, s_0, b_1)$$
$$+ V(d_0, b_1, c_2)P(c_2 \mid d_0, t_0, s_0, b_1)$$

where E stands for expected value. The needed values of V we obtain directly from the information stored in the influence diagram:

$$V(d_0, b_1, c_1) = 0 + 100 - 40 = 60$$
$$V(d_0, b_1, c_2) = 0 + 100 - 200 = -100.$$

The needed conditional probabilities can be obtained in the following way. Given that $D = d_0$ and $B = b_1$, the three chance nodes comprise a causal network as illustrated above. Thus we can compute the probability of c_1 given (t_0, s_0) and the probability of c_2 given (t_0, s_0) in that network using the techniques in chapters 6–8. After obtaining these conditional probabilities, we can compute the expected value of V given (d_0, t_0, s_0, b_1). That value is

$$E(V \mid d_0, t_0, s_0, b_1) = 28.$$

Since we pass maximums to decision nodes, we simply mark node B with a value of 28 and repeat the computation with B having a value of b_2. That is, we compute

$$E(V \mid d_0, t_0, s_0, b_2) = 0.$$

If this value were larger, it would become the new value marked at B. Having exhausted all B's alternatives, we backtrack to node S and T and add

$$28 \times P(s_0, t_0 \mid d_0)$$

to the 0 stored at S and T. Again, $P(s_0, t_0 \mid d_0)$ can be computed using the techniques in chapters 6–8. (Since decision B is not an ancestor of S, T, or C, only a decision at D is necessary in order for these chance nodes to constitute a causal network.) $P(s_0, t_0 \mid d_0)$ turns out to be equal to 1. We then let S and T take their second pair of values, s_0 and t_1 respectively, and repeat the same computations with B equal to first b_1 and then b_2. The larger of the two resultant values is then multiplied by $P(s_0, t_1 \mid d_0)$ and that product is added to the 28 stored at S and T. This procedure is repeated for all combinations of values of S and T. In this case all the remaining terms turn out to be 0 and thus the final value stored at S and T is 28. Again since maximums are simply passed to decision nodes, this value of 28 is then passed to node D (we also mark the node with decision d_0). We then let $D = d_1$ and repeat the entire procedure. If we obtain a larger expected value, that value will become the new value marked at D (along with d_1 becoming the new decision). Finally we let $D = d_2$ and again repeat the procedure. In this way we actually perform the same computations as those in the decision tree in Figure 10.6. The difference is that the tree is built on the fly by the program. The model builder need never concern himself with any details other than the natural relationships in the influence diagram.

The used car example is a very convenient textbook example. Before ending this chapter, we give a more meaningful example concerning a company's expected profit. (This example was also used in a different context in Chapter 2.)

Example 9.4.

A firm is setting its price for a product which is also sold by one other company. The other firm will announce its price first and then our customers will wait for our firm to set its price before choosing their orders. Some of these customers will be placing their orders early, before we make our production lot size decision. Our goal is to maximize our expected profit which depends on our sales, our price, and the lot size.

—Shachter [1988]

If we let

$$C = \text{competitor price}$$
$$P = \text{price}$$
$$S = \text{sales}$$
$$E = \text{early order}$$
$$L = \text{lot size}$$
$$V = \text{profit},$$

then the influence diagram in Figure 9.9 represents the relationships among these variables.

This section has been a brief introduction to the use of influence diagrams in decision analysis. For more details on this subject, the reader is referred to Howard and Matheson [1984] and Shachter [1986, 1988]. A discussion of influence diagrams in which the variables are continuous can be found in Shachter and Kenley [1989].

PROBLEMS 9.2

1. Create the influence diagram for the problem described in Problem 9.1.

2. Create the decision tree which corresponds to the influence diagram in Example 9.4.

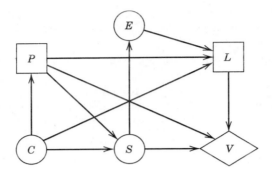

FIGURE 9.9   The influence diagram for the problem in Example 9.4.

COMPUTER PROBLEM 9.2

Write a program which performs decision analysis using an influence diagram. The specifications for building the influence diagram are the same as those for a program which builds an arbitrary causal network (Computer Problem 5.6) except for the modifications needed to build an influence diagram. The output of the program is the best decision and its expected value.

9.3. THE UTILITY NEED NOT BE MONEY

In many practical problems the utility which we wish to maximize need not be money as in the convenient textbook examples in the previous sections. Consider the following fictitious example. Suppose that a patient has an illness which causes him to be bedridden and that there is an operation which could cure him. However, a possible side effect of the operation is blindness. Let

t_1 = the patient decides to have the operation

t_2 = the patient decides not to have the operation

w_1 = the patient gets well

w_2 = the patient does not get well

b_1 = the patient ends up blind

b_2 = the patient does not end up blind,

and suppose that

$$P(w_1 \mid t_1) = .9 \qquad P(b_1 \mid t_1) = .3$$
$$P(w_1 \mid t_2) = .2 \qquad P(b_1 \mid t_2) = 0.$$

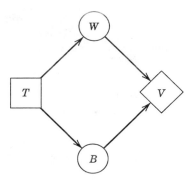

FIGURE 9.10 The influence diagram which represents the situation in which a patient must decide on a treatment.

Then the simple influence diagram in Figure 9.10 represents the relationships in this problem. (We are assuming that the side effect of being blind is independent of getting well given that the patient has the treatment.)

The possible outcomes are

$$(b_1, w_1), \qquad (b_1, w_2), \qquad (b_2, w_1), \qquad (b_2, w_2),$$

and each of these outcomes has a different utility to the patient. Clearly, the best outcome is (b_2, w_1) and the worst is (b_1, w_2). However, it is not immediately obvious whether (b_1, w_1) or (b_2, w_2) is worse. We must assume that the decision maker is able to order all the outcomes according to desirability (two outcomes could be ranked to be equivalent). Suppose, after careful deliberation, the patient decides that

$$V(b_1, w_2) < V(b_1, w_1) < V(b_2, w_2) < V(b_2, w_1).$$

It is not sufficient merely to rank the outcomes. To reach a decision we must also determine how much more desirable one outcome is than another. This is accomplished in the following way. An arbitrary value, which is usually taken to be 0, is assigned to the worst outcome, and another arbitrary value, usually taken to be 1, is assigned to the best outcome. In the present example, if we let

$$r_1 = (b_1, w_2), \qquad r_2 = (b_1, w_1), \qquad r_3 = (b_2, w_2), \qquad r_4 = (b_2, w_1),$$

then

$$V(r_1) = 0 \qquad \text{and} \qquad V(r_4) = 1.$$

In order to determine the utility of r_2, the decision maker determines an α_2, where $0 < \alpha_2 < 1$, such that he feels that the utility of r_2 is given by

$$V(r_2) = \alpha_2 V(r_1) + (1 - \alpha_2)V(r_4).$$

The value of α_2 is determined by considering a "gamble" in which he gets outcome r_1 with probability α and outcome r_4 with probability $1 - \alpha$. The α, which yields a gamble which he feels is of the same utility as simply being assured of outcome r_2, is the value of α_2. Suppose that, after careful deliberation, the patient decides that α_2 equals .9. Then

$$V(r_2) = .9(0) + .1(1) = .1.$$

In the same way the patient must decide on a value of α_3 to determine the utility of r_3. If he decides that α_3 equals .7, then

$$V(r_3) = .7(0) + .3(1) = .3.$$

When a new utility is determined, it should be compared directly to the previous utilities to see if one feels that they are consistent. That is, there should be some gamble in which one will receive outcome r_2 with probability α and outcome r_4 with probability $1 - \alpha$ which is equivalent to being guaranteed outcome r_3. In symbols

$$V(r_3) = \alpha V(r_2) + (1 - \alpha)V(r_4).$$

Suppose that the patient decides that the value of this α is .8. However, based on the utilities already computed, $V(r_2) = .1$, $V(r_3) = .3$, and $V(r_4) = 1$. Thus we must have that

$$.3 = \alpha(.1) + (1 - \alpha)1,$$

which implies that $\alpha = .78$. The patient must then decide which value of α really represents his feelings. We will assume that he decides that this α is equal to .78. In other words, he stays with his original assessments.

Once utilities are assigned to all possible outcomes, the decision can be determined using the technique outlined in the previous section. In the present example, we can easily compute the expected utilities directly:

$$E(V \mid t_1) = V(r_1)P(r_1 \mid t_1) + V(r_2)P(r_2 \mid t_1) + V(r_3)P(r_3 \mid t_1) + V(r_4)P(r_4 \mid t_1)$$

$$= (0)(.3)(.1) + (.1)(.3)(.9) + (.3)(.7)(.1) + (1)(.7)(.9)$$

$$= .678$$

$$E(V \mid t_2) = V(r_1)P(r_1 \mid t_2) + V(r_2)P(r_2 \mid t_2) + V(r_3)P(r_3 \mid t_2) + V(r_4)P(r_4 \mid t_2)$$

$$= (0)(0)(.8) + (.1)(0)(.2) + (.3)(1)(.8) + (1)(1)(.2)$$

$$= .440.$$

Therefore the patient should take the treatment.

The primary purpose of this chapter has been to show how problems in decision theory can be solved using influence diagrams. The discussion of decision theory and utility theory has thus been very sketchy. There exist many more considerations, both of a theoretical and of a practical nature. For instance, the utility of money declines as more money is received. If a person would originally perform an unpleasant task for $10,000, he may no longer perform that task after winning $1,000,000 in the state lottery. Consequently,

$$V(\$1,010,000) \neq V(\$1,000,000) + V(\$10,000).$$

As another example, if given the choice of being given $5000 or the opportunity to receive $0 if a coin turns up heads and $11,000 if it turns up tails, one may choose the definite $5000. Hence the "expected value" of the gamble is not equal to

$$\$5500 = .5(0) + .5(\$11,000).$$

There exist excellent texts on decision theory and utility theory which discuss these issues and much more. Specifically, the interested reader is referred to Berger [1985].

9.4. THE RELATIONSHIP BETWEEN EXPERT SYSTEMS AND DECISION THEORY

The field of expert systems is young and therefore its relationships to other disciplines has not been firmly established. Traditionally, decision theory investigates problems containing relationships which exist on a one-time basis and also problems containing relationships which are repeatable. For example, Joe can use decision theory to decide whether to buy the Starfire. The relationships which Joe analyzes in order to solve his problem may never exist in another problem. A physician can also use decision theory to help him determine the diseases and decide on the treatments for a particular patient. However, this problem is repeatable. That is, he will treat many patients with a similar symptoms. The computer program which assists him is called an expert system. A particular expert system, which engages in probabilistic reasoning, may or may not include decision nodes and a value node; however, it will always at least include chance nodes. We see then that probabilistic reasoning in expert systems is a subfield of decision theory. Of course, all the methods developed in this text are also applicable to problems that contain relationships which exist on a one-time basis.

CHAPTER 10

VARIABILITY AND UNCERTAINTY IN PROBABILITIES; OBTAINING PROBABILITIES

Aside from the considerations in chapter 2, we have thus far treated probabilities as abstract entities, associated with the values of propositional variables, which are changed as information is passed through a causal network. That is, we have given little consideration to the meaning of probabilities. In this chapter we consider again this meaning. Specifically, we discuss the meaning of the change in the probability of a proposition as information is gathered and the meaning of uncertainty in probability values themselves. Furthermore, we show how to augment probabilities obtained from an expert with information in a data base and how to obtain probabilities solely from a data base.

Consider first the following example:

Example 10.1. Suppose that we have an urn filled with 100 coins, and the compositions of the coins are such that the probability of a heads on a toss of coin 1 is .01, the probability of a heads on a toss of coin 2 is .02,..., and the probability of a heads on a toss of coin 100 is 1. That is, we would expect a heads to come about roughly ten times in 1000 tosses of coin 1. The frequentist would say that .01 is the propensity or objective probability associated with coin 1, whereas the subjectivist need only say that the sequence is exchangeable and .01 represents his belief. Throughout this chapter we will use the frequentist's expression "collective," realizing that, for practical purposes, a collective is the same as an infinitely exchangeable sequence. Thus .01 is the probability associated with the collective of all tosses of coin 1 and the event that a heads turns up.

If we choose a coin at random from the urn and toss that coin, the probability of a heads on that toss is equal to

$$\sum_{i=1}^{100} P(\text{heads} \mid \text{coin } i)P(\text{coin } i) = \sum_{i=1}^{100} \left(\frac{i}{100} \times \frac{1}{100} \right) = \frac{100 \times 101}{100 \times 100 \times 2} \simeq .5.$$

This value is the probability of a heads given the information that we picked a coin at random from the urn and tossed it. To repeat this experiment (i.e., to obtain another member of the same collective), we must replace the coin, randomly pick another coin, and toss it. If we repeated this experiment many times, we would expect that a heads would come up approximately 50% of the time.

Suppose now that we are given the coin, which has a probability of heads equal to .5, and we are told that the probability is indeed .5. If we tossed that coin many times, we would also expect that a heads would come up about 50% of the time. Each repetition of this experiment is a member of the collective of all tosses of this particular coin. Hence we see that we have two different collectives (one being many repetitions of picking a coin from the urn and tossing it, the other many repetitions of tossing the coin with probability of heads equal to .5), each with the probability of heads coming up equal to .5.

The situation in these two experiments is different even though the probabilities are equal. The toss of a coin picked at random from the urn is a member of another collective, and there exists a method with which we could approximately determine that collective. This other collective is the collective consisting of many tosses of the particular coin picked from the urn. We could determine, with high certainty, this collective by tossing the coin many times. If we had that option and we found that heads came up about .8 fraction of the time, we would bet on heads on the next toss according to a value about equal to .8, rather than according to the value .5. Therefore, by obtaining more information, we can determine another collective to which a particular toss belongs, and we can thereby improve on our predictive accuracy for the outcome of that toss. On the other hand, if we know that we are tossing the coin with probability of heads equal to .5, there is no other collective which we can determine (assuming that we are unable to compute the effects of atmospheric conditions, placement of the coin in the hand, etc.). Thus we cannot, by obtaining additional information, improve on our predictive accuracy for the outcome of a toss of that coin.

Next we show how this example pertains to expert systems.

Example 10.2. Suppose that we have three propositional variables, A, B, and C, each with two alternatives, where

$$a_1 = \text{patient has a particular disease}$$

$$b_1 = \text{patient has a particular symptom}$$

$$c_1 = \text{patient tests positive on a particular test},$$

and that these variables are related by the causal network in Figure 10.1.

Either by probability propagation or by direct computation, it is easy to determine that

$$P(a_1 \mid b_1) = .85.$$

Therefore, if a patient had the particular symptom, we would conclude that the probability of his having the disease is very high. If this particular disease called for a serious operation, we may decide to operate if it were not possible to obtain any additional information. Such a situation would be equivalent to the situation in which we know that we are holding the coin with probability of heads equal to .5. However, in this example, we do have the option of obtaining additional information. That is, we can conduct the test. Suppose that we elect to do this and the outcome of test is negative. It is easy to compute that

$$P(a_1 \mid b_1, c_2) = .006.$$

Thus, by running this test, we have learned that it is very improbable that the patient has the disease. It is important to realize the meaning of these different probabilities. The particular patient either does or does not have the disease. Ideally, we would like to determine which is true. However, assuming that the only facts which we can possibly obtain are the values of variables B and C, we cannot accomplish this. We can only determine collectives of which the patient is a member. First we know that he is in the collective of all patients considered by this system. Given this collective, the probability of his having the disease is .1. That is to say, if we operated on all the patients ever considered by the system, we should be choosing the correct alternative about 10% of the time. When we learn that the patient has the symptom, we know that he is in the collective of all people considered by this system who also have this particular symptom. If we operated on all such people, we should be choosing the correct alternative approximately 85% of the time. This percentage may be good; however, it is little consolation to the patients on whom we operate who do not have the disease. If it were not possible to obtain additional information, it may indeed be wise to operate based only on the information that the symptom is present. However, in the current example, we can run the test which has the possibility of determining that the patient is in the collective of all people considered by this system who have this particular symptom and who test negative on the test. Given this collective, the probability of the disease is only .006. Accordingly, given that the patient is in this collective, we would most likely choose not to operate. We would be remiss if we operated owing to the large probability value of 85% without performing the test. Hence we see that we are not only interested in whether the disease is highly probable, but also in whether that probability can change significantly in the light of future information.

In Section 10.1 we show how to analyze the possible change in probability values due to future information. In Section 10.2 we show how uncertainty in

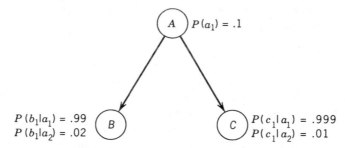

FIGURE 10.1   The causal network for Example 10.2.

the a priori and conditional probability values which are assigned to the network can be represented, obtained, and incorporated into the structure of a causal network. Moreover, in that section, we show how to augment probabilities obtained from an expert with information in a data base and how to obtain probabilities solely from a data base. Finally, in Section 10.3 we obtain a formula which is useful when we are using an approximation technique. Many of the results in Sections 10.1 and 10.2.2 are obtained from Spiegelhalter [1988].

10.1. POSSIBLE CHANGES IN PROBABILITY VALUES DUE TO FUTURE INFORMATION

We develop the concepts in this section in stages. Consider first the causal network in Figure 10.2. The a priori probability of a_1 in that network is given by

$$P(a_1) = P(a_1 \mid c_1)P(c_1) + P(a_1 \mid c_2)P(c_2)$$

$$= (.9)(.2) + (.4)(.8) = .5.$$

If a_1 stands for a patient having a particular disease, this would mean that approximately 50% of all patients ever considered by this system should have this disease. However, we see directly that this probability can change to .9 or .4 depending on whether c_1 or c_2 occurs. Perhaps c_1 stands for a patient having a particular risk factor for the disease. Thus, in this simple network, we see immediately that the possible future value of a_1, based on information obtained elsewhere in the network, is in the interval [.4,.9]. We have an interval because C may not become known for certain. That is, virtual evidence, E, may arrive for C. It is possible to show that $P(a_1 \mid E) \in [.4, .9]$. This is a special case of Theorem 10.1, which will be stated shortly. Note that this range does not include possible values due to virtual evidence which arrives directly for A. Aside from the range of possible future values, we are also interested in the variance of the future values, since a small variance would indicate that it is unlikely that the future values will deviate significantly from the expected

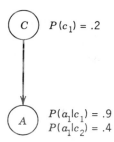

FIGURE 10.2   The first causal network discussed in Section 10.1.

value. An expected value and a variance are always relative to a particular random variable on a specific probability space. In the present considerations, the random variable is the $P(a_1 \mid C)$, while the probability distribution is $P(C)$. We will denote an expected value in this space by E_C and a variance by V_C. It is easy to see that the expected value of $P(a_1 \mid C)$ is given by

$$E_C(P(a_1 \mid C)) = P(a_1 \mid c_1)P(c_1) + P(a_1 \mid c_2)P(c_2) = P(a_1) = .5.$$

Therefore the variance of $P(a_1 \mid C)$ is obtained as follows:

$$
\begin{aligned}
V_C(P(a_1 \mid C)) &= E_C((P(a_1 \mid C) - P(a_1))^2) \\
&= (P(a_1 \mid c_1) - P(a_1))^2 P(c_1) + (P(a_1 \mid c_2) - P(a_1))^2 P(c_2) \\
&= .04.
\end{aligned}
$$

Both the range, [.4,.9], and the variance, .04, are useful in determining whether we should try to obtain the information concerning c_1. Of course, the cost of obtaining this information is also a consideration. To incorporate this cost into the decision we would use the techniques discussed in chapter 9.

Next consider the causal network in Figure 10.3. This network is the same as the one in Figure 10.2 except A now has a child B. B might stand for the result of a particular test for the disease, where b_1 means the test is positive. We see then that roughly 70% of the patients with the disease test positive, while about 10% of those without the disease test positive. In this case we can determine the range of possible future values of the probability of a_1 due to C alone, (i.e., the possible values due to C becoming known for certain or due to virtual evidence arriving for C) due to B alone, or due to both C and B. The range due to both C and B is the maximum range of future values. We have already computed the range due to C alone. To determine the range due to B alone, we compute

$$P(a_1 \mid b_1) = .875$$

$$P(a_1 \mid b_2) = .25.$$

In this simple example, these values can be computed directly from the definition of conditional probability. In practice they can be determined by probability propagation, as discussed in chapters 6 and 7. The variance, with respect to B, of $P(a_1 \mid B)$ is then given by

$$V_B(P(a_1 \mid B)) = E_B\left((P(a_1 \mid B) - P(a_1))^2\right)$$

$$= (P(a_1 \mid b_1) - P(a_1))^2 P(b_1) + (P(a_1 \mid b_2) - P(a_1))^2 P(b_2)$$

$$= .094.$$

Hence the range in possible future values of the probability of a_1, due to information received about B alone, is $[.25, .875]$ and the variance is .094. We can use the ranges and variances due to C and B individually to determine whether we should investigate the values of either of these variables. On the other hand, we can compute the total range of possible future values of the probability of a_1 by considering C and B together:

$$P(a_1 \mid b_1, c_1) = .984$$

$$P(a_1 \mid b_1, c_2) = .824$$

$$P(a_1 \mid b_2, c_1) = .75$$

$$P(a_1 \mid b_2, c_2) = .182.$$

Again, these values can be obtained by probability propagation. Thus the range of all possible future values of the probability of a_1 is $[.182, .984]$. The variance, relative to both B and C, is given by

$$V_{B,C}(P(a_1 \mid B,C)) = E\left((P(a_1 \mid B,C) - P(a_1))^2\right)$$

$$= (P(a_1 \mid b_1,c_1) - P(a_1))^2 P(b_1,c_1) + (P(a_1 \mid b_1,c_2) - P(a_1))^2 P(b_1,c_2)$$

$$+ (P(a_1 \mid b_2,c_1) - P(a_1))^2 P(b_2,c_1) + (P(a_1 \mid b_2,c_2) - P(a_1))^2 P(b_2,c_2)$$

$$= .116.$$

Next consider the more complex network in Figure 10.4. This figure is meant to represent the situation in which the only parents of A are C and D, the only child of A is B, the only parent of B is E, and these vertices are embedded in a larger network. Using the method outlined in this section, we can compute the range of possible future values of the probability of a_1 due to C, D, B, and E. That is, we can compute for all i, j, k, m the values of

$$P(a_1 \mid c_i, d_j, b_k, e_m).$$

The minimum and maximum values determine the range. We can also compute

$$V_{C,D,E,F}(P(a_1 \mid C,D,E,F)).$$

The resultant range is the range of possible future values of a_1 due to local influences. The following lemma and theorem show that this range is the max-

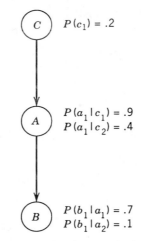

$P(c_1) = .2$

$P(a_1 | c_1) = .9$
$P(a_1 | c_2) = .4$

$P(b_1 | a_1) = .7$
$P(b_1 | a_2) = .1$

FIGURE 10.3 The second causal network discussed in Section 10.1.

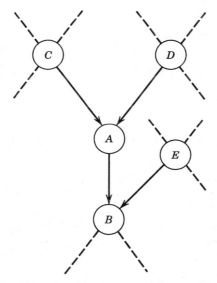

FIGURE 10.4 The third causal network discussed in Section 10.1.

imum range of possible future values of a_1 due to information received from any vertices in the network:

Lemma 10.1. Let a vertex A be given. Then the set $m(A)$, which contains all parents of A, children of A, and parents of children of A, d-separates A from the set $V - (\{A\} \cup m(A))$; that is, $m(A)$ d-separates A from all other vertices in the network. The set $m(A)$, is called the "Markov blanket" around A [Pearl, 1986b].

Proof. The proof follows immediately from the definition of d-separation and is left as an exercise. $\quad\square$

Theorem 10.1. If $m(A)$ is the Markov blanket around A, and W is any subset of $V - (\{A\} \cup m(A))$, then for each i there is no combination of values W' of W such that

$$P(a_i \mid W') < P(a_i \mid m(A))$$

for all combinations of values of variables in $m(A)$. The theorem also holds when $>$ is replaced by $<$.

Proof. Suppose W' is such a set of instantiated values. By Lemma 10.1 and Theorem 6.2, for all combinations of values of $m(A)$

$$P(a_i \mid m(A) \cup W') = P(a_i \mid m(A)).$$

Therefore, for all combinations of values of $m(A)$,

$$P(a_i \mid W') < P(a_i \mid m(A) \cup W'))$$

$$\frac{P(\{a_1\} \cup W')}{P(W')} < \frac{P(\{a_i\} \cup m(A) \cup W')}{P(m(A) \cup W')}$$

$$P(\{a_i\} \cup W')P(m(A) \cup W') < P(\{a_i\} \cup m(A) \cup W')P(W')$$

$$\sum_{m(A)} (P(\{a_i\} \cup W')P(m(A) \cup W')) < \sum_{m(A)} (P(\{a_i\} \cup m(A) \cup W')P(W'))$$

$$P(\{a_i\} \cup W')P(W') < P(\{a_i\} \cup W')P(W')$$

$$1 < 1,$$

a contradiction. The case for $>$ is proven in the same fashion. $\quad\square$

Owing to Theorem 10.1, we can determine the minimum and maximum possible future values of the probability of a value of A by looking only at the variables in the Markov blanket around A. This fact is very important because it enables us to determine locally the maximum possible change in the probability of A due to new information received anywhere in the network. In practice, however, we may still investigate the impact of variables which are not in the Markov blanket. For instance, there may be a test which is far removed from A whose impact we may wish to investigate relative to A in order to determine whether we should conduct that particular test.

The discussion in this section has centered around how the probability of a variable A may change from its a priori value to conditional values on account of possible future information. There is nothing in our development, though, which limits us to the a priori value. That is, we can determine the conditional probability $P'(A)$ based on certain information and use the method outlined here to determine the range of possible new conditional probabilities (along with the variance) based on future information. Indeed, it is in this manner that we would ordinarily apply the method. For example, we would determine the probability of a disease based on some observed manifestations and then ascertain the possible influence that a particular test could have on that probability in order to determine whether we should conduct the test.

Before ending this section, we return to the causal network in Figure 10.3 to illustrate a convenient computational technique. Previously, we computed that

$$V_C(P(a_1 \mid C)) = .04$$
$$V_{B,C}(P(a_1 \mid B,C)) = .116.$$

We determined each of these values separately. Spiegelhalter [1988] has proved that

$$V_{B,C}(P(a_1 \mid B,C)) = V_C(P(a_1 \mid C)) + E_{B,C}\left((P(a_1 \mid B,C) - P(a_1 \mid C))^2\right).$$

Therefore, if we first investigate the possible impact on $P(a_1)$ due to C, we can examine the possible further impact due to B as follows:

$$E_{B,C}\left((P(a_1 \mid B,C) - P(a_1 \mid C))^2\right)$$

$$= (P(a_1 \mid b_1,c_1) - P(a_1 \mid c_1))^2 P(b_1,c_1) + (P(a_1 \mid b_1,c_2) - P(a_1 \mid c_2))^2 P(b_1,c_2)$$

$$+ (P(a_1 \mid b_2,c_1) - P(a_1 \mid c_1))^2 P(b_2,c_1) + (P(a_1 \mid b_2,c_2) - P(a_1 \mid c_2))^2 P(b_2,c_2)$$

$$= .076.$$

We then have that
$$V_{B,C}(P(a_1)) = .04 + .076 = .116.$$

This procedure can be used iteratively. For example, if there is another variable D in the network and if we wish to compute possible further impact due to D, we would determine the value of

$$E_{B,C,D}\left((P(a_1 \mid B,C,D) - P(a_1 \mid B,C))^2\right).$$

PROBLEMS 10.1

1. Prove Lemma 10.1.

2. Consider Example 7.5. In chapter 7, we determined the new probabilities of the other variables in the network given that a patient had visited Asia and had dyspnea. Given this information, compute the maximum range of future possible values and the variance of the probability that the patient has lung cancer. Second, compute the possible impact that a chest X ray could have on the probability of the patient having lung cancer.

10.2. REPRESENTING THE UNCERTAINTY IN PROBABILITY VALUES

As discussed in chapter 2, if we take either a frequentistic or subjectivistic approach to probability, we are never certain of any probability values. Only a pure logical approach (e.g., applications of the principle of indifference) claims to ever determine probability values for certain. For example, if a coin were tossed 1000 times and 511 tosses came up heads, the frequentist would obtain a confidence interval for the probability of heads, while a subjectivist would obtain a posterior probability interval. The subjectivist could also obtain a Beta posterior distribution for the probability value (the methods for doing this will be discussed in the Subsection 10.2.1; the extreme frequentist does not accept the validity of probability distributions for probability values). If an individual says that the probability of the Lakers winning the NBA championship is .4, and if this value represents the individual's betting belief, at best this would mean that he feels that the value is somewhere between .39 and .41.

Thus we see that whether probabilities are obtained from data or from an individual, we cannot be absolutely certain of their values. In Subsection 10.2.1 we show how to augment the probabilities obtained from an expert with information in a data base and how to obtain the uncertainty in probability values, whereas in Subsection 10.2.2 we show how to incorporate this uncertainty into the structure of a causal network.

10.2.1. Obtaining Probabilities and the Uncertainty in Probabilities

In Section 2.5, when we discussed obtaining probability values from a data base, we made the simplification that the relative frequencies simply represent the probabilities. But we also mentioned that the subjective probabilities of an expert could be augmented with frequency data. In this subsection, we show how to accomplish this augmentation and further how to obtain the uncertainty in probability values.

Consider first the case where a propositional variable, say D, has precisely two alternatives. For example, d_1 might represent the presence of a particular disease. If we let x be a variable which represents a possible value of $P(d_1)$, then the general formula for the beta distribution is given by

$$\beta(a,b) = \frac{(a+b+1)!}{a!\,b!} x^a(1-x)^b \qquad \text{where} \quad a,b \geq 0.$$

This function is a probability density function $\mu(x)$ for the possible values of $P(d_1)$. If $P(d_1)$ is thought of as an objective probability, then this distribution is a subjective probability of the *true* value of $P(d_1)$. Hardy [1889] and Whitworth [1897] both suggested using the beta distribution for quantification of prior knowledge. It is left as an exercise to show that for all $a,b \geq 0$, if

$$\mu(x) = \beta(a,b),$$

then

$$\int_0^1 \mu(x)\,dx = 1.$$

The expected value of x represents the assessed value of $P(d_1)$. It is also left as an exercise to show that

$$P(d_1) = \int_0^1 x\mu(x)\,dx = \frac{a+1}{a+b+2}.$$

The values of a and b can be obtained as follows: We could ask the expert to specify the value of $P(d_1)$. If he felt that this value were .4, then we would set

$$.4 = \frac{a+1}{a+b+2}.$$

The values of a and b are based on how confident he is in the probability value. We can liken his confidence to a sample of size $a + b$. Thus, if he were not very confident, we might assign a value of 8 to $a + b$, while if he were very confident we might assign a value of 100, and if he were extremely confident, we might assign a value of 1000. Once we fix this hypothetical sample size, we can solve for a and b. For instance, if the estimate is .4 and the hypothetical sample size $a + b$ is 8, then $a = 3$ and $b = 5$.

Alternatively, we could ask the expert to specify a and $a + b$ such that his experience is approximately equivalent to having seen a successes in $a + b$ trials. His certainty is again relative to the size of $a + b$. We would then solve for $P(d_1)$.

It is also left as an exercise to show that the distribution $\beta(a,b)$ reaches its maximum at $a/(a+b)$ and that the distribution is highly peaked around that value if $a + b$ is large. The reason that $P(d_1)$ is not equal to $a/(a+b)$ is that the expected value of the probability in general is not equal to the maximizing value of the distribution. If $a = b = 0$, the distribution is uniform. These values are often used when we have no input from an expert and we are "equally undecided" about the value of $P(d_1)$ before obtaining any information from the data base. Figure 10.5 depicts $\beta(3,5)$ and $\beta(19,29)$.

Example 10.3. Assume that the expert represents his beliefs by $\beta(3,5)$. This means that he believes that $P(d_1)$ is equal to .4 and his probability distribution for $P(d_1)$ is the one in Figure 10.5(a). If we did not have a data base, then this value and this distribution would be used in the causal network, as described in the following subsection.

Suppose that the experiment is repeatable and that, in addition to the prior distribution obtained from the expert, we have data on n repetitions and k of them have value of D equal to d_1. If we assume that the sequence is infinitely exchangeable, then it can be shown [Jeffreys, 1939] that, given this additional

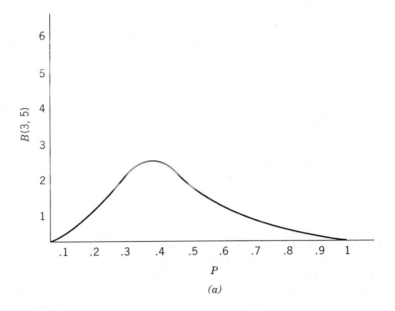

(a)

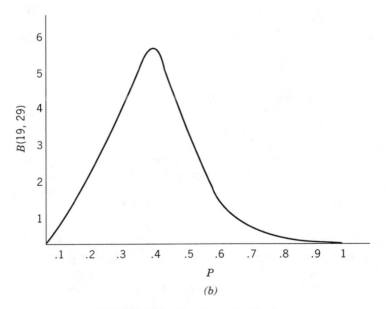

(b)

FIGURE 10.5   Two beta distributions.

information, the new value of $P(d_1)$ is given by

$$P(d_1) = \frac{k + a + 1}{n + a + b + 2}$$

and the posterior beta distribution is equal to

$$\beta(a + k, b + (n - k)).$$

In Example 10.3, if $n = 20$ and $k = 12$, then

$$P(d_1) = \frac{12 + 3 + 1}{20 + 3 + 5 + 2} = .53$$

and the posterior beta distribution is

$$\beta(3 + 12, 5 + (20 - 12)) = \beta(15, 13).$$

This value and this distribution would then be used in the causal network to represent $P(d_1)$ and our uncertainty in its value.

Example 10.4. Assume that the expert represents his belief by $\beta(m, m)$ for some constant $m \geq 0$. Then

$$P(d_1) = \frac{m + 1}{m + m + 2} = .5.$$

That is, if a and b are equal, the $P(d_1)$ is equal to .5 regardless of the size of a and b. Although the $P(d_1)$ is the same for all values of m, the size of m is very important because it represents the expert's uncertainty in the probability value. A value of $m = 0$ is an expression of ignorance. That is to say, the expert feels that d_1 and d_2 are equally likely because he is totally ignorant as to their relative likelihoods. As discussed in chapter 2, an example of such a situation would be when two individuals are to race and we have no information as to their relative speeds. Assigning $m = 0$ is an application of the second interpretation of the principle of indifference (as discussed in chapter 2). If in ten trials we observed eight successes, we would then have that

$$P(d_1) = \frac{k + m + 1}{n + m + m + 2} = \frac{8 + 0 + 1}{10 + 0 + 0 + 2} = .75.$$

Hence we see that an expression of ignorance is quickly overwhelmed by data. On the other hand, suppose we were to toss a coin just received in change at the store. Since we have good reason to believe an arbitrary coin is fair, we might assign a value of 100 to m. If in ten tosses we observed eight successes, we would then have that

$$P(d_1) = \frac{k + m + 1}{n + m + m + 2} = \frac{8 + 100 + 1}{10 + 100 + 100 + 2} = .514.$$

We see that the probability is hardly changed by the evidence.

Next assume that the propositional variable D can have t possible values. The distribution which we shall use is a generalization of the beta distribution, called the Dirichlet distribution. The expert is asked to specify values, $a_1, a_2, \ldots$, and a_t, which are all ≥ 0, such that his experience is approximately equivalent to having seen d_1 occur a_1 times, d_2 occur a_2 times, $\ldots$, and d_t occur a_t times in $a_1 + a_2 + \cdots + a_t$ total occurrences. The larger the value of $a_1 + a_2 + \cdots + a_t$, the more confident he is in his probabilities. The density function for the possible value of $P(d_i)$ is then given by

$$\mathrm{Dir}_i(a_1, a_2, \ldots, a_t) = \frac{(b_i + a_i + t - 1)!}{a_i!(b_i + t - 2)!} x^{a_i}(1 - x)^{(b_i + t - 2)},$$

where

$$b_i = \sum_{j=1}^{t} a_j - a_i.$$

It is left as an exercise to show that if $\mu(x) = \mathrm{Dir}_i(a_1, a_2, \ldots, a_t)$, then

$$\int_0^1 \mu(x)\,dx = 1.$$

The expected value of x represents the assessed value of $P(d_i)$. It is also left as an exercise to show that

$$P(d_i) = \int_0^1 x\mu(x)\,dx = \frac{a_i + 1}{a_1 + a_2 + \cdots + a_t + t}.$$

Notice that if $t = 2$, the Dirichlet distribution for $P(d_1)$ is

$$\mu(x) = \frac{(a_2 + a_1 + 2 - 1)!}{a_1!(a_2 + 2 - 2)!} x^{a_1}(1 - x)^{(a_2 + 2 - 2)},$$

which is the same as the beta distribution with $a = a_1$ and $b = a_2$.

The concept of exchangeability and the de Finetti representation theorem can be generalized to the case where there are more than two alternatives [Hewitt & Savage, 1955]. If we assume that each of the sequences is infinitely exchangeable, it can be shown [Jeffreys, 1939] that if in n occurrences we find that d_1 occurs k_1 times, d_2 occurs k_2 times, etc., and d_t occurs k_t times, then the $P(d_i)$ based on this additional information is given by

$$P(d_i) = \frac{a_i + k_i + 1}{a_1 + a_2 + \cdots + a_t + n + t}$$

and the Dirichlet posterior distribution for $P(d_i)$ is equal to

$$\mathrm{Dir}_i(a_1 + k_1, a_2 + k_2, \ldots, a_t + k_t).$$

Example 10.5. Suppose $t = 3$ and the expert says that

$$a_1 = 4, \qquad a_2 = 6, \qquad a_3 = 10.$$

Then

$$b_1 = 6 + 10 = 16$$

$$\text{Dir}_1(4,6,10) = \frac{(16+4+3-1)!}{4!(16+3-2)!}x^4(1-x)^{(16+3-2)}$$

$$b_2 = 4 + 10 = 14$$

$$\text{Dir}_2(4,6,10) = \frac{(14+6+3-1)!}{6!(14+3-2)!}x^6(1-x)^{(14+3-2)}$$

$$b_3 = 4 + 6 = 10$$

$$\text{Dir}_3(4,6,10) = \frac{(10+10+3-1)!}{10!(10+3-2)!}x^{10}(1-x)^{(10+3-2)}$$

$$P(d_1) = \frac{4+1}{4+6+10+3} = .217$$

$$P(d_2) = \frac{6+1}{4+6+10+3} = .304$$

$$P(d_3) = \frac{10+1}{4+6+10+3} = .478.$$

If there is no data base, then these probabilities and distributions would be used in the causal network. Suppose, however, that in 30 cases d_1 occurs eight times, d_2 occurs seven times, and d_3 occurs 15 times. Then

$$P(d_1) = \frac{4+8+1}{4+6+10+30+3} = .245$$

$$P(d_2) = \frac{6+7+1}{4+6+10+30+3} = .264$$

$$P(d_3) = \frac{10+15+1}{4+6+10+30+3} = .491.$$

It is left as an exercise to compute the posterior Dirichlet distributions.

If the expert says that, for all i, $a_i = m$ for some constant $m \geq 0$, then all the probabilities are initially the same. In this case the distribution is called the symmetric Dirichlet. Again, the larger the value of m, the more certain the expert is that they are the same.

We have assumed that the expert represents his beliefs by a Dirichlet distribution. However, what if he feels that such a distribution cannot represent his beliefs? Zabell [1982] has obtained the result that for a large class of repeatable experiments the prior distributions must be Dirichlet. In the case of three or more alternatives, Zabell's result is based on Johnson's *sufficientness postulate*, which is as follows [Johnson, 1932]:

Johnson's Sufficientness Postulate: Suppose $t \geq 3$, there are t alternatives $d_1, d_2, \ldots,$ and d_t, and in n repetitions of the experiment, d_i occurs k_i times for $1 \leq i \leq t$. Then the new probability $P(d_i)$ based on these n trials depends only on k_i and n and not on where the k_i occurrences occur or on the values of k_j for $j \neq i$.

For example, if $t = 3$, $n = 10$, and $k_1 = 3$, then the $P(d_1)$ on the 11th trial is the same regardless of how many times d_2 or d_3 occurred in the first ten trials or on where the three occurrences of d_1 occurred. Notice that this postulate implies exchangeability for any one of the alternatives. In the case where (1) the expert says that the outcome of the trials can change his assessed probabilities in any way (if the expert says that the outcome of the trials cannot affect his assessed probabilities, then the new probability $P(d_i)$ on the $n + 1$st trial is independent of what happens on the n previous trials and therefore we would have no reason to use the information in the data base); (2) the expert says that the sequence is infinitely exchangeable; and (3) the expert says that the sufficientness postulate represents his beliefs, Zabell [1982] proves that the prior distributions must be Dirichlet. Johnson [1932] originally sketched an incomplete version of this proof.

Notice that the assumption in the sufficientness postulate is vacuous when $t = 2$. In this case, if the expert says that the outcomes of the trials can change his assessed probabilities, if the sequence is infinitely exchangeable, and if we assume that there exist values $C(n)$, $E_1(n)$, and $E_2(n)$ such that the new values of $P(d_1)$ and $P(d_2)$, based on n repetitions of the experiment, are given by

$$P(d_1) = E_1(n) + C(n)k_1$$
$$P(d_2) = E_2(n) + C(n)k_2,$$

then again the prior distributions must be Dirichlet (k_i is again the number of occurrences of d_i). This result is also obtained in Zabell [1982].

Johnson's sufficientness postulate makes a somewhat minimal assumption. That is, that absence of knowledge about different types means that information about the frequency of one type yields no information about the likelihood of other types occurring. However, in some cases, such as cryptanalytic work, there is information in the frequencies of the frequencies (the frequency of a frequency k is the number of alternatives which occur k times in the n trials) and the postulate is not justified [Good, 1953, 1965]. Good [1967] shows how to mix symmetric Dirichlets in such cases.

Zabell [1982] obtains another interesting result. Suppose the expert believes that the new probability $P(d_i)$, based on n repetitions of the experiment, is given by

$$P(d_i) = f(k_i, n),$$

where k_i is again the number of times d_i occurs and f is a function which is independent of i. Given this assumption (along with the other assumptions stated above), Zabell proves that the prior Dirichlet distribution must be symmetrical. That is, if we feel that we should be able to compute the posterior probabilities from such a function, then we must assign the same prior probability to all of the alternatives.

PROBLEMS 10.2.1

1. Show that for all $a, b \geq 0$, if $\mu(x) = \beta(a, b)$, then

$$\int_0^1 \mu(x)\,dx = 1 \qquad \text{and} \qquad \int_0^1 x\mu(x)\,dx = \frac{a+1}{a+b+2}.$$

Hint:

$$\int_0^1 x^a(1-x)^b\,dx = \frac{a!\,b!}{(a+b+1)!}.$$

2. Show that $\beta(a, b)$ reaches its maximizing value at $a/(a+b)$ and that it is highly peaked around that value if $a + b$ is large.

3. Show that if $\mu(x) = \text{Dir}_i(a_1, a_2, \ldots, a_t)$ where $a_i \geq 0$ for all i, then

$$\int_0^1 \mu(x)\,dx = 1 \qquad \text{and} \qquad \int_0^1 x\mu(x)\,dx = \frac{a_i + 1}{a_1 + a_2 \ldots a_t + t}.$$

Hint: Use the hint in Problem 1.

4. Determine the posterior Dirichlet distributions in Example 10.5.

10.2.2. Representing Uncertainty in Probability Values in a Causal Network

We will show how the uncertainty in probability values can be incorporated into the structure of a causal network with a series of examples:

Example 10.6. Suppose that in the simple causal network in Figure 10.6, a_1 represents the presence of a certain disease and b_1 represents the presence of a certain symptom. In addition, suppose that we essentially know the probability values for certain. For example, we obtained the $P(b_1 \mid a_1)$ from ten different large samples and they all approximately yielded the value .9. If we then learned for certain that the patient has symptom b_1, we would determine that the new probability of a_1 is equal to .86. We would conclude that of all the patients having this symptom, about 86% should have this disease.

Example 10.7. Suppose now that we have very accurate data for the conditional probability of B given A, but not very good data for the a priori probability of A. Furthermore, suppose that our expert is not very confident in his assessed value of $P(A)$. We would then obtain a beta distribution for $P(a_1)$

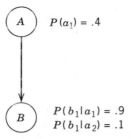

FIGURE 10.6   The causal network for Example 10.6.

which is not highly peaked. We could discretize the beta distribution possibly as follows:

$$P(P(a_1) = .1) = .2$$
$$P(P(a_1) = .4) = .6$$
$$P(P(a_1) = .7) = .2.$$

Of course, by including more possible values for $P(a_1)$, we could represent the distribution more accurately. We can represent this information in a causal network by adding a parent C to A. This situation is depicted in Figure 10.7. If $P(a_1)$ is thought of as an objective probability, then $P(c_1)$ is the subjective probability that $P(a_1)$ is equal to .1. Of course, with the assumption that $P(a_1)$ is an objective probability, $P(a_1)$ must have one unique value. The situation is no different than when we say that the probability of heads on the next toss is equal to .5. A specific toss will always be either heads or tails. Computing the probability of a_1, we obtain that

$$P(a_1) = \sum_{i=1}^{3} = P(a_1 \mid c_i)P(c_i) = .4,$$

which is the same as the value of $P(a_1)$ in Example 10.6. If we then instantiate B for b_1, we obtain that

$$P(a_1 \mid b_1) = .86,$$

which is again the same value as in Example 10.6. One may wonder what we have gained by including our uncertainty in the a priori value of $P(A)$ in the network, since, when we obtain information, the new probability of a_1 is the same as it is in Example 10.6. There is a difference, however, in the information which we can supply to the user. In Example 10.6 we could inform the user that approximately 86% of all patients with this disease have this illness. Thus, if the decision is made to perform a serious operation, it should be the correct decision about 86% of the time. In the current example we can determine the remaining variability in the probability of a_1 due to C. That is, if we let P' denote a conditional probability based on the instantiation of B

for b_1, we can compute the following probabilities either by propagation or direct computation:

$$P'(a_1) = P(a_1 \mid b_1) = .86$$

$$P'(c_1) = P(c_1 \mid b_1) = .086$$

$$P'(c_2) = P(c_2 \mid b_1) = .6$$

$$P'(c_3) = P(c_3 \mid b_1) = .314$$

$$P'(a_1 \mid c_1) = P(a_1 \mid b_1, c_1) = .5$$

$$P'(a_1 \mid c_2) = P(a_1 \mid b_1, c_2) = .86$$

$$P'(a_1 \mid c_3) = P(a_1 \mid b_1, c_3) = .95$$

$$V_C(P'(a_1 \mid C)) = (.5 - .86)^2(.086) + (.86 - .86)^2(.6) + (.95 - .86)^2(.314)$$

$$= .0143.$$

These values can be interpreted as follows: The program can only determine for certain that $P'(a_1)$ is a value in the interval [.5,.95], where $P'(a_1)$ is the probability of a patient, from the collective of all patients with this symptom, having this illness. Consequently, if a serious operation were to be performed, the user could only conclude that the operation is the correct course of action, based on the available information, approximately between 50% and 95% of the time. Given this variability, the user might decide to obtain information from sources other than the expert system before making a decision, whereas if the user knew for certain that it would be the correct course of action about 86% of the time, he may not. On the other hand, given the small variance and large expected value, he may decide to operate without obtaining additional information. The point is that by supplying the user with the information concerning the variability, he is better able to make a decision.

Uncertainty in the conditional probabilities in the network (i.e., probabilities associated with vertices which are not roots) can be incorporated into the structure of the network in the same way. The example that follows illustrates the method.

Example 10.8. Suppose that we are able to obtain a fairly accurate value for the conditional probability of symptom b_1 given disease a_1 is present, and a substantially less accurate value for the conditional probability of the symptom given the disease is not present. Specifically, suppose we ascertain that

$$P(P(b_1 \mid a_1) = .8) = .1 \qquad P(P(b_1 \mid a_2) = 0) = .3$$

$$P(P(b_1 \mid a_1) = .9) = .8 \qquad P(P(b_1 \mid a_2) = .1) = .4$$

$$P(P(b_1 \mid a_1) = 1) = .1 \qquad P(P(b_1 \mid a_2) = .2) = .3.$$

We can represent this uncertainty by the causal network in Figure 10.8.

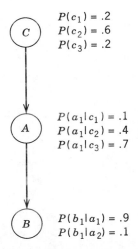

$$P(c_1) = .2$$
$$P(c_2) = .6$$
$$P(c_3) = .2$$

$$P(a_1|c_1) = .1$$
$$P(a_1|c_2) = .4$$
$$P(a_1|c_3) = .7$$

$$P(b_1|a_1) = .9$$
$$P(b_1|a_2) = .1$$

FIGURE 10.7   The causal network for Example 10.7.

Note in Example 10.8 that if B is instantiated for b_1 and we wish to determine the variability in the probability of a_1 due to the uncertainty in the probabilities stored in the network, then we must compute the variability in the probability of a_1 due to C, D, and E. This is due to the fact that D and E are not d-separated from A by $\{B\}$. That is, if we let P' again denote a conditional probability based on the instantiation of B for b_1, we must compute

$$\min_{i,j,k} P'(a_1 \mid c_i, d_j, e_k), \quad \max_{i,j,k} P'(c_i, d_j, e_k), \quad \text{and} \quad V_{C,D,E}(P'(a_1 \mid C, D, E)).$$

If there is a longer path from A to B, it will be necessary to also include the uncertainty in the probabilities stored for the nodes on this path in the computation since none of the parent nodes, which represent this uncertainty, are d-separated from A by $\{B\}$. Thus the number of necessary computations can grow exponentially with the distance in the DAG from A to B. Clearly, when A and B are embedded in a network, there are many configurations in which an unfeasible number of computations are necessary even when A and B are close together. We can, however, determine a minimum on the variability due to the uncertainty in the stored probabilities by computing the variability due only to C (that is, the uncertainty in the a priori probability of A). The variability is a minimum in the sense that the interval obtained for $P'(a_1)$ is a subset of the true interval and the variance obtained is less than or equal to the true variance. The minimum could be improved on by including any feasible number of uncertainties in other stored probabilities in the computation. A fairly loose maximum for the variability in the probability of a_1 due to the uncertainty in the probabilities stored in the network can be obtained by computing the variability due to all variables in the Markov blanket around A (by

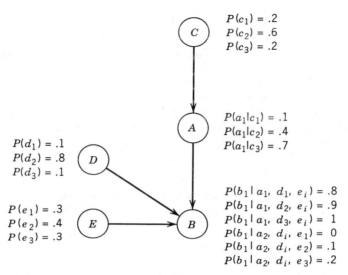

$$P(c_1) = .2$$
$$P(c_2) = .6$$
$$P(c_3) = .2$$

$$P(a_1|c_1) = .1$$
$$P(a_1|c_2) = .4$$
$$P(a_1|c_3) = .7$$

$$P(d_1) = .1$$
$$P(d_2) = .8$$
$$P(d_3) = .1$$

$$P(b_1 \mid a_1, d_1, e_i) = .8$$
$$P(b_1 \mid a_1, d_2, e_i) = .9$$
$$P(b_1 \mid a_1, d_3, e_i) = 1$$
$$P(b_1 \mid a_2, d_i, e_1) = 0$$
$$P(b_1 \mid a_2, d_i, e_2) = .1$$
$$P(b_1 \mid a_2, d_i, e_3) = .2$$

$$P(e_1) = .3$$
$$P(e_2) = .4$$
$$P(e_3) = .3$$

FIGURE 10.8   The causal network for Example 10.8.

Theorem 10.1). Note that some of the variables in the Markov blanket will now be variables which represent uncertainties in probabilities.

The uncertainty in a probability value is also useful when we are determining whether we should obtain additional evidence. Suppose that the network in Figure 10.8 is part of a larger network and there are other arcs emanating from A. Further, suppose that we have gathered certain information and have computed $P'(a_1)$, the probability of a_1 based on that evidence. We could compute the variance in the future probability of a_1 based on all the variables in the Markov blanket around A, excluding C. Then we could compute the additional variance due to C (using the method outlined at the end of the Section 10.1). If we find that the additional variance due to C is unacceptably large compared to the variance due to the other variables, we may decide not to bother to obtain information elsewhere in the network. Of course, this decision is also related to the cost of obtaining that information.

The examples in this chapter have related probabilities to relative frequencies because this author is most comfortable expressing ideas in this framework. However, except for the augmentation of probabilities with information in a data base, the methods in this chapter are also applicable to problems in which the probabilities solely represent beliefs and are not related to relative frequencies.

PROBLEM 10.2.2

Suppose that a_1 represents the presence of a certain disease, c_1 represents the presence of a certain risk factor for the disease, and b_1 represents a positive result on a certain test for the presence of the disease. Moreover, suppose that we determine that exactly 70% of the patients with the disease test positive, while it is equally likely that either 5% or 15% of those without the disease test positive. Finally, suppose that the disease is present in exactly 90% of those with the risk factor, whereas the probability is .25 that the disease is present in 70% of those without the risk factor and the probability is .75 that the disease is present in 30% of those without the risk factor. Construct the causal network which represents this situation and determine the range of possible future values and the variance of the probability of a_1.

10.3. THE PRINCIPLE OF INTERVAL CONSTRAINTS

Recall from the end of chapter 7 that probability propagation in causal networks is NP-hard. At that time it was mentioned that in some cases the only solution may be to use approximation techniques. Such techniques were originally used in NESTOR [Cooper, 1984] (in our discussion of NESTOR we replaced them by probability propagation) and in Peng and Reggia's method for obtaining the most probable explanations in the case of two-level networks [Peng & Reggia, 1987a and b].

When using these approximation techniques, we obtain interval constraints for the probabilities rather than point values. For example, we might obtain that the three most probable explanations are E_1, E_2, and E_3 and that

$$P'(E_1) \in [.2, .5], \qquad P'(E_2) \in [.2, .4], \qquad P'(E_3) \in [.1, .6],$$

where $P'(E_i)$ is the probability of E_i conditional on the information given to the system. When we obtain such intervals, we are left with the opposite problem from the one encountered when we fail to include our uncertainty in probabilities in the network (as discussed in section 10.2). That is, we can supply the user with our uncertainties in the probabilities, but we cannot supply him with expected point values. As can be seen in the present example, we cannot even provide him with a ranking of the explanations. By looking only at the interval constraints on the probabilities of E_1 and E_3, it is not possible to rank them relative to each other. In this section, we will show how to obtain expected values for the point probabilities from interval constraints on the probabilities. These results are taken from Neapolitan and Kenevan [1988] and [Neapolitan, 1989c].

The method is for the case where there are t mutually exclusive and exhaustive alternatives and we know that the probability of each alternative is in some interval. First we illustrate the method by considering the case where there are precisely two alternatives. Suppose that p_1 is the probability of the first alternative and p_2 is the probability of the second alternative and that

$$p_1 \in [0, .5] \qquad \text{and} \qquad p_2 \in [.5, 1].$$

Let x_i be a variable which represents the possible values that p_i can have. Then, since

$$p_1 + p_2 = 1,$$

it is easy to see that x_1 can take each value in the interval $[0,.5]$ in precisely one way, that is, with $x_2 = 1 - x_1$, and that x_2 can take each value in the interval $[.5,1]$ in precisely one way. Therefore, if we apply the principle of indifference to the possible values that p_1 could have, we obtain the uniform distribution in $[0,.5]$ as the probability distribution for p_1. Similarly, we obtain the uniform distribution in $[.5,1]$ as the the probability distribution for p_2. The expected values of p_1 and p_2 are then given by

$$E(p_1) = \int_0^{.5} \frac{x_1}{.5 - 0} dx_1 = .25$$

$$E(p_2) = \int_{.5}^1 \frac{x_1}{1 - .5} dx_1 = .75.$$

The assumption here is that probability values are distributed in nature according to the principle of indifference. That is to say, if we take all distributions which have precisely two alternatives in which the probability of the first alternative is in the interval $[0,.5]$ and the probability of the second alternative is in the interval $[.5,1]$, then the collective which contains the probabilities of the first alternative has the uniform distribution. There is no experimental evidence for this assumption. Rather, like many applications of the principle of indifference, it is only a reasonable symmetry argument.

Next we extend this result to the case where there are t mutually exclusive and exhaustive alternatives. First we need the following lemma:

Lemma 10.2. If $\sum_{i=1}^t p_i = 1$ and $p_i \in [a_i, b_i]$, then for $j = 1, 2, \dots, t$,

$$p_j \geq 1 - \sum_{i \neq j} b_i$$

$$p_j \leq 1 - \sum_{i \neq j} a_i.$$

Proof. Since $\sum_{i=1}^t p_i = 1$,

$$p_j + \sum_{i \neq j} p_i = 1,$$

and since $b_i \geq p_i$,

$$p_j \geq 1 - \sum_{i \neq j} b_i.$$

The second inequality is obtained in a similar manner. $\square$

Using this lemma, it is possible to obtain tighter interval constraints by setting, for $j = 1, 2, \ldots, t$,

$$\text{New } b_j = \min\left(b_j, 1 - \sum_{i \neq j} a_i\right)$$

$$\text{New } a_j = \max\left(a_j, 1 - \sum_{i \neq j} b_i\right).$$

It is left as an exercise to show that it does not matter in which order these new values are computed.

Example 10.9. Suppose that we determine that

$$p_1 \in [0, .4], \qquad p_2 \in [.1, .3], \qquad p_3 \in [.2, .6].$$

Then

$$\text{New } a_1 = \max(0, 1 - (.3 + .6)) = .1$$

$$\text{New } a_2 = \max(.1, 1 - (.4 + .6)) = .1$$

$$\text{New } a_3 = \max(.2, 1 - (.4 + .3)) = .3.$$

In the exercises you will be asked to prove that if the tightening method changes any upper bound then it is not possible that it changes any lower bounds. Thus we need not compute the new values of b_i.

Example 10.10. Suppose that we have four mutually exclusive and exhaustive explanations, $\mathcal{E}_1$, $\mathcal{E}_2$, $\mathcal{E}_3$, and $\mathcal{E}_4$, and we know that

$$P'(\mathcal{E}_1) \in [.2, .5] \qquad P'(\mathcal{E}_2) \in [.2, .4] \qquad P'(\mathcal{E}_3) \in [.1, .6],$$

where $P'(\mathcal{E}_i)$ is the probability of $\mathcal{E}_i$ conditional on the information given to the system. Suppose further that we have no information concerning $P'(\mathcal{E}_4)$. Then the initial interval constraint on $P'(\mathcal{E}_4)$ is $[0,1]$. However, we can tighten that constraint using the constraints on the other probabilities as follows:

$$\text{New } b_4 = \min\left(1, 1 - \sum_{i=1}^{3} \min(P'(\mathcal{E}_i))\right) = \min(1, 1 - (.2 + .2 + .1)) = .5.$$

Therefore we obtain that $P'(\mathcal{E}_4) \in [0, .5]$.

Next we prove the theorem which obtains expected values of the probabilities of t mutually exclusive and exhaustive alternatives from interval constraints on the probabilities. It is assumed in this theorem that Lemma 10.2 has been used to obtain the tighter intervals.

Theorem 10.2. *Principle of Interval Constraints.* If there are t mutually exclusive and exhaustive alternatives each with probability equal to p_i, and all that is known is that for $i = 1, 2, \ldots, t$,

$$p_i \in [a_i, b_i],$$

then there exist t density functions, $\mu_i(x_i)$, each defined on $[a_i, b_i]$, such that

$$\int_{a_i}^{b_i} \mu_i(x_i) \, dx_i = 1$$

and

$$E(p_i) = \int_{a_i}^{b_i} x_i \mu_i(x_i) \, dx_i,$$

where $\mu_i(x_i)$ is obtained by applying the principle of indifference to the probability values. The principle is applied by determining the way the intervals $[a_j, b_j]$ for $j \neq i$ constrain the measure of the ways that, for each $x_i \in [a_i, b_i]$, there exist $x_j \in [a_j, b_j]$ such that $x_i + \sum_{j \neq i} x_j = 1$.

When $t \geq 2$, the function $\mu_i(x_i)$ and the value $E(p_i)$ are obtained from the other intervals as follows (i is fixed at 1 to simplify notation):

1. If $t \geq 3$, define

$$\underline{x}_2(x_1) = \max(a_2, 1 - x_1 - b_3 - b_4 - \cdots - b_t)$$

$$\overline{x}_2(x_1) = \min(b_2, 1 - x_1 - a_3 - a_4 - \cdots - a_t)$$

$$\underline{x}_3(x_1, x_2) = \max(a_3, 1 - x_1 - x_2 - b_4 - \cdots - b_t)$$

$$\overline{x}_3(x_1, x_2) = \min(b_3, 1 - x_1 - x_2 - a_4 - \cdots - a_t)$$

$$\vdots$$

$$\underline{x}_{t-1}(x_1, x_2, \ldots, x_{t-2}) = \max(a_{t-1}, 1 - x_1 - x_2 - \cdots - x_{t-2} - b_t)$$

$$\overline{x}_{t-1}(x_1, x_2, \ldots, x_{t-2}) = \min(b_{t-1}, 1 - x_1 - x_2 - \cdots - x_{t-2} - a_t).$$

2. $\mu_1(x_1)$ is given by

$$\mu_1(x_1) = \frac{\int_{\underline{x}_2}^{\overline{x}_2} \int_{\underline{x}_3}^{\overline{x}_3} \cdots \int_{\underline{x}_{t-2}}^{\overline{x}_{t-2}} \int_{\underline{x}_{t-1}}^{\overline{x}_{t-1}} dx_{t-1} \, dx_{t-2} \, dx_{t-3} \ldots dx_2}{\int_{a_1}^{b_1} \int_{\underline{x}_2}^{\overline{x}_2} \int_{\underline{x}_3}^{\overline{x}_3} \cdots \int_{\underline{x}_{t-2}}^{\overline{x}_{t-2}} \int_{\underline{x}_{t-1}}^{\overline{x}_{t-1}} dx_{t-1} \, dx_{t-2} \, dx_{t-3} \ldots dx_2 \, dx_1}.$$

Note that the arguments to the functions $\overline{x}_i$ and $\underline{x}_i$ have not been included in the integrals for the sake of clarity, and if $t = 2$ the numerator is 1 while the denominator is $b_1 - a_1$.

3. $E(p_1)$, the expected value of the probability of the first alternative, is given by

$$E(p_1) = \int_{a_1}^{b_1} x_1 \mu_1(x_1) \, dx_1.$$

Proof. The theorem will be proved for the case where $t = 3$; the proof for an arbitrary t is a straightforward generalization. Create n equally spaced points in each of the intervals $[a_i, b_i]$. Label the points in $[a_1, b_1]$ as $x_{11}, x_{12}, \ldots, x_{1n}$. Since we are interested only in the limit as $n \to \infty$, we may assume that, for a given $x_1 \in [a_1, b_1]$, if there is a point x_2 among the n points in $[a_2, b_2]$ and a point x_3 among the n points in $[a_3, b_3]$, such that there exists an $\hat{x}_2$ in a small neighborhood of x_2 and an $\hat{x}_3$ in a small neighborhood of x_3 such that

$$x_1 + \hat{x}_2 + \hat{x}_3 = 1,$$

then the points $\hat{x}_2$ and $\hat{x}_3$ are the points x_2 and x_3, and that the probability that p_1 is equal to x_1 is approximately equal to

$$P_n(x_1) = \frac{r_n(x_1)}{\sum_{k=1}^{n} r_n(x_{1k})},$$

where $r_n(x_1)$ is defined to be the number of ways that it is possible to obtain $x_1 + x_2 + x_3 = 1$, where x_2 is one of the n points from $[a_2, b_2]$ and x_3 is one of the n points from $[a_3, b_3]$. We see that we are computing the probability that p_1 has a particular value by applying the principle of indifference to the possible values of p_1.

The density function $\mu_1(x_1)$ is approximately equal to $P_n(x_1)/\Delta x_1$ or $P_n(x_1)n/(b_1 - a_1)$, and therefore, taking the limit, we have

$$\mu_1(x_1) = \lim_{n \to \infty} \frac{P_n(x_1)n}{(b_1 - a_1)}.$$

We need to determine the function $r_n(x_1)$. If we define

$$\underline{x}_2(x_1) = \max(a_2, 1 - x_1 - b_3)$$
$$\overline{x}_2(x_1) = \min(b_2, 1 - x_1 - a_3),$$

then $\underline{x}_2(x_1)$ is the smallest point in the interval $[a_2, b_2]$ such that $x_1 + x_2 + x_3 = 1$ for some $x_3 \in [a_3, b_3]$, and $\overline{x}_2(x_1)$ is the largest such point. These points are depicted in Figure 10.9. Clearly, if $x_2 \in [\underline{x}_2(x_1), \overline{x}_2(x_1)]$, then there is exactly one point $x_3 \in [a_3, b_3]$ such that $x_1 + x_2 + x_3 = 1$, and if x_2 is not in $[\underline{x}_2(x_1), \overline{x}_2(x_1)]$, there is no such x_3. If we set

$$f(x_1) = \overline{x}_2(x_1) - \underline{x}_2(x_1),$$

then we have that

$$r_n(x_1) = \frac{f(x_1)n}{(b_2 - a_2)}$$

$$P_n(x_1) = \frac{f(x_1)n/(b_2 - a_2)}{\sum_{k=1}^{n} f(x_{1k})n/(b_2 - a_2)} = \frac{f(x_1)}{\sum_{k=1}^{n} f(x_{1k})}$$

$$\mu_1(x_1) = \lim_{n \to \infty} \frac{f(x_1)}{\sum_{k=1}^{n} f(x_{1k})(b_1 - a_1)/n}. \quad \blacksquare$$

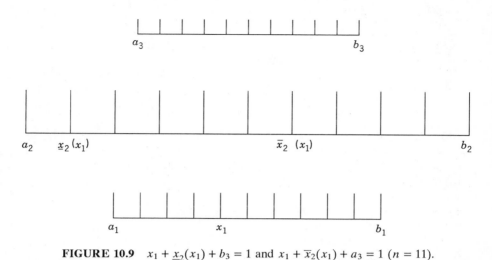

FIGURE 10.9 $x_1 + \underline{x}_2(x_1) + b_3 = 1$ and $x_1 + \overline{x}_2(x_1) + a_3 = 1$ $(n = 11)$.

If we set

$$C = \lim_{n \to \infty} \left(\sum_{k=1}^{n} f(x_{1k}) \right) (b_1 - a_1)/n,$$

we then have that

$$1 = \int_{a_1}^{b_1} \mu_1(x_1) \, dx_1 = \frac{\int_{a_1}^{b_1} f(x_1) \, dx_1}{C},$$

which implies that

$$C = \int_{a_1}^{b_1} f(x_1) \, dx_1,$$

and therefore

$$\mu_1(x_1) = \frac{f(x_1)}{\int_{a_1}^{b_1} f(x_1) \, dx_1}. \quad \square$$

Note that the expected probability, which is obtained using the above theorem, is a coordinate for the centroid of a convex region inside an $t - 1$ dimensional cube, but that it is a different region for each probability.

It can be shown [Jeffreys, 1939] that if all the interval constraints are equal to [0,1], then the density function in the principle of interval constraints is equal to the symmetric Dirichlet distribution in which $a_i = 0$ for all i (here we mean the a_i's in the Dirichlet distribution, not the ones in the interval constraints).

The integrals which yield the expected probabilities can be computed by numerical integration. Recall Example 10.10 in which we had four explanations,

$\mathcal{E}_1$, $\mathcal{E}_2$, $\mathcal{E}_3$, and $\mathcal{E}_4$ such that

$$P(\mathcal{E}_1) \in [.2,.5], \qquad P(\mathcal{E}_2) \in [.2,.4], \qquad P(\mathcal{E}_3) \in [.1,.6], \qquad P(\mathcal{E}_4) \in [0,.5].$$

When we compute the expected probabilities based on these interval constraints, we obtain that

$$E(P(\mathcal{E}_1)) = .320, \quad E(P(\mathcal{E}_2)) = .287, \quad E(P(\mathcal{E}_3)) = .246, \quad E(P(\mathcal{E}_4)) = .146.$$

Notice that the expected value of $P(\mathcal{E}_1)$ is significantly higher than the expected value of $P(\mathcal{E}_3)$, even though an inspection of the interval constraints would not enable us to conclude that either would be higher.

We offer a final example:

Example 10.11. $[a_i, b_i] = [0, b]$ for $i = 1, \ldots, N$ where $1/N \le b \le 1$. When Theorem 10.2 is used to compute the expected values of the probabilities, each expected value is determined to be $1/N$, as expected, since all probabilities have the same interval constraint. These expected values are thus equal to the probabilities which would be obtained if we applied the principle of indifference directly to the alternatives.

Recall the case discussed at the beginning of this section in which we had interval constraints on the probabilities of the three most probable explanations. It may be thought that we can group all the remaining explanations into one explanation, assign an initial interval constraint of [0,1] to the probability of that explanation, and apply the principle of interval constraints to the probabilities of four explanations. Philosophically, this is not a sound practice. It would be similar to doing the following: we know only that there are t mutually exclusive and exhaustive alternatives. We then group the last $(t-3)$ of them into one alternative and apply the principle of indifference to conclude that the probability of each of the first three alternatives is .25. Clearly, we can obtain different probabilities by arbitrarily grouping alternatives. In the case of the principle of interval constraints we must assign an initial interval constraint of [0,1] to the probability of each of the $(t-3)$ remaining alternatives. If the alternatives are explanations, then t is ordinarily a forbiddingly large number. However, if we believed almost for certain that the top three explanations were the only possible ones, then we could apply the principle of interval constraints to the probabilities of only those three explanations.

PROBLEMS 10.3

1. Show that when we tighten the interval constraints, as discussed in the text, it does not matter in which order we compute the new values.

2. Show that when we tighten the interval constraints, as discussed in the text, if any of the upper bounds are changed then it is not possible that any of the lower bounds are changed.

COMPUTER PROBLEM 10.3

Using numerical integration, write a program which determines the expected
values of the probabilities of a set of mutually exclusive and exhaustive alternatives from interval constraints on the probabilities. Hint: Recursively use the
Romberg method [Cheney and Kincaid, 1980].

REFERENCES

Andreassen, S., M. Woldbye, B. Falck, and S. K. Andersen [1987], "MUNIN—A Causal Probabilistic Network for Interpretation of Electromyographic Findings," *Proceedings of 10th International Joint Conference on Artificial Intelligence*, Milan, Italy.

Arrnborg, S., D. G. Corneil, and A. Proskurowski [1987], "Complexity of Finding Embeddings in a *K*-tree," *SIAM Journal of Algebraic and Discrete Methods*, Vol. 8.

Ash, R. B. [1970], *Basic Probability Theory*, Wiley, New York.

Balcazar, J. L., J. Diaz, and J. Gabarro [1988], *Structural Complexity I*, Springer-Verlag, New York.

Belinfante, F. J. [1970], "Experiments to Disprove That Nature Would Be Deterministic," mimeographed, Purdue University, West Lafayette, Indiana.

Bell J. S. [1964], "On the Einstein Podolsky Rosen Paradox," *Physics*, Vol. 1, No. 3.

Ben-Bassat, M., R. W. Carlson, V. K. Puri, M. D. Davenport, J. A. Schriver, M. Latif, R. Smith, L. D. Portigal, E. H. Lipnick, and M. H. Weil [1980], "Pattern-Based Interactive Diagnosis of Multiple Disorders: The Medas System," *IEEE Transactions on Pattern Analysis and Machine Intelligence*, Vol. PAMI-2, No. 2.

Ben-Bassat, M., K. L. Klove, and M. H. Weil [1980], "Sensitivity Analysis in Bayesian Classification Models: Multiplicative Deviations," *IEEE Transactions on Pattern Analysis and Machine Intelligence*, Vol. PAMI-2, No. 3.

Berger, J. O. [1985], *Statistical Decision Theory and Bayesian Analysis*, Springer-Verlag, New York.

Bhatnagar, R. K., and L. N. Kanal [1986], "Handling Uncertain Information," in L. N. Kanal and J. F. Lemmer, Eds., *Uncertainty in Artificial Intelligence*, North-Holland, Amsterdam.

Binford, T. O., T. S. Levitt, and W. Mann [1987], "Bayesian Inference in Model-Based Machine Vision," *Proceedings of the Third AAAI Workshop on Uncertainty in Artificial Intelligence*, Seattle, Washington.

Bohm, D. [1952], "A Suggested Interpretation of the Quantum Theory in Terms of 'Hidden Variables,' Part I," *Physical Review*, Vol. 85.

Bondy, J. A., and U. S. R. Murty [1982], *Graph Theory with Applications*, North-Holland, New York.

Brooks, R. A. [1981], "Symbolic Reasoning Among 3-D Models and 3-D Images," *Artificial Intelligence*, Vol. 17.

Buchanan, B. G., and E. H. Shortliffe [1984], *Rule-Based Expert Systems*, Addison-Wesley, Reading, Massachusetts.

Carnap, R. [1950], "The Two Concepts of Probability," in *Logical Foundations of Probability*, University of Chicago Press, Chicago, Illinois.

Carnap, R. [1952], *The Continuum of Inductive Methods*, University of Chicago Press, Chicago, Illinois.

Charniak, E. [1983], "The Bayesian Basis of Common Sense Medical Diagnosis," *Proceedings of AAAI*, Washington, D.C.

Chavez, R. M. [1989], "A Fully Polynomial Randomized Approximation Scheme for the Bayesian Inference Problem," Technical Report KSL-88-72, Knowledge Systems Laboratory, Stanford University, Stanford, California.

Cheeseman, P. [1984], "A Method of Computing Generalized Bayesian Probability Values for Expert Systems," *Proceedings of IEEE Workshop on Principles of Knowledge-Based Systems*, Karlsruhe, West Germany.

Cheney, W., and D. Kincaid [1980], *Numerical Mathematics and Computing*, Brooks/Cole, Monterey, California.

Chin, H. L., and G. F. Cooper [1987], "Stochastic Simulation of Bayesian Belief Networks," *Proceedings of the Third AAAI Workshop on Uncertainty in Artificial Intelligence*, Seattle, Washington.

Church, A. [1940], "On the Concept of a Random Sequence," *Bulletin of the American Mathematical Society*, Vol. 46.

Cohen, P. R. [1985], *Heuristic Reasoning About Uncertainty: An Artificial Intelligence Approach*, Pitman, Boston, Massachusetts.

Cooper, G. F. [1984], "'NESTOR': A Computer-Based Medical Diagnostic That Integrates Causal and Probabilistic Knowledge," Technical Report HPP-84-48, Stanford University, Stanford, California.

Cooper, G. F. [1987], "An Algorithm for Computing Probabilistic Propositions," *Proceedings of the Third AAAI Workshop on Uncertainty in Artificial Intelligence*, Seattle, Washington.

Cooper, G. F. [1988], "Probabilistic Inference Using Belief Networks Is NP-Hard," Technical Report KSL-87-27, Stanford University, Stanford, California.

Cooper, G. F. [1989], Private correspondence.

Cox, R. T. [1946], "Probability, Frequency, and Reasonable Expectation," *The American Journal of Physics*, Vol. 14, No. 1.

Cox, R. T. [1979], "Of Inference and Inquiry, An Essay in Inductive Logic," in R. D. Levine and M. Tribus, Eds., *The Maximum Entropy Formalism*, MIT Press, Cambridge, Massachusetts.

Darroch, J. N., S. L. Lauritzen, and T. P. Speed [1980], "Markov Fields and Log-Linear Interaction Models for Contingency Tables," *The Annals of Statistics*, Vol. 8, No. 3.

Dias, P. M. C., and A. Shimony [1981], "A Critique of Jaynes' Maximum Entropy Principle," *Advances in Applied Mathematics*, Vol. 2.

Dombel, F. T. de, D. J. Leaper, J. R. Staniland, A. P. McCann, and J. C. Horricks [1972], "Computer-Aided Diagnosis of Acute Abdominal Pain," *British Medical Journal*, Vol. 2.

Doob, J. L. [1936], "Note on Probability," *Annals of Mathematics*, Vol. 37, No. 2.

Doyle, J. A. [1979], "A Truth Maintenance System," *Artificial Intelligence*, Vol. 12.

Duda, R. O., P. E. Hart, and N. J. Nilsson [1976], "Subjective Bayesian Methods for Rule-Based Inference Systems," Technical Report 124, Stanford Research Institute, Menlo Park, California.

Duda, R. O., P. E. Hart, N. J. Nilsson, and G. L. Sutherland [1978], "Semantic Network Representation in Rule-Based Inference Systems," in D. A. Waterman and F. Hayes-Roth, Eds., *Pattern-Directed Inference Systems*, Academic Press, New York.

Earman, J. [1986], *A Primer on Determinism*, Kluwer Academic, Netherlands.

Einstein, A., B. Podolsky, and N. Rosen [1935], "Can Quantum-Mechanical Description of Physical Reality Be Considered Complete?" *Physical Review*, Vol. 47.

Feinstein, A. R. [1977], "The Haze of Bayes, the Aerial Palaces of Decision Analysis, and the Computerized Ouija Board," *Clinical Pharmacology and Therapeutics*, Vol. 21, No. 4.

Feller, W. [1968], *Introduction to Probability Theory and Its Applications*, Vol. I, Wiley, New York.

Fine, T. L. [1973], *Theories of Probability*, Academic Press, New York.

Finetti, B. de [1931], "La prévision: ses lois logiques, ses sources subjectives," *Annales de l'Institute Henri Poincaré*, Vol. 7.

Finetti, B. de [1964], "Foresight: Its Logical Laws, Its Subjective Sources," in H. E. Kyburg, Jr. and H. E. Smokler, Eds., *Studies in Subjective Probability*, Wiley, New York.

Finetti, B. de [1972], *Probability, Induction, and Statistics*, Wiley, New York.

Garey, M. R., and D. S. Johnson [1979], *Computers and Intractability: A Guide to the Theory of NP-Completeness*, W. H. Freeman, San Francisco, California.

Geiger, D., and J. Pearl [1988], "On the Logic of Causal Models," *Proceedings of the Fourth AAAI Workshop on Uncertainty in Artificial Intelligence*, University of Minnesota, Minneapolis, Minnesota.

Geiger, D., T. Verma, and J. Pearl [1989], "d-Separation: From Theorems to Algorithms," *Proceedings of the Fifth Workshop on Uncertainty in Artificial Intelligence*, University of Windsor, Windsor, Ontario, Canada.

Golumbic, M. C. [1980], *Graph Theory and Perfect Graphs*, Academic Press, New York.

Good, I. J. [1952], "Rational Decisions," *Journal of the Royal Statistical Society B*, Vol. 14.

Good, I. J. [1953], "The Population Frequencies of Species and the Estimation of Population Parameters," *Biometrika*, Vol. 40.

Good, I. J. [1965], *The Estimation of Probabilities: An Essay on Modern Bayesian Methods*, Research Monograph No. 30, MIT Press, Cambridge, Massachusetts.

Good, I. J. [1967], "A Bayesian Significance Test for Multinomial Distributions," *Journal of the Royal Statistical Society B*, Vol. 29.

Hardy, G. F. [1889], Letter, *Insurance Record* (reprinted in *Transactions of Faculty of Actuaries*, Vol. 8, 1920).

Harré, R. [1970], "Probability and Confirmation," in *The Principles of Scientific Thinking*, University of Chicago Press, Chicago, Illinois.

Harris, J. M. [1981], "The Hazards of Bedside Bayes," *Journal of the American Medical Association*, Vol. 246, No. 22.

Heckerman, D. [1986], "Probabilistic Interpretations for MYCIN's Certainty Factors," in L. N. Kanal and J. F. Lemmer, Eds., *Uncertainty in Artificial Intelligence*, North-Holland, Amsterdam.

Heckerman, D. [1989], "A Tractable Inference Algorithm for Diagnosing Multiple Diseases," *Proceedings of the Fifth Workshop on Uncertainty in Artificial Intelligence*, University of Windsor, Windsor, Ontario, Canada.

Heckerman, D., and E. J. Horvitz [1987], "On the Expressiveness of Rule-Based Systems for Reasoning with Uncertainty," *Proceedings of AAAI*, Seattle, Washington.

Heisenberg, W. [1971], *Physics and Beyond*, Harper & Row, New York.

Hempel, C. G. [1965], "Studies in the Logic of Confirmation," in *Aspects of Scientific Explanation and Other Essays in the Philosophy of Science*, Free Press, New York.

Henrion, M. [1986], "Uncertainty in Artificial Intelligence: Is Probability Epistemologically and Heuristically Adequate?" in J. Mumpower and O. Renn, Eds., *Expert Systems and Expert Judgement*, Springer-Verlag, New York.

Hewitt, E., and L. J. Savage [1955], "Symmetric Measures on Cartesian Product," *Transactions of the American Mathematical Society*, Vol. 80.

Heyde, J. E. [1957], *Entwertung der Kausalität?*, Kohlhammer, Stuttgart; Europa Verlag, Zurich, Vienna.

Horvitz, E. J. [1987], "Reasoning About Beliefs and Actions Under Computational Resource Constraints," *Proceedings of the Third AAAI Workshop on Uncertainty in Artificial Intelligence*, Seattle, Washington.

Horvitz, E. J., D. E. Heckerman, and C. P. Langlotz [1986], "A Framework for Comparing Alternative Formalisms for Plausible Reasoning," *Proceedings of 5th National Conference on AI (AAAI-86)*, Philadelphia, Pennsylvania.

Horvitz, E. J., H. J. Suermondt, and G. F. Cooper [1989], "Bounded Conditioning: Flexible Inference for Decisions Under Scarce Resources," *Proceedings of the Fifth Workshop on Uncertainty in Artificial Intelligence*, University of Windsor, Windsor, Ontario, Canada.

Howard, R. A. [1984], "The Used Car Buyer," in R. A. Howard and J. E. Matheson, Eds., *Applications of Decision Analysis*, Vol. II, Strategic Decisions Group, Menlo Park, California.

Howard, R. A., and J. E. Matheson [1984], "Influence Diagrams," in R. A. Howard and J. E. Matheson, Eds., *Applications of Decision Analysis*, Vol. II, Strategic Decisions Group, Menlo Park, California.

Iversen, G. R., W. H. Longcor, F. Mosteller, J. P. Gilbert, and C. Youtz [1971], "Bias and Runs in Dice Throwing and Recording: A Few Million Throws," *Psychometrika*, Vol. 36.

Jackson, P. [1986], *Introduction to Expert Systems*, Addison-Wesley, Reading, Massachusetts.

Jammer, M. [1974], *The Philosophy of Quantum Mechanics*, Wiley, New York.

Jaynes, E. T. [1979], "Where Do We Stand on Maximum Entropy," in R. D. Levine and M. Tribus, Eds., *The Maximum Entropy Formalism*, MIT Press, Cambridge, Massachusetts.

Jeffreys, H. [1939], *Theory of Probability*, Clarendon Press, Oxford.

Johnson, W. E. [1932], "Probability: the Inductive and Deductive Problems," *Mind*, Vol. 49.

Keynes, J. M. [1948], *A Treatise on Probability*, Macmillan, London (originally published 1921).

Kolmogorov, A. N. [1929], "Das Gesetz der geiterierten Logarithmus," *Math. Ann.*, Vol. 101.

Kolmogorov, A. N. [1950], *Foundations of the Theory of Probability*, Chelsea, New York (originally appeared in the *Ergebnisse der Wahrscheinlichkeitrechnung* in 1933).

Kolmogorov, A. N. [1963], "On Tables of Random Numbers," *Sankhya, the Indian Journal of Statistics*, Series A 25.

Kulikowski, C. A., and S. H. Weiss [1982], "Representation of Expert Knowledge for Consultant," in P. Szolovits, Ed., *AI in Medicine*, Westview Press, Boulder, Colorado.

Kyburg, H. E. [1970], *Probability and Inductive Logic*, Macmillan, London.

Lambalgen, M. van [1987], *Random Sequences*, Ph.D. Thesis, University of Amsterdam.

Laplace, P. S. de [1951], *A Philosophical Essay and Probabilities*, Dover, New York (originally published in 1820).

Lauritzen, S. L., and D. J. Spiegelhalter [1988], "Local Computation with Probabilities in Graphical Structures and Their Applications to Expert Systems," *Journal of the Royal Statistical Society B*, Vol. 50, No. 2.

Lemmer, J. F. [1983], "Generalized Updating of Incompletely Specified Distributions," *Large Scale Systems*, No. 5.

Lesser V. R., and L. D. Erman [1977], "A Retrospective View of Hearsay II Architecture," *Proceedings of 5th International Joint Conference on Artificial Intelligence*, Cambridge, Massachusetts.

Lewis, C. I. [1962], *An Analysis of Knowledge and Valuation*, Open Court Press, LaSalle, Illinois.

Li, M., and P. Vitanyi [1988], *Two Decades of Applied Kolmogorov Complexity*, to be published.

Lindsay, R., B. G. Buchanan, E. A. Feigenbaum, and J. Lederberg [1980], *Applications of AI for Chemical Inference: The DENDRAL Project*, McGraw-Hill, New York.

Marbe, K. [1916], *Die Gleichformigkeit in der Welt*, Munich.

McDermott, J. [1982], "R1: A Rule-Based Configurer of Computer Systems," *Artificial Intelligence*, Vol. 19, No. 1.

Miller, A. C., M. M. Merkhofer, R. A. Howard, J. E. Matheson, and T. R. Rice [1976], *Development of Automated Aids for Decision Analysis*, Stanford Research Institute, Menlo Park, California.

Moore, E. C. [1985], "Semantic Considerations on Non-Monotonic Logic," *Artificial Intelligence*, Vol. 25.

Nagel, E. [1939], *Principles of the Theory of Probability*, University of Chicago Press, Chicago, Illinois.

Neapolitan, R. E. [1986], Private lecture notes, CS 335, Northeastern Illinois University, Chicago, Illinois.

Neapolitan, R. E. [1987], "A Note of Caution on Combining Certainty," *International Journal of Pattern Recognition and Artificial Intelligence*, Vol. 1, No. 3 and 4.

Neapolitan, R. E. [1989a], Private lecture notes, CS 435, Northeastern Illinois University, Chicago, Illinois.

Neapolitan, R. E. [1989b], "A Limiting Frequency Approach to Probability Based on the Weak Law of Large Numbers," submitted to the *British Journal for the Philosophy of Science*.

Neapolitan, R. E. [1989c], "The Principle of Interval Constraints—A Generalization of the Symmetric Dirichlet Prior," submitted to *Mathematical Biosciences*.

Neapolitan, R. E., and J. R. Kenevan [1988], "Justifying the Principle of Interval Constraints," *Proceedings of the Fourth AAAI Workshop on Uncertainty in Artificial Intelligence*, University of Minnesota, Minneapolis, Minnesota.

Neapolitan, R. E., C. Georgakis, M. Evens, J. R. Kenevan, H. Jiwani, and D. B. Hier [1987], "Using Set Covering and Uncertain Reasoning to Rank Explanations," *Proceedings of SCAMC*, Washington, D.C.

Pearl, J. [1986a], "Fusion, Propagation, and Structuring in Belief Networks," *Artificial Intelligence*, Vol. 29.

Pearl, J. [1986b], "Markov and Bayes Networks," UCLA Technical Report CSD 860024 R-46-II, UCLA, Los Angeles.

Pearl, J. [1986c], "A Constraint-Propagation Approach to Probabilistic Reasoning," in L. N. Kanal and J. F. Lemmer, Eds., *Uncertainty in Artificial Intelligence*, North-Holland, Amsterdam.

Pearl, J. [1987a], "Evidential Reasoning Using Stochastic Simulation of Causal Models," *Artificial Intelligence*, Vol. 32.

Pearl, J. [1987b], "Distributed Revision of Composite Beliefs," *Artificial Intelligence*, Vol. 33.

Pearl, J. [1988], *Probabilistic Reasoning in Intelligent Systems*, Morgan Kaufmann, San Mateo, California.

Pearl, J. [1989], Private correspondence.

Peng, Y., and J. A. Reggia [1987a], "A Probabilistic Causal Model for Diagnostic Problem Solving—Parts I and II," *IEEE Transactions on Systems, Man and Cybernetics*, Vol. SMC-17.

Peng, Y., and J. A. Reggia [1987b], "Being Comfortable with Plausible Diagnostic Hypotheses," Computer Science Technical Report 1753, University of Maryland, College Park, Maryland.

Pople, H. E. [1982], "Heuristic Methods for Imposing Structure on Ill-Structured Problems: The Structuring of Medical Diagnosis," in P. Szolovits, Ed., *AI in Medicine*, Westview Press, Boulder, Colorado.

Popper, K. R. [1975], *Logic of Scientific Discovery*, Hutchinson & Co. (originally published in 1935).

Popper, K. R. [1983], *Realism and the Aim of Science*, Rowman & Littlefield, Totowa, New Jersey.

Putnam, H. [1963], "'Degree of Confirmation' and Inductive Logic," in P. A. Schilpp, Ed., *The Philosophy of Rudolf Carnap*, Open Court Press, LaSalle, Illinois.

Reggia, J. A., D. S. Nau, and P. Y. Wang [1983], "A Theory of Abductive Inference in Expert Systems," Computer Science Technical Report 1338, University of Maryland, College Park, Maryland.

Salmon, W. C. [1966], *The Foundations of Scientific Inference*, University of Pittsburgh Press, Pittsburgh, Pennsylvania.

Savage, L. J. [1954], *The Foundations of Statistics*, Wiley, New York.

Seidenfeld, T. [1986], "Entropy and Uncertainty," *Philosophy of Science*, Vol. 53.

Shachter, R. D. [1986], "Evaluating Influence Diagrams," *Operations Research*, Vol. 34, No. 6.

Shachter, R. D. [1988], "Probabilistic Inference and Influence Diagrams," *Operations Research*, Vol. 36, No. 4.

Shachter, R. D. [1989], "Evidence Absorption and Propagation Through Evidence Reversals," *Proceedings of the Fifth Workshop on Uncertainty in Artificial Intelligence*, University of Windsor, Windsor, Ontario, Canada.

Shachter, R. D., and D. Heckerman [1987], "A Backwards View for Assessments," *AI Magazine*, Fall.

Shachter, R. D., and C. R. Kenley [1989], "Gaussian Influence Diagrams," *Management Science*, Vol. 35, No. 5.

Shafer, G. [1976], *A Mathematical Theory of Evidence*, Princeton University Press, Princeton, New Jersey.

Shafer, G. [1986], "Probabilistic Judgement in Artificial Intelligence," in L. N. Kanal and J. F. Lemmer, Eds., *Uncertainty in Artificial Intelligence*, North-Holland, Amsterdam.

Shastri, L., and J. A. Feldman [1984], "Semantic Networks and Neural Nets," Technical Report 131, Computer Science Department, University of Rochester, Rochester, New York.

Shore, J. E. [1986], "Relative Entropy, Probabilistic Inference, and AI," in L. N. Kanal and J. F. Lemmer, Eds., *Uncertainty in Artificial Intelligence*, North-Holland, Amsterdam.

Spiegelhalter, D. J. [1986a], "'A Statistical View of Uncertainty in Expert Systems," in W. Gale, Ed., *Artificial Intelligence and Statistics*, Addison-Wesley, Reading, Massachusetts.

Spiegelhalter, D. J. [1986b], "Probabilistic Reasoning in Expert Systems," in L. N. Kanal and J. F. Lemmer, Eds., *Uncertainty in Artificial Intelligence*, North-Holland, Amsterdam.

Spiegelhalter, D. J. [1987], "Coherent Evidence Propagation in Expert Systems," *The Statistician*, Vol. 36.

Spiegelhalter, D. J. [1988], "Analysis of Softness," in L. N. Kanal and T. S. Levitt, Eds., *Uncertainty in Artificial Intelligence III*, North-Holland, Amsterdam.

Srinivas, S., S. Russell, and A. Alice [1989], "Automated Construction of Sparse Bayesian Networks From Unstructured Probabilistic Models and Domain Information," *Proceedings of the Fifth Workshop on Uncertainty in Artificial Intelligence*, University of Windsor, Windsor, Ontario, Canada.

Suermondt, H. J., and G. F. Cooper [1988], "Updating Probabilities in Multiply-Connected Belief Networks," *Proceedings of the Fourth AAAI Workshop on Uncertainty in Artificial Intelligence*, University of Minnesota, Minneapolis, Minnesota.

Suermondt, H. J., and G. F. Cooper [1989], Private correspondence.

Szolovits, P., and S. G. Pauker [1978], "Categorical and Probabilistic Reasoning in Medical Diagnosis," *Artificial Intelligence*, Vol. 11.

Tarjan, R. E. [1976], "Maximum Cardinality Search and Chordal Graphs," Unpublished lecture notes in CS 259, Stanford University, Stanford, California.

Tarjan, R. E., and M. Yannakakis [1984], "Simple Linear-Time Algorithms to Test Chordality of Graphs, Test Acyclicity of Hypergraphs, and Selectively Reduce Hypergraphs," *SIAM Journal Computing*, Vol. 13, No. 3.

Thompson, B. A., and W. A. Thompson [1985], "Inside an Expert System," *BYTE Magazine*, April.

Verma, T., and J. Pearl [1988], "Causal Networks: Semantics and Expressiveness," *Proceedings of the Fourth AAAI Workshop on Uncertainty in Artificial Intelligence*, University of Minnesota, Minneapolis, Minnesota.

Von Mises, R. [1957], *Probability, Statistics, and Truth*, George Allen & Unwin, London (originally published in Vienna, 1928).

Wald, J., M. Farach, M. Tagamets, and J. A. Reggia [1989], "Generating Plausible Diagnostic Hypotheses With Self-Processing Causal Networks," *Journal of Experimental and Theoretical Artificial Intelligence*, in press.

Weatherford, R. [1982], *Philosophical Foundations of Probability Theory*, Routledge & Kegan Paul, London.

Wen, W. X. [1988], "MCE Reasoning in Recursive Causal Networks," *Proceedings of the Fourth AAAI Workshop on Uncertainty in Artificial Intelligence*, University of Minnesota, Minneapolis, Minnesota.

Whitworth, W. A. [1897], *DCC Exercises in Choice and Chance* (reprinted 1965, Hafner, New York).

Winston, P. [1984], *Artificial Intelligence*, Addison-Wesley, Reading, Massachusetts.

Yager, R. R., S. Ovchinnikov, R. M. Yong, and H. T. Nguyen, Eds., [1987], *Fuzzy Sets and Applications: Selected Papers by L. A. Zadeh*, Wiley, New York.

Yannakakis, M. [1981], "Computing the Minimum Fill-in Is NP-Complete," *SIAM Journal of Algebraic and Discrete Methods*, Vol. 2, No. 1.

Zabell, S. L. [1982], "W. E. Johnson's 'Sufficientness' Postulate," *The Annals of Statistics*, Vol. 10, No. 4.

Zabell, S. L. [1988], "Symmetry and Its Discontents," in B. Skyrms and W. L. Harper, Eds., *Causation, Chance, and Credence*, Kluwer Academic, Netherlands.

INDEX

429